Lecture Notes in Computer Science 7615

Commenced Publication in 1973
Founding and Former Series Editors:
Gerhard Goos, Juris Hartmanis, and Jan van Leeuwen

Maria Serna (Ed.)

Algorithmic Game Theory

5th International Symposium, SAGT 2012
Barcelona, Spain, October 22-23, 2012
Proceedings

Volume Editor

Maria Serna
Universitat Politècnica de Catalunya
ALBCOM Research Group
Departament de Llenguatges i Sistemes Informàtics
Jordi Girona 1-3
08034 Barcelona, Spain
E-mail: mjserna@lsi.upc.edu

ISSN 0302-9743 e-ISSN 1611-3349
ISBN 978-3-642-33995-0 e-ISBN 978-3-642-33996-7
DOI 10.1007/978-3-642-33996-7
Springer Heidelberg Dordrecht London New York

Library of Congress Control Number: 2012948192

CR Subject Classification (1998): I.6, H.5.3, J.1, K.6.0, H.3.5, J.4, K.4.4, G.1.2, F.2.2

LNCS Sublibrary: SL 3 – Information Systems and Application, incl. Internet/Web and HCI

Typesetting: Camera-ready by author, data conversion by Scientific Publishing Services, Chennai, India

Printed on acid-free paper

Springer is part of Springer Science+Business Media (www.springer.com)

Preface

The present volume is devoted to the Fifth International Symposium on Algorithmic Game Theory (SAGT), an interdisciplinary event intended to provide a forum for researchers and practitioners to exchange innovative ideas and to be aware of each other's approaches and findings. The main focus of SAGT is on the study of the algorithmic aspects of game theory; typical questions include how scarce computational resources affect the way games between selfish agents are played and the impact of selfishness on the quality of the outcome of a multiplayer system. This is a departure from traditional algorithmic theory in which players are supposed to be cooperative.

The algorithmic approach to game theory has been applied primarily to problems from economics and computer science (e.g., auctions, network, and routing problems). We believe that this approach can be used to pose new questions and to give answers to problems in other fields like physics and biology and hope SAGT will be one of the fora that make this convergence happen.

SAGT 2012 took place in the Technical University of Catalonia, Barcelona (Spain), during October 22–23, 2012. The present volume contains all contributed papers that were accepted at SAGT 2012 in alphabetical order by author.

In response to the call for papers, the Program Committee received 44 submissions of which 22 were selected for inclusion in the scientific program of the symposium after a detailed evaluation (each submission was read by at least three Program Committee members) and electronic discussion. Papers co-authored by a Program Committee member were handled by a special sub-committee.

We would like to thank all of the authors who submitted papers and the members of the Program Committees, the external reviewers, and the Organizing Committee.

We gratefully acknowledge the support from the Software Department and the ALBCOM research group from Universitat Politècnica de Catalunya, as well as the financial support from Universitat Politècnica de Catalunya and the Generalitat de Catalunya.

We wish to thank the creator of the EasyChair System, a free conference management system, which was very helpful in the selection of the scientific program and in the preparation of this volume

August 2012 Maria Serna

Organization

Program Committee

Carme Àlvarez	Universitat Politècnica de Catalunya, Spain
Peter Bro Miltersen	Aarhus University, Denmark
Ioannis Caragiannis	Patras University, Greece
Piotr Faliszewski	AGH University of Science and Technology, Poland
Michele Flammini	University of L'Aquila, Italy
Dimitris Fotakis	National Technical University of Athens, Greece
Martin Hoefer	RWTH Aachen, Germany
Annamaria Kovacs	Goethe University Frankfurt am Main, Germany
Katrina Ligett	California Institute of Technology, USA
Igal Milchtaich	Bar Ilan University, Israel
Vahab S. Mirrokni	GOOGLE Research New York, USA
Paolo Penna	Università di Salerno, Italy
Maria Polukarov	University of Southampton, UK
Maria Serna (Chair)	Universitat Politècnica de Catalunya, Spain

Steering Committee

Elias Koutsoupias	National and Kapodistrian University of Athens, Greece
Marios Mavronicolas	University of Cyprus
Dov Monderer	The Technion - Israel Institute of Technology, Israel
Burkhard Monien	Paderborn University, Germany
Christos Papadimitriou	University of California at Berkeley, USA
Giuseppe Persiano	University of Salerno, Italy
Paul Spirakis (Chair)	RACTI and University of Patras, Greece
Berthold Vöcking	RWTH Aachen University, Germany

Additional Reviewers

Alam, Muddasser
Anshelevich, Elliot
Apt, Krzysztof
Barman, Siddharth
Bhaskar, Umang
Bhattacharya, Sayan
Bilò, Davide
Biló, Vittorio

Bodine-Baron, Elizabeth
Busch, Costas
Cavallo, Ruggiero
Chen, Ning
Christodoulou, George
Dams, Johannes
Díaz, Josep
Elkind, Edith
Erdelyi, Gabor
Fanelli, Angelo
Ferraioli, Diodato
Fu, Hu
Gairing, Martin
Giotis, Ioannis
Gkatzelis, Vasilis
Haghpanah, Nima
Huang, Chien-Chung
Kanellopoulos, Panagiotis
Koczy, Laszlo
Kontogiannis, Spyros
Kyropoulou, Maria
Lu, Pinyan
Maneva, Elitza
Monaco, Gianpiero
Moscardelli, Luca
Nikolova, Evdokia
Nunnari, Salvatore
Oren, Sigal
Pasquale, Francesco
Piliouras, Georgios
Röglin, Heiko
Samet, Dov
Schapira, Michael
Skopalik, Alexander
Skowron, Piotr
Sundararajan, Mukund
Ventre, Carmine
Vöcking, Berthold
Voice, Thomas
Wagner, Lisa
Zick, Yair
Zohar, Aviv

Table of Contents

A Classification of Weakly Acyclic Games*

Krzysztof R. Apt[1,2] and Sunil Simon[1]

[1] CWI, Amsterdam, The Netherlands
[2] ILLC, University of Amsterdam

Abstract. Weakly acyclic games form a natural generalization of the class of games that have the finite improvement property (FIP). In such games one stipulates that from any initial joint strategy some finite improvement path exists. We classify weakly acyclic games using the concept of a scheduler recently introduced in [1].

1 Introduction

1.1 Background

Given a strategic game, when we allow the players to improve their choices on a unilateral basis, we are naturally brought to the concept of an *improvement path*, in which at each stage a single player who did not select a best response is allowed to select a better strategy. By definition every finite improvement path terminates in a Nash equilibrium. This suggests the *finite improvement property* (FIP), introduced in [2], according to which every improvement path is finite. This is obviously a desired property of a game that in particular is satisfied by the congestion games.

However, the FIP is a very strong property and many natural games do not satisfy it. In particular, [3] studied the congestion games in which the payoff functions are players specific. These games do not have the FIP. Milchtaich proved that such games belong to a larger class of games (essentially introduced in [4]), called *weakly acyclic games*, in which one only stipulates that from any initial joint strategy some finite improvement path exists.

Weakly acyclic games have a natural appeal because the concept of an improvement path captures the idea of a possible 'interaction' resulting from players' preference for better strategies and hence provides a link with distributed computing. In particular, [5] introduced a natural class of weakly acyclic games, which model the routing aspects on the Internet. In turn, [6] showed that for weakly acyclic games, a modification of the traditional no-regret algorithm yields almost sure convergence to a pure Nash equilibrium. Further, [7] proved that the existence of a unique (pure) Nash equilibrium in every subgame implies that the game is weakly acyclic.

* A full version with all the proofs is available at the authors' homepages.

M. Serna (Ed.): SAGT 2012, LNCS 7615, pp. 1–12, 2012.

1.2 Our Work

If we view a strategic game as a distributed system in which the players attempt to find a Nash equilibrium by means of a 'better response (respectively, 'best response) dynamics', then the property of being weakly acyclic only guarantees that finding a Nash equilibrium is always possible. However, such an existence guarantee does not help the players to find it. By adding to the game a ***scheduler***, a concept recently introduced in [1], we ensure that the players always reach a Nash equilibrium, by repeatedly interacting with it. A scheduler is simply a function that given a finite sequence of joint strategies selects a player who can improve his payoff in the last joint strategy. Each player interacts with the scheduler by submitting to it a strategy he selected. Subsequently the scheduler again selects a player who did not select a best response. This interaction process leaves open how each player selects his better (respectively, best) strategy.

In the presence of a scheduler for a strategic game G we can view the resulting interaction as a 'supergame' between the central authority represented by the scheduler and the players of G. The aim of the central authority is to reach a Nash equilibrium in spite of a limited guarantee on the behaviour of the players: all it can be sure of is that each selected player will select a better response (respectively a best response). The resulting interaction results in an improvement path (respectively a best response improvement path). If all so generated improvement paths are finite, we say that the game *respects* the scheduler.

By providing a classification of the schedulers we obtain a natural classification of weakly acyclic games. An advantage of such a classification is that given a weakly acyclic game we can determine under what adverse circumstances a Nash equilibrium still can be reached. Consequently some existing results can be improved. In particular, we show in Section 7 how we can strengthen our recent result from [1] concerning a class of social network games. In turn, [8] recently strengthened the above mentioned theorem of [3] by showing that congestion games with player specific payoff functions respect every local best response scheduler, defined below in Section 3.

In what follows we introduce eight natural categories of schedulers. They yield nine classes of finite weakly acyclic games that for two player games collapse into five classes.

2 Preliminaries

Assume a set $N := \{1, \ldots, n\}$ of players, where $n > 1$. A ***strategic game*** for n players, written as $(S_1, \ldots, S_n, p_1, \ldots, p_n)$, consists of a non-empty set S_i of ***strategies*** and a ***payoff function*** $p_i : S_1 \times \cdots \times S_n \to \mathbb{R}$, for each player i.

Fix a strategic game $G := (S_1, \ldots, S_n, p_1, \ldots, p_n)$. We denote $S_1 \times \cdots \times S_n$ by S, call each element $s \in S$ a ***joint strategy***, denote the ith element of s by s_i, and abbreviate the sequence $(s_j)_{j \neq i}$ to s_{-i}. Occasionally we write (s_i, s_{-i}) instead of s.

We call a strategy s_i of player i a ***best response*** to a joint strategy s_{-i} of his opponents if $\forall s'_i \in S_i \; p_i(s_i, s_{-i}) \geq p_i(s'_i, s_{-i})$. If s_i is (not) a best response to s_{-i}, we say that player i ***selected*** (***did not select***) ***a best response in*** s. Next, we call a joint strategy s a ***Nash equilibrium*** if each s_i is a best response to s_{-i}, that is, if $\forall i \in N\ \forall s'_i \in S_i\ p_i(s_i, s_{-i}) \geq p_i(s'_i, s_{-i})$. We also define

$$BR(s) := \{i \mid \text{player } i \text{ selected a best response in } s\},$$

$$NBR(s) := \{i \mid \text{player } i \text{ did not select a best response in } s\}.$$

Further, we call a strategy s'_i of player i a ***better response*** given a joint strategy s if $p_i(s'_i, s_{-i}) > p_i(s_i, s_{-i})$. Following [2] a ***path*** in S is a sequence $(s^1, s^2, \ldots)$ of joint strategies such that for every $k > 1$ there is a player i such that $s^k = (s'_i, s^{k-1}_{-i})$ for some $s'_i \neq s^{k-1}_i$. A path is called an ***improvement path*** (respectively, a ***best response improvement path***, in short a ***BR-improvement path***) if it is maximal and for all $k > 1$, $p_i(s^k) > p_i(s^{k-1})$ (respectively, s^k_i is a best response to s^{k-1}_{-i}), where i is the player who deviated from s^{k-1}. So in an improvement path each deviating player selects a better response, while in a BR-improvement path each deviating player selects a best response.

The ***better response graph*** (respectively, the ***best response graph***) associated with the game G is defined as $(S, \rightarrow)$, where $s \rightarrow s'$ if (s, s') is a step in an improvement path (respectively, in an BR-improvement path).

Given joint strategies $s, s' \in S$ and a player i we define

$$s \xrightarrow{i} s' \text{ iff } s_{-i} = s'_{-i} \text{ and } p_i(s') > p_i(s),$$

$$s \xRightarrow{i} s' \text{ iff } s \xrightarrow{i} s' \text{ and } s'_i \text{ is a best response to } s'_{-i}.$$

Recall that G has the ***finite improvement property*** (***FIP***), (respectively, the ***finite best response property*** (***FBRP***)) if every improvement path (respectively, every BR-improvement path) is finite. Obviously, if a game has the FIP or the FBRP, then it has a Nash equilibrium – it is the last element of each path. Following [4,3] we call a strategic game ***weakly acyclic*** (respectively, ***BR-weakly acyclic***) if for any joint strategy there exists a finite improvement path (respectively, BR-improvement path) that starts at it.

3 Schedulers

In what follows we introduce some classes of weakly acyclic games. They are defined in terms of schedulers. By a ***scheduler*** we mean a function f that given finite sequence $s^1, \ldots, s^k$ of joint strategies that does not end in a Nash equilibrium selects a player who did not select in s^k a best response. In practice schedulers will be applied only to initial prefixes of improvement paths.

Consider an improvement path $\rho = (s^1, s^2, \ldots)$. We say that ρ ***respects*** a scheduler f if for all k smaller than the length of ρ we have $s^{k+1} = (s'_i, s^k_{-i})$, where $f(s^1, \ldots, s^k) = i$. We say that a strategic game ***respects a scheduler*** f if all improvement paths ρ which respect f are finite. Further, we say that a strategic game ***respects a BR-scheduler*** f if all BR-improvement paths ρ which respect f are finite.

In what follows we study various types of schedulers. We say that a scheduler f is ***state-based*** if for some function $g : S \rightarrow \mathbb{R}$ we have

$$f(s^1, \ldots, s^k) = g(s^k).$$

We say that a function $g : \mathcal{P}(N) \rightarrow N$ is a ***choice function*** if for all $A \neq \emptyset$ we have $g(A) \in A$. Next, we say that a scheduler f is ***set-based*** if for some choice function $g : \mathcal{P}(N) \rightarrow N$

$$f(s^1, \ldots, s^k) = g(NBR(s^k)).$$

Finally, we say that a set-based scheduler f is ***local*** if the above choice function g satisfies the following property:

$$\text{if } g(A) \in B \subseteq A \text{ then } g(A) = g(B). \tag{1}$$

A simple way of producing choice functions $g : \mathcal{P}(N) \rightarrow N$ that satisfy (1) is the following. Take a permutation π of $1, \ldots, n$ and define for $A \neq \emptyset$ $[\pi](A) := \pi(k)$, where k is the smallest element of N such that $\pi(k) \in A$. That is, $[\pi](A)$ is the first element on the list $\pi(1), \ldots, \pi(n)$ that belongs to A.

In Section 7 we shall need the following characterization result.

Proposition 1. *A choice function $g : \mathcal{P}(N) \rightarrow N$ satisfies (1) iff it is of the form $[\pi]$ for some permutation π of $1, \ldots, n$.*

Proof. Suppose a choice function $g : \mathcal{P}(N) \rightarrow N$ satisfies (1). Define a permutation π of $1, \ldots, n$ inductively as follows:

$$\pi(1) := g(N), \ \pi(2) := g(N \setminus \{\pi(1)\}), \ldots, \ \pi(n) := g(N \setminus \{\pi(1), \ldots, \pi(n-1)\}).$$

Take now a nonempty subset A of N. Let $\pi(k) = [\pi](A)$. By definition $\{\pi(1), \ldots, \pi(k-1)\} \cap A = \emptyset$ and $\pi(k) \in A$. Let $B := N \setminus \{\pi(1), \ldots, \pi(k-1)\}$. By definition $g(B) = \pi(k)$. Further, $A \subseteq B$ and $\pi(k) \in A$, so by property (1) we have $g(A) = g(B) = [\pi](A)$.

Next, it is straightforward to check that each function $[\pi]$ satisfies (1). □

The games that respect schedulers fall into various categories. In what follows *FIP* (respectively, *FBRP*) stands for the class of games that have the FIP (respectively, FBRP), *WA* (respectively, *BRWA*) for the class of weakly acyclic games (respectively, BR-weakly acyclic games), *Sched* (respectively, $Sched_{BR}$) stands for the class of games that respect a scheduler (respectively, a BR-scheduler), etc.

4 Schedulers versus State-Based Schedulers

We prove here three implications which show that the classes of games $Sched_{BR}$, *Sched*, *State* and $State_{BR}$ coincide.

Theorem 1 (*Sched* ⇒ *State*). *If a game respects a scheduler, then it respects a state-based scheduler.*

Proof. Fix a strategic game $G = (S_1, \ldots, S_n, p_1, \ldots, p_n)$. Let $Y := \cup_{k \in \mathbb{N}} Y_k$, where

- $Y_0 := \{s \in S \mid s \text{ is a Nash equilibrium}\}$,
- $Y_{k+1} := Y_k \cup \{s \mid \exists i \, \forall s' (s \xrightarrow{i} s' \rightarrow s' \in Y_k)\}$.

For each $s \in Y_{k+1} \setminus Y_k$, let $f_{State}(s) := i$, where i is such that $\forall s' (s \xrightarrow{i} s' \rightarrow s' \in Y_k)$.

If $Y = S$, then we can view f_{State} as a state-based scheduler. We now prove two claims concerning the set Y and the scheduler f_{State}.

Claim 1. *If a strategic game G respects a scheduler, then $Y = S$.*

Proof. Suppose that G respects a scheduler f. Assume by contradiction that $Y \neq S$. Take $s^0 \in S \setminus Y$. Suppose $f(s^0) = i_1$. By the definition of Y there exists a joint strategy s^1 such that $s^0 \xrightarrow{i_1} s^1$ and $s^1 \in S \setminus Y$. Suppose $f((s^0, s^1)) = i_2$. Again, by the definition of Y there exists a joint strategy s^2 such that $s^1 \xrightarrow{i_2} s^2$ and $s^2 \in S \setminus Y$. Iterating this argument we construct an infinite improvement path which respects f, which yields a contradiction. □

Claim 2. *If for a strategic game G we have $Y = S$, then G respects f_{State}.*

Proof. We prove by induction on k that all improvements paths that start in a joint strategy from Y_k and respect f_{State} are finite.

The claim holds vacuously for $k = 0$. Suppose it holds for some $k \geq 0$. Take some $s \in Y_{k+1} \setminus Y_k$ and an improvement path ξ that respects f_{State} and starts in s. Suppose that $f_{State}(s) := i$. Then for some s', $s \xrightarrow{i} s'$ is the first step in ξ. By the definition of f_{State}, $s' \in Y_k$, so by the induction hypothesis ξ is finite. □

Suppose now that a game G respects a scheduler. By Claim 1 $Y = S$, so f_{State} is a state-based scheduler. By Claim 2, G respects f_{State}. □

The above proof uses a construction similar to the one used to compute the winning regions of reachability games, see, e.g., [9, page 104].

Theorem 2 (*Sched$_{BR}$* ⇒ *State$_{BR}$*). *If a game respects a BR-scheduler, then it respects a state-based BR-scheduler.*

Proof. The proof is the same as that of Theorem 1 with the relation $\xRightarrow{i}$ used instead of $\xrightarrow{i}$. □

Theorem 3 (*Sched$_{BR}$* ⇒ *Sched*). *If a finite game respects a BR-scheduler, then it respects a scheduler.*

Proof. The idea of the proof is as follows. Suppose that a game respects a BR-scheduler f_{BR}. We construct then a scheduler f inductively by repeatedly scheduling the same player until he plays a best response, and subsequently scheduling the same player as f_{BR} does.

To make it precise we need some notation. We call an initial prefix of an improvement path an ***improvement sequence***. To indicate the deviating players at each step of an improvement sequence $(s^0, \ldots, s^k)$ we shall write it alternatively as $s^0 \stackrel{i_1}{\rightarrow} s^1 \stackrel{i_2}{\rightarrow} \ldots \stackrel{i_k}{\rightarrow} s^k$.

Given an improvement sequence ξ we denote by $[\xi]_{BR}$ the subsequence of it obtained by deleting the joint strategies that do not result from a selection of a best response. In general $[\xi]_{BR}$ is not a improvement sequence (for example, it does not need to be a maximal sequence), but it is if every maximal subsequence of it of the form $s^0 \stackrel{i}{\rightarrow} s^1 \stackrel{i}{\rightarrow} \ldots \stackrel{i}{\rightarrow} s^m$ ends with a selection of a best response.

Given a finite sequence of joint strategies ξ we denote its last element by $last(\xi)$ and denote the extension of ξ by a joint strategy s by ξ, s. We define the desired scheduler f inductively by the length of the sequences. For a sequence of length 1, so a joint strategy that is not a Nash equilibrium, we put $f(s) := f_{BR}(s)$.

Suppose now that we defined f on all sequences of length k. Consider a sequence ξ, s of length $k+1$. If ξ, s is not an improvement path or $last(\xi) \stackrel{f(\xi)}{\rightarrow} s$ does not hold, then we define $f(\xi, s)$ arbitrarily. Otherwise we put

$$f(\xi, s) := \begin{cases} f_{BR}([\xi, s]_{BR}) & \text{if } s_i \text{ is a best response to } s_{-i} \\ f(\xi) & \text{otherwise} \end{cases}$$

We claim that G respects the scheduler f. To see it take an improvement path ξ that respects f. By the definition of f, $[\xi]_{BR}$ is an improvement sequence that respects f_{BR}. By assumption $[\xi]_{BR}$ is finite, so ξ is finite, as well. □

Note that the above theorem fails to hold for infinite games. Indeed, consider a two player game $(\{0\}, [0,1], p_1, p_2)$, where $[0,1]$ stands for the real interval $\{r \mid 0 \leq r \leq 1\}$ and $p_1(0, s_2) = p_2(0, s_2) := s_2$. Then 1 is a unique best response of player 2 to the strategy 0, so this game respects the unique BR-scheduler. However, it does not respect the unique scheduler.

5 Other Implications

For two player games another implication holds.

Proposition 2 (*Sched* ⇒ *FBRP*). *If a two player game respects a scheduler, then every BR-improvement path is finite.*

Proof. Suppose that a two player game G respects a scheduler. Note that the best response graph of G has the property that for every node s that is not a source node, the set $NBR(s)$ has at most one element. Take a BR-improvement path ξ. Suppose that (s, s') is the first step in ξ and that η is the suffix of ξ that starts at s'. Then every element s'' of η is such that the set $NBR(s'')$ has at most one element. Hence η respects any scheduler and consequently is finite. So ξ is finite, as well. □

By definition if a two player game respects a set-based scheduler, then it respects a local scheduler. Thus for two player games we get the implications and equivalences depicted in Figure 1. All implications can be shown to be proper by example games which are omitted due to lack of space.

$$\begin{array}{ccccccccccc}
\mathit{FIP} & \Longrightarrow & \mathit{Local} & \Longleftrightarrow & \mathit{Set} & \Longrightarrow & \mathit{State} & \Longleftrightarrow & \mathit{Sched} & \Longrightarrow & \mathit{WA} \\
\Downarrow & & \Downarrow & & \Downarrow & & \Updownarrow & & \Updownarrow & & \Uparrow \\
\mathit{FBRP} & \Leftrightarrow & \mathit{Local}_{BR} & \Leftrightarrow & \mathit{Set}_{BR} & \Leftrightarrow & \mathit{State}_{BR} & \Leftrightarrow & \mathit{Sched}_{BR} & \Rightarrow & \mathit{BRWA}
\end{array}$$

Fig. 1. Classification of two player finite weakly acyclic games

For arbitrary finite games we have the implications and equivalences depicted in Figure 2. Again, all implications can be shown to be proper.

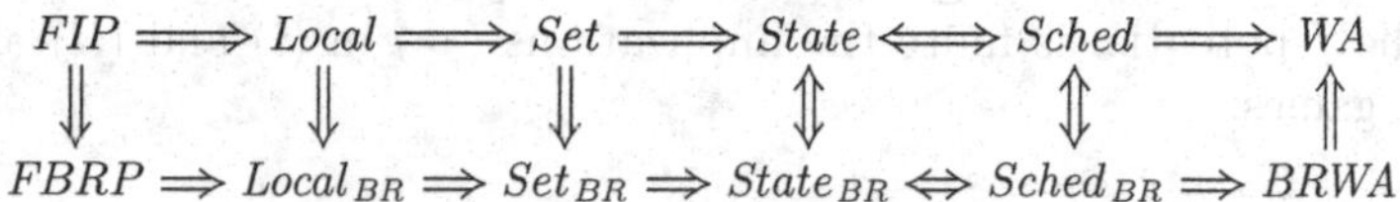

Fig. 2. Classification of finite weakly acyclic games

6 Potentials

To characterize the finite games that have the FIP [2] introduced the concept of a (generalized ordinal) ***potential***. We now introduce an appropriately modified notion to characterize the games that respect a scheduler. We shall use it in the next section to reason about a natural class of games.

Consider a game $(S_1, \ldots, S_n, p_1, \ldots, p_n)$ and a scheduler f. We say that a function $F : S \rightarrow \mathbb{R}$ is an ***f-potential*** (respectively, an ***f-BR-potential***) if for every initial prefix of an improvement path (respectively, an BR-improvement path) $(s^1, \ldots, s^k, s^{k+1})$ that respects f we have $F(s^{k+1}) > F(s^k)$.

Note that when f is a state-based scheduler, then a function F is an f-potential iff for all i, s'_i and s

$$\text{if } f(s) = i \text{ and } p_i(s'_i, s_{-i}) > p_i(s_i, s_{-i}), \text{ then } F(s'_i, s_{-i}) > F(s_i, s_{-i}),$$

and similarly for the f-BR-potentials. In the proof below we use the following classic result of [10].

Lemma 1 (König's Lemma). *Any finitely branching tree is either finite or it has an infinite path.*

Theorem 4. *Consider a finite game G.*

(i) G respects a scheduler f iff an f-potential exists.
(ii) G respects a BR-scheduler f iff an f-BR-potential exists.

Proof. (i) ($\Rightarrow$) Consider a branching tree the root of which has all joint strategies as successors, the non-root elements of which are joint strategies, and whose branches are the improvement paths that respect f. Because the game is finite this tree is finitely branching. By König's Lemma this tree is finite, so we conclude that the number of improvement paths that respect f is finite. Given a joint strategy s define $F(s)$ to be the number of improvement sequences (in the sense of the proof of Theorem 3) that respect f and that terminate in s. Clearly F is an f-potential.
($\Leftarrow$) Let F be an f-potential. Suppose by contradiction that an infinite improvement path that respects f exists. Then the corresponding values of F form a strictly increasing infinite sequence. This is a contradiction, since there are only finitely many joint strategies.
The proof of (ii) is analogous. □

The argument given in (i)($\Rightarrow$) follows the proof of [3] of the fact that every game that has the FIP has a generalized ordinal potential. Note that when the range of the f-potential is finite the implications ($\Leftarrow$) in (i) and (ii) also hold for infinite games.

7 An Application: Cyclic Coordination Games

In coordination games the players need to coordinate their strategies in order to choose among multiple pure Nash equilibria. Here we consider a natural set up according to which the players are arranged in a directed simple cycle and the payoff functions can yield three values: 0 if one chooses a 'noncommitting' strategy, 1 if one coordinates with the neighbour and -1 otherwise. We call such games ***cyclic coordination games***. They are special cases of strategic games introduced in [1] that are naturally associated with social networks built over arbitrary weighted directed graphs. We showed there that a similar game respects a local scheduler. We now show how using the concept of an f-potential we can strengthen this result.

More precisely, let $G_{coord} = (S_1, \ldots, S_n, p_1, \ldots, p_n)$ be a (possibly infinite) strategic game in which there is a special strategy $t_0 \in \bigcap_{i \in N} S_i$ common to all the players. For $i \in N$, let $i \oplus 1$ and $i \ominus 1$ denote the increment and decrement operations done in cyclic order within $\{1, \ldots, n\}$. That is, for $i \in \{1, \ldots, n-1\}$, $i \oplus 1 = i + 1$, $n \oplus 1 = 1$, for $i \in \{2, \ldots, n\}$, $i \ominus 1 = i - 1$, and $1 \ominus 1 = n$. The payoff functions are defined as,

$$p_i(s) := \begin{cases} 0 & \text{if } s_i = t_0, \\ 1 & \text{if } s_i = s_{i \ominus 1} \text{ and } s_i \neq t_0, \\ -1 & \text{otherwise.} \end{cases}$$

Theorem 5. *Each coordination game G_{coord} respects every local scheduler.*

Proof. For $n = 2$, it is easy to see that G_{coord} has the FIP and hence the result follows. Therefore, assume that $n > 2$. We prove the result by showing that for every local scheduler f, it is possible to associate an f-potential with the game G_{coord}.

Let f be a local scheduler. By Proposition 1, the choice function g associated with f is of the form $[\pi]$ for some permutation π of $1, \ldots, n$. Let $l = \pi(n)$ be the last element in the permutation π (this will be the only information about π that we shall rely upon). Let $U := \{-1, 0, 1\}^n$ and let $F : S \to U$ be a function defined by $F(s) := (p_l(s), p_{l\oplus 1}(s), p_{l\oplus 2}(s), \ldots, p_{l\oplus(n-1)}(s))$.

For $x \in U$ and $i \in \{1, \ldots, n\}$, x_i denotes the i-th entry in x and as before, $x_{-1} = (x_2, \ldots, x_n)$. We also use the notation $F(s)[i]$ to denote the i-th entry in the n-tuple $F(s)$.

Let $\prec_L$ be the strict counterpart of the lexicographic ordering over the $(n-1)$-tuples of $-1, 0, 1$, where $-1 \prec_L 0 \prec_L 1$. We extend $\prec_L$ to a relation $\lhd \subseteq U \times U$. For $x, y \in U$ such that $x \neq y$, $x \lhd y$ if one of the following mutually exclusive conditions holds:

C1 $x_1 \in \{-1, 1\}$ and $y_1 = 0$,
C2 $x_1 = y_1 = 0$ and $x_{-1} \prec_L y_{-1}$,
C3 $x_1, y_1 \in \{-1, 1\}$ and $x_{-1} \prec_L y_{-1}$,
C4 $x_1, y_1 \in \{-1, 1\}$, $x_{-1} = y_{-1}$ and $x_1 \prec_L y_1$.

In other words, if the first entry of y is 0 and that of x is not 0, then $x \lhd y$. If the first entry of both x and y is 0, then to order x and y we use the lexicographic ordering over the $(n-1)$-tuples x_{-1} and y_{-1}. If the first entry of both x and y is not 0, then again to order x and y we use the lexicographic ordering over x_{-1} and y_{-1}, the exception being when $x_{-1} = y_{-1}$. In this case, to determine the ordering we use the lexicographic ordering over x_1 and y_1.

Claim 3. *The relation $\lhd$ is a strict total ordering over U.*

Assuming Claim 3, consider an initial prefix $\xi_{k+1} = (s^1, \ldots, s^k, s^{k+1})$ of an improvement path ξ that respects f. We claim that $F(s^k) \lhd F(s^{k+1})$. We have the following cases:

- $f(s^k) = l \oplus i$ where $i \in \{1, \ldots, n-1\}$. Since ξ respects f, we have $p_{l\oplus i}(s^k) < p_{l\oplus i}(s^{k+1})$, so $F(s^k)[i+1] \prec_L F(s^{k+1})[i+1]$. Since $i \neq n$, if $i > 1$, then by the definition of the payoff functions, for all $j \in \{1, \ldots, i-1\}$, $p_{l\oplus j}(s^k) = p_{l\oplus j}(s^{k+1})$. If $i \neq n-1$, then $p_l(s^k) = p_l(s^{k+1})$ and it immediately follows that $F(s^k) \lhd F(s^{k+1})$. Therefore, the interesting case is when $i = n-1$. Here we show that the first entry in $F(s^{k+1})$ remains 0 after the update by player $n-1$ iff the first entry in $F(s^k)$ is 0.
    - If $F(s^k)[1] = 0$, then $F(s^{k+1})[1] = 0$.
      Indeed, suppose $F(s^k)[1] = 0$. Since $f(s^k) \neq l$, we have $s_l^k = s_l^{k+1}$. By the definition of the payoff function, for any joint strategy s, $p_l(s) = 0$ iff $s_l = t_0$. Thus irrespective of the choice of $l \oplus (n-1)$ we have $p_l(s^{k+1}) = 0$, so $F(s^{k+1})[1] = 0$.

   - If $F(s^k)[1] \neq 0$, then $F(s^{k+1})[1] \neq 0$.
     Suppose $F(s^k)[1] \neq 0$. By the definition of the payoff functions, $s^k_l \neq t_0$. Since $f(s^k) \neq l$, we have $s^k_l = s^{k+1}_l$. Therefore irrespective of the choice of $l \oplus (n-1)$ we have $p_l(s^{k-1}) \neq 0$, so $F(s^{k+1})[1] \neq 0$.

  Thus by conditions C2 and C3 in the definition of $\lhd$, and the fact that $(F(s^k))_{-1} \prec_L (F(s^{k+1}))_{-1}$, we indeed have $F(s^k) \lhd F(s^{k+1})$.

- $f(s^k) = l$. Since ξ respects f, for all $i \in \{1, \ldots, n-1\}$ we have $l \oplus i \in BR(s^k)$. We claim that in this case, $s^k_l \neq t_0$ and $s^k_{l\ominus 1} = t_0$. Suppose not. If $s^k_l = t_0$, then for all $i \in \{1, \ldots, n-1\}$, $l \oplus i \in BR(s^k)$ implies that $s^k_{l\oplus i} = t_0$. This in turn implies that $l \in BR(s^k)$, which is a contradiction. If $s^k_{l\ominus 1} \neq t_0$, then we have the following two possibilities:
   - $s^k_{l\ominus 1} = s^k_l$. This implies $l \in BR(s^k)$, which is a contradiction.
   - $s^k_{l\ominus 1} \neq s^k_l$. Then there exists $j \in \{1, \ldots, n-1\}$ such that $s^k_{l\oplus j} = s^k_{l\ominus 1}$ and $s^k_{l\oplus(j-1)} \neq s^k_{l\oplus j}$. Since $s^k_{l\oplus j} \neq t_0$, this implies that $l \oplus j \notin BR(s^k)$, which is a contradiction.

  Now, if $s^k_l \neq t_0$, $s^k_{l\ominus 1} = t_0$ and $p_l(s^k) < p_l(s^{k+1})$, then $s^{k+1}_l = t_0$. By C1 in the definition of $\lhd$, it then follows that $F(s^k) \lhd F(s^{k+1})$.

Finally, since the set U which is the range of the function F is finite and $\lhd$ is a strict total order on U, we can use an appropriate encoding $e : U \to \mathbb{R}$ such that $u_1 \lhd u_2$ iff $e(u_1) < e(u_2)$. Then $e(F(s^k)) < e(F(s^{k+1}))$. So $e \circ F$ is an f-potential. By the remark following Theorem 4 the result follows. □

Proof of Claim 3. Let $x, y \in U$ such that $x \neq y$. We have the following cases.

- $x_1 \in \{-1, 1\}$ and $y_1 = 0$. Then by C1, $x \lhd y$.
- $x_1 = 0$ and $y_1 = 0$. Then by C2, if $x_{-1} \prec_L y_{-1}$ then $x \lhd y$ else $y \lhd x$.
- $x_1 = 0$ and $y_1 \in \{-1, 1\}$. Then by C1, $y \lhd x$.
- $x_1, y_1 \in \{-1, 1\}$ and $x_{-1} \neq y_{-1}$. Then by C3, if $x_{-1} \prec_L y_{-1}$ then $x \lhd y$ else $y \lhd x$.
- $x_1, y_1 \in \{-1, 1\}$ and $x_{-1} = y_{-1}$. Then by C4, if $x_1 \prec_L y_1$ then $x \lhd y$ else $y \lhd x$.

Further, it can be verified that the relation $\lhd$ is transitive by a straightforward case analysis. □

Note that Theorem 5 cannot be extended to set-based schedulers. Indeed, suppose that $n > 2$, and for some $t \neq t_0$ we have $t \in \cap_{i\in N} S_i$. Consider the joint strategy $s := (t, t_0, \ldots, t_0)$ and a set-based scheduler f such that for all $k \in \{1, \ldots, n\}$, $f(\{k, k \oplus 1\}) := k \oplus 1$, $f(\{k, k \oplus 2\}) := k$, with arbitrary values for other inputs. Then the following infinite improvement path respects this scheduler. For the sake of readability we underlined the strategies that are not best responses.

$$(\underline{t}, \underline{t_0}, \ldots, t_0),\ (\underline{t}, t, \underline{t_0}, \ldots, t_0),\ (t_0, \underline{t}, \underline{t_0}, \ldots, t_0), \ldots$$

Finally, observe the following properties of the coordination games.

Theorem 6.

(i) The game G_{coord} has the FIP iff $n = 2$ or $\cap_{i \in N} S_i = \{t_0\}$.
(ii) In G_{coord}, starting from each joint strategy there exists an improvement path of length $\leq n$ and a BR-improvement path of length $\leq 2n - 2$.

Note that s is a Nash equilibrium in the game G_{coord} iff it is is of the form $(t, \ldots, t)$. So we can alternatively state item (i) as: The game G_{coord} has the FIP iff $n = 2$ or it has exactly one Nash equilibrium.

Proof. (i) $(\Rightarrow)$ As already mentioned when $n = 2$, G_{coord} has the FIP. If $n > 2$, then the above example implies that $\cap_{i \in N} S_i = \{t_0\}$.

$(\Leftarrow)$ Suppose that G_{coord} does not have the FIP. Consider an infinite improvement path ξ. Some player, say i, is selected in ξ infinitely often. This means that player i selects in ξ some strategy $t \neq t_0$ infinitely often. Indeed, otherwise from some moment on in each joint strategy in ξ his strategy would be t_0, which is not the case.

Each time player i switches in ξ to the strategy t, the strategy of his predecessor $i \ominus 1$ is necessarily t, as well. So also player $i \ominus 1$ switches in ξ to t infinitely often. Iterating this reasoning we conclude that each player selects in ξ the strategy t infinitely often. In particular $t \in \cap_{i \in N} S_i$.

(ii) Take a joint strategy s. Note that if all payoffs in s are ≥ 0, then s is a Nash equilibrium. Suppose that some payoff in s is < 0. Then repeatedly select the first player in the cyclic order whose payoff is negative and let him switch to t_0. After at most n steps the Nash equilibrium $(t_0, \ldots, t_0)$ is reached.

For the BR-improvement path we use the local scheduler f associated with the identity permutation, i.e., we repeatedly schedule the first player in the cyclic order who did not select a best response.

Consider a joint strategy s taken from a BR-improvement path. Observe that for all k if $s_k \neq t_0$ and $p_k(s) \geq 0$ (so in particular if s_k is a best response to s_{-k}), then $s_k = s_{k \ominus 1}$. So for all $i > 1$, the following property holds:

$Z(i)$: if $f(s) = i$ and $s_{i-1} \neq t_0$ then for all $j \in \{n, 1, \ldots, i-1\}$, $s_j = s_{i-1}$.

In words: if i is the first player who did not select a best response and player $i-1$ strategy is not t_0, then this is a strategy of every earlier player and of player n.

Along each BR-improvement path that respects f the value of $f(s)$ strictly increases until the path terminates or at certain stage $f(s) = n$. Note that then $s_{n-1} = t_0$ since otherwise on the account of property $Z(n)$ all players' strategies are equal, so s is a Nash equilibrium and hence $f(s)$ is undefined. So the unique best response for player n is t_0. This switch begins a new round with player 1 as the next scheduled player. Player 1 also switches to t_0 and from now on every consecutive player switches to t_0, as well. The resulting path terminates once player $n - 2$ switches to t_0. Consequently the length of the generated BR-improvement path is $\leq 2n - 2$. □

We can naturally extend the notion of a scheduler to one that chooses a non-empty set of players. These players then simultaneously select a better (respectively, best) response. Such a set-valued scheduler models a controlled concurrent better (respectively, best) response dynamics. We have then the following extension of Theorem 5. For every player l the game G_{coord} respects every set-valued scheduler $\bar{g}$ such that for every $A \subseteq N$ such that $|A| > 1$, $\bar{g}(A) \subseteq A \backslash \{l\}$. One can check that if (s, s') is a step of a 'concurrent' improvement path that respects $\bar{g}$, then $F(s) \lhd F(s')$. We plan to study set-valued schedulers in another paper.

Acknowledgement. We thank Ruben Brokkelkamp and Mees de Vries for helpful discussions.

References

1. Simon, S., Apt, K.R.: Choosing products in social networks. Manuscript, CWI, Amsterdam, The Netherlands, Computing Research Repository, CoRR (2012), `http://arxiv.org/abs/1202.2209`
2. Monderer, D., Shapley, L.S.: Potential games. Games and Economic Behaviour 14, 124–143 (1996)
3. Milchtaich, I.: Congestion games with player-specific payoff functions. Games and Economic Behaviour 13, 111–124 (1996)
4. Young, H.P.: The evolution of conventions. Econometrica 61(1), 57–84 (1993)
5. Engelberg, R., Schapira, M.: Weakly-Acyclic (Internet) Routing Games. In: Persiano, G. (ed.) SAGT 2011. LNCS, vol. 6982, pp. 290–301. Springer, Heidelberg (2011)
6. Marden, J., Arslan, G., Shamma, J.: Regret based dynamics: convergence in weakly acyclic games. In: Proceedings of the Sixth International Joint Conference on Autonomous Agents and Multiagent Systems (AAMAS 2007), pp. 194–201. IFAAMAS (2007)
7. Fabrikant, A., Jaggard, A.D., Schapira, M.: On the Structure of Weakly Acyclic Games. In: Kontogiannis, S., Koutsoupias, E., Spirakis, P.G. (eds.) SAGT 2010. LNCS, vol. 6386, pp. 126–137. Springer, Heidelberg (2010)
8. Brokkelkamp, K.R., de Vries, M.J.: Convergence of Ordered Improvement Paths in Generalized Congestion Games. In: Serna, M. (ed.) SAGT 2012. LNCS, vol. 7615, pp. 61–71. Springer, Heidelberg (2012)
9. Grädel, E.: Back and forth between logic and games. In: Apt, K.R., Grädel, E. (eds.) Lectures in Game Theory for Computer Scientists, pp. 99–145. Cambridge University Press (2011)
10. König, D.: Über eine Schlußweise aus dem Endlichen ins Unendliche. Acta Litt. Ac. Sci. 3, 121–130 (1927)

Selfishness Level of Strategic Games*

Krzysztof R. Apt[1,2] and Guido Schäfer[1,3]

[1] Centrum Wiskunde & Informatica (CWI), Amsterdam, The Netherlands
[2] University of Amsterdam, The Netherlands
[3] VU University Amsterdam, The Netherlands

Abstract. We introduce a new measure of the discrepancy in strategic games between the social welfare in a Nash equilibrium and in a social optimum, that we call ***selfishness level***. It is the smallest fraction of the social welfare that needs to be offered to each player to achieve that a social optimum is realized in a pure Nash equilibrium. The selfishness level is unrelated to the price of stability and the price of anarchy and in contrast to these notions is invariant under positive linear transformations of the payoff functions. Also, it naturally applies to other solution concepts and other forms of games.

We study the selfishness level of several well-known strategic games. This allows us to quantify the implicit tension within a game between players' individual interests and the impact of their decisions on the society as a whole. Our analysis reveals that the selfishness level often provides more refined insights into the game than other measures of inefficiency, such as the price of stability or the price of anarchy.

1 Introduction

The discrepancy in strategic games between the social welfare in a Nash equilibrium and in a social optimum has been long recognized by the economists. One of the flagship examples is Cournot competition, a strategic game involving firms that simultaneously choose the production levels of a homogeneous product. The payoff functions in this game describe the firms' profit in the presence of some production costs, under the assumption that the price of the product depends negatively on the total output. It is well-known, see, e.g., [1, pages 174–175], that the price in the social optimum is strictly higher than in the Nash equilibrium, which shows that the competition between the producers of a product drives its price down.

In computer science the above discrepancy led to the introduction of the notions of the *price of anarchy* [2] and the *price of stability* [3] that measure the ratio between the social welfare in a worst and, respectively, a best Nash equilibrium and a social optimum. This originated a huge research effort aiming at determining both ratios for specific strategic games that possess (pure) Nash equilibria.

* A full version with all proofs is available at the authors' homepages.

M. Serna (Ed.): SAGT 2012, LNCS 7615, pp. 13–24, 2012.

These two notions are *descriptive* in the sense that they refer to an existing situation. In contrast, we propose a notion that measures the discrepancy between the social welfare in a Nash equilibrium and a social optimum, which is *normative*, in the sense that it refers to a modified situation. On an abstract level, the approach that we propose here is discussed in [4], in chapter "How to Promote Cooperation", from where we cite (see page 134): "An excellent way to promote cooperation in a society is to teach people to care about the welfare of others."

Our approach draws on the concept of *altruistic games* (see, e.g., [5] and more recent [6]). In these games each player's payoff is modified by adding a positive fraction α of the social welfare in the considered joint strategy to the original payoff. The ***selfishness level*** of a game is defined as the infimum over all $\alpha \geq 0$ for which such a modification yields that a social optimum is realized in a pure Nash equilibrium.

Intuitively, the selfishness level of a game can be viewed as a measure of the players' willingness to cooperate. A low selfishness level indicates that the players are open to align their interests in the sense that a small share of the social welfare is sufficient to motivate them to choose a social optimum. In contrast, a high selfishness level suggests that the players are reluctant to cooperate and a large share of the social welfare is needed to stimulate cooperation among them. An infinite selfishness level means that cooperation cannot be achieved through such means.

Often the selfishness level of a strategic game provides better insights into the game under consideration than other measures of inefficiency, such as the price of stability or the price of anarchy. To illustrate this point, we elaborate on our findings for the public goods game with n players. In this game, every player i chooses an amount $s_i \in [0, b]$ that he wants to contribute to a public good. A central authority collects all individual contributions, multiplies their sum by $c > 1$ (here we assume for simplicity that $n \geq c$) and distributes the resulting amount evenly among all players. The payoff of player i is thus $p_i(s) := b - s_i + \frac{c}{n} \sum_j s_j$.

In the (unique) Nash equilibrium, every player attempts to "free ride" by contributing 0 to the public good (which is a dominant strategy), while in the social optimum every player contributes the full amount of b. As we will show, the selfishness level of this game is $(1 - \frac{c}{n})/(c - 1)$. This bound suggests that the temptation to free ride (i) increases as the number of players grows and (ii) decreases as the parameter c increases. Both phenomena were observed by experimental economists, see, e.g., [5, Section III.C.2]. In contrast, the price of stability (which coincides with the price of anarchy) for this game is c, which is rather uninformative.

In this paper, we define the selfishness level by taking pure Nash equilibrium as the solution concept. This is in line with how the price of anarchy and price of stability were defined originally [2, 3]. However, the definition applies equally well to other solution concepts and other forms of games.

Our Contributions. In this paper, we study the selfishness level of some selected classical and fundamental strategic games. These games are often used to illustrate the consequences of selfish behaviour and the effects of competition. To this aim, we first derive a characterization result that allows us to determine the selfishness level of a strategic game. Our characterization shows that the selfishness level is determined by the maximum *appeal factor* of unilateral profitable deviations from specific social optima, which we call *stable*. Intuitively, the appeal factor of a single player deviation refers to the ratio of the gain in his payoff over the resulting loss in social welfare.

We show that the selfishness level of a finite game can be an arbitrary real number that is unrelated to the price of stability. A nice property of our selfishness level notion is that, unlike the price of stability and the price of anarchy, it is invariant under positive linear transformations of the payoff functions.

We then use the above characterization result to analyze the selfishness level of several strategic games. In particular, we show that the selfishness level of finite ordinal potential games is finite. We also derive explicit bounds on the selfishness level of fair cost sharing games and congestion games with linear delay functions. These bounds depend on the specific parameters of the underlying game, but are independent of the number of players. Moreover, our bounds are tight.

Further, we show that the selfishness level of the Prisoner's Dilemma with n players is $1/(2n-3)$ and that of the public goods game with n players is $\max\{0, (1-\frac{c}{n})(c-1)\}$. Finally, the selfishness level of Cournot competition (an example of an infinite ordinal potential game), Tragedy of the Commons, and Bertrand competition turns out to be infinite.

Related Work. There are only few articles in the algorithmic game theory literature that study the influence of altruism in strategic games [7–11]. In these works, altruistic player behavior is modeled by altering each player's perceived payoff in order to account also for the welfare of others. The models differ in the way they combine the player's individual payoff with the payoffs of the other players. All these studies are descriptive in the sense that they aim at understanding the impact of altruistic behavior on specific strategic games.

Closest to our work are the articles [10] and [8]. Elias et al. [10] study the inefficiency of equilibria in network design games (which constitute a special case of the cost sharing games considered here) with altruistic (or, as they call it, socially-aware) players. As we do here, they define each player's cost function as his individual cost plus α times the social cost. They derive lower and upper bounds on the price of anarchy and the price of stability, respectively, of the modified game. In particular, they show that the price of stability is at most $(H_n + \alpha)/(1 + \alpha)$, where n is the number of players.

Chen et al. [8] introduce a framework to study the *robust price of anarchy*, which refers to the worst-case inefficiency of other solution concepts such as coarse correlated equilibria (see [12]) of altruistic extensions of strategic games. In their model, player i's perceived cost is a convex combination of $(1-\bar{\alpha}_i)$ times his individual cost plus $\bar{\alpha}_i$ times the social cost, where $\bar{\alpha}_i \in [0, 1]$ is the altruism level of i. If all players have a uniform altruism level $\bar{\alpha}_i = \bar{\alpha}$, this model relates to the one we consider here by setting $\alpha = \bar{\alpha}/(1-\bar{\alpha})$ for $\bar{\alpha} \in [0,1)$. Although not being the main focus of the paper, the authors also provide upper bounds of $2/(1+\bar{\alpha})$ and $(1-\bar{\alpha})H_n + \bar{\alpha}$ on the price of stability for linear congestion games and fair cost sharing games, respectively.

Note that in all three cases the price of stability approaches 1 as α goes to ∞. This seems to suggest that the selfishness level of these games is ∞. However, this is not the case as outlined above.

Other models of altruism were proposed in [7, 9]. Chen and Kempe [9] define the perceived cost of a player as $(1-\beta)$ times his individual cost plus β/n times the social cost, where $\beta \in [0,1]$. Caragiannis et al. [7] define the perceived cost of player i as $(1-\xi)$ times his individual cost plus ξ times the sum of the costs of all other players (i.e., excluding player i), where $\xi \in [0,1]$. Both models are equivalent to the model the we consider here by using the transformations $\alpha = \beta/((1-\beta)n)$ for $\beta \in [0,1)$ and $\alpha = \xi/(1-2\xi)$ for $\xi \in [0, \frac{1}{2})$.

In network congestion games, researchers studied the effect of imposing tolls on the edges of the network in order to reduce the inefficiency of Nash equilibria; see, e.g., [13]. From a high-level perspective, these approaches can also be regarded as being normative. Conceptually, our selfishness level notion is related to the *Stackelberg threshold* introduced by Sharma and Williamson [14]. The authors consider network routing games in which a fraction $\beta \in [0,1]$ of the flow is first routed centrally and the remaining flow is then routed selfishly. The Stackelberg threshold refers to the smallest value β that is needed to improve upon the social cost of a Nash equilibrium flow.

2 Selfishness Level

A ***strategic game*** (in short, a game) $G = (N, \{S_i\}_{i \in N}, \{p_i\}_{i \in N})$ is given by a set $N = \{1, \ldots, n\}$ of $n > 1$ players, a non-empty set of ***strategies*** S_i for every player $i \in N$, and a ***payoff function*** p_i for every player $i \in N$ with $p_i : S_1 \times \cdots \times S_n \to \mathbb{R}$. The players choose their strategies simultaneously and every player $i \in N$ aims at choosing a strategy $s_i \in S_i$ so as to maximize his individual payoff $p_i(s)$, where $s = (s_1, \ldots, s_n)$.

We call $s \in S_1 \times \cdots \times S_n$ a ***joint strategy***, denote its ith element by s_i, denote $(s_1, \ldots, s_{i-1}, s_{i+1}, \ldots, s_n)$ by s_{-i} and similarly with S_{-i}. Further, we write (s'_i, s_{-i}) for $(s_1, \ldots, s_{i-1}, s'_i, s_{i+1}, \ldots, s_n)$, where we assume that $s'_i \in S_i$. Sometimes, when focussing on player i we write (s_i, s_{-i}) instead of s.

A joint strategy s a ***Nash equilibrium*** if for all $i \in \{1, \ldots, n\}$ and $s'_i \in S_i$, $p_i(s_i, s_{-i}) \geq p_i(s'_i, s_{-i})$. Further, given a joint strategy s we call the sum $SW(s) := \sum_{i=1}^{n} p_i(s)$ the ***social welfare*** of s. When the social welfare of s is maximal we call s a ***social optimum***.

Given a strategic game $G := (N, \{S_i\}_{i \in N}, \{p_i\}_{i \in N})$ and $\alpha \geq 0$ we define the game $G(\alpha) := (N, \{S_i\}_{i \in N}, \{r_i\}_{i \in N})$ by putting $r_i(s) := p_i(s) + \alpha SW(s)$. So when $\alpha > 0$ the payoff of each player in the $G(\alpha)$ game depends on the social welfare of the players. $G(\alpha)$ is then an altruistic version of the game G.

Suppose now that for some $\alpha \geq 0$ a pure Nash equilibrium of $G(\alpha)$ is a social optimum of $G(\alpha)$. Then we say that G is ***α-selfish***. We define the ***selfishness level*** of G as

$$\inf\{\alpha \in \mathbb{R}_+ \mid G \text{ is } \alpha\text{-selfish}\}. \tag{1}$$

Here we adopt the convention that the infimum of an empty set is ∞. Further, we stipulate that the selfishness level of G is denoted by α^+ iff the selfishness level of G is $\alpha \in \mathbb{R}_+$ but G is *not* α-selfish (equivalently, the infimum does not belong to the set). We show below (Theorem 2) that pathological infinite games exist for which the selfishness level is of this kind; none of the other studied games is of this type.

The above definitions refer to strategic games in which each player i maximizes his payoff function p_i and the social welfare of a joint strategy s is given by $SW(s)$. These definitions obviously apply to strategic games in which every player i minimizes his cost function c_i and the social cost of a joint strategy s is defined as $SC(s) := \sum_{i=1}^{n} c_i(s)$. The definition also extends in the obvious way to other solution concepts (e.g., mixed or correlated equilibria) and other forms of games (e.g., subgame perfect equilibria in extensive games).

Note that the social welfare of a joint strategy s in $G(\alpha)$ equals $(1+\alpha n)SW(s)$, so the social optima of G and $G(\alpha)$ coincide. Hence we can replace in the above definition the reference to a social optimum of $G(\alpha)$ by one to a social optimum of G.

Intuitively, a low selfishness level means that the share of the social welfare needed to induce the players to choose a social optimum is small. This share can be viewed as an 'incentive' needed to realize a social optimum. Let us illustrate this definition on three simple examples.

Example 1. **Prisoner's Dilemma**

	C	D
C	2, 2	0, 3
D	3, 0	1, 1

	C	D
C	6, 6	3, 6
D	6, 3	3, 3

Consider the Prisoner's Dilemma game G (on the left) and the resulting game $G(\alpha)$ for $\alpha = 1$ (on the right). In the latter game the social optimum, (C, C), is also a Nash equilibrium. One can easily check that for $\alpha < 1$, (C, C) is also a social optimum of $G(\alpha)$ but not a Nash equilibrium. So the selfishness level of this game is 1.

Example 2. **Battle of the Sexes**

	F	B
F	2, 1	0, 0
B	0, 0	1, 2

Here each Nash equilibrium is also a social optimum, so the selfishness level of this game is 0.

Example 3. **Game with a bad Nash equilibrium**
The following game results from equipping each player in the Matching Pennies game with a third strategy E (for edge):

	H	T	E
H	$1,-1$	$-1,\ 1$	$-1,-1$
T	$-1,\ 1$	$1,-1$	$-1,-1$
E	$-1,-1$	$-1,-1$	$-1,-1$

Its unique Nash equilibrium is (E, E). It is easy to check that the selfishness level of this game is ∞.

Recall that, given a finite game G that has a Nash equilibrium, its ***price of stability*** is the ratio $SW(s)/SW(s')$ where s is a social optimum and s' is a Nash equilibrium with the highest social welfare in G. So the price of stability of G is 1 iff its selfishness level is 0. However, in general there is no relation between these two notions.

Theorem 1. *For every finite $\alpha > 0$ and $\beta > 1$ there is a finite game whose selfishness level is α and whose price of stability is β.*

Further, in contrast to the price of stability (and to the ***price of anarchy***, defined as the ratio $SW(s)/SW(s')$ where s is a social optimum and s' is a Nash equilibrium with the lowest social welfare in G) the notion of the selfishness level is invariant under simple uniform payoff transformations. Given a game G and a value a we denote by $G + a$ (respectively, aG) the game obtained from G by adding to each payoff function the value a (respectively, by multiplying each payoff function by a).

Proposition 1. *Consider a game G and $\alpha \geq 0$.*

1. *For every a, G is α-selfish iff $G + a$ is α-selfish,*
2. *For every $a > 0$, G is α-selfish iff aG is α-selfish.*

This result allows us to better frame the notion of selfishness level. Namely, suppose that the original n-players game G was set up by a designer who has a fixed budget $SW(s)$ for each joint strategy s and that the selfishness level of G is $\alpha < \infty$. Then we should scale $G(\alpha)$ by the factor $a := 1/(1 + \alpha n)$ so that for each joint strategy s its social welfare in the original game G and $aG(\alpha)$ is the same.

By the above proposition, α is the smallest non-negative real such that $aG(\alpha)$ has a Nash equilibrium that is a social optimum. The game $aG(\alpha)$ can then be viewed as the intended transformation of G. That is, each payoff function p_i of the game G is transformed into the payoff function

$$r_i(s) := \frac{1}{1+\alpha n} p_i(s) + \frac{\alpha}{1+\alpha n} SW(s).$$

Note that the selfishness level is not invariant under a multiplication of the payoff functions by a value $a \leq 0$. Indeed, for $a = 0$ each game aG has the selfishness level 0. For $a < 0$ take the game G from Example 3 whose selfishness level is ∞. In the game aG the joint strategy (E, E) is both a Nash equilibrium and a social optimum, so the selfishness level of aG is 0.

Theorem 2. *There exists a game whose selfishness level is 0^+, i.e., it is α-selfish for every $\alpha > 0$, but it is not 0-selfish.*

3 A Characterization Result

We now characterize the games with a finite selfishness level. To this end we shall need the following notion. We call a social optimum s ***stable*** if for all $i \in N$ and $s'_i \in S_i$ the following holds: if (s'_i, s_{-i}) is a social optimum, then $p_i(s_i, s_{-i}) \geq p_i(s'_i, s_{-i})$. In other words, a social optimum is stable if no player is better off by unilaterally deviating to another social optimum.

It will turn out that to determine the selfishness level of a game we need to consider deviations from its stable social optima. Consider a deviation s'_i of player i from a social stable optimum s. If player i is better off by deviating to s'_i, then by definition the social welfare decreases, i.e., $SW(s_i, s_{-i}) - SW(s'_i, s_{-i}) > 0$. If this decrease is small, while the gain for player i is large, then strategy s'_i is an attractive and socially acceptable option for player i. We define player i's ***appeal factor*** of strategy s'_i given the social optimum s as

$$AF_i(s'_i, s) := \frac{p_i(s'_i, s_{-i}) - p_i(s_i, s_{-i})}{SW(s_i, s_{-i}) - SW(s'_i, s_{-i})}.$$

In what follows we shall characterize the selfishness level in terms of bounds on the appeal factors of profitable deviations from a stable social optimum.

Theorem 3. *Consider a strategic game $G := (N, \{S_i\}_{i \in N}, \{p_i\}_{i \in N})$.*

1. *The selfishness level of G is finite iff a stable social optimum s exists for which $\alpha(s) := \max_{i \in N, s'_i \in U_i(s)} AF_i(s'_i, s)$ is finite, where $U_i(s) := \{s'_i \in S_i \mid p_i(s'_i, s_{-i}) > p_i(s_i, s_{-i})\}$.*
2. *If the selfishness level of G is finite, then it equals $\min_{s \in SSO} \alpha(s)$, where SSO is the set of stable social optima.*
3. *If G is finite, then its selfishness level is finite iff it has a stable social optimum. In particular, if G has a unique social optimum, then its selfishness level is finite.*
4. *If $\beta > \alpha \geq 0$ and G is α-selfish, then G is β-selfish.*

4 Examples

We now use the above characterization result to determine or compute an upper bound on the selfishness level of some selected games. First, we exhibit a well-known class of games (see [15]) for which the selfishness level is finite.

Potential Games. Given a game $G := (N, \{S_i\}_{i\in N}, \{p_i\}_{i\in N})$, a function $P : S_1\times\cdots\times S_n \to \mathbb{R}$ is called an ***ordinal potential function*** for G if for all $i \in N$, $s_{-i} \in S_{-i}$ and $s_i, s_i' \in S_i$, $p_i(s_i, s_{-i}) > p_i(s_i', s_{-i})$ iff $P(s_i, s_{-i}) > P(s_i', s_{-i})$. A game that possesses an ordinal potential function is called an ***ordinal potential game***.

Theorem 4. *Every finite ordinal potential game has a finite selfishness level.*

In particular, every finite congestion game (see [16]) has a finite selfishness level. We derive below explicit bounds for two special cases of these games.

Fair Cost Sharing Games. In a fair cost sharing game, see, e.g., [17], players allocate facilities and share the cost of the used facilities in a fair manner. Formally, a fair cost sharing game is given by $G = (N, E, \{S_i\}_{i\in N}, \{c_e\}_{e\in E})$, where $N = \{1, \ldots, n\}$ is the set of players, E is the set of facilities, $S_i \subseteq 2^E$ is the set of facility subsets available to player i, and $c_e \in \mathbb{R}_+$ is the cost of facility $e \in E$. It is called a *singleton* cost sharing game if for every $i \in N$ and for every $s_i \in S_i$: $|s_i| = 1$. For a joint strategy $s \in S_1 \times \cdots \times S_n$ let $x_e(s)$ be the number of players using facility $e \in E$, i.e., $x_e(s) = |\{i \in N \mid e \in s_i\}|$. The cost of a facility $e \in E$ is evenly shared among the players using it. That is, the cost of player i is defined as $c_i(s) = \sum_{e\in s_i} c_e/x_e(s)$. The social cost function is given by $SC(s) = \sum_{i\in N} c_i(s)$.

We first consider singleton cost sharing games. Let $c_{\max} = \max_{e\in E} c_e$ and $c_{\min} = \min_{e\in E} c_e$ refer to the maximum and minimum costs of the facilities, respectively.

Proposition 2. *The selfishness level of a singleton cost sharing game is at most* $\max\{0, \frac{1}{2}c_{\max}/c_{\min} - 1\}$. *Moreover, this bound is tight.*

This result should be contrasted with the price of stability of H_n and the price of anarchy of n for cost sharing games [17]. Cost sharing games admit an exact potential function and thus by Theorem 4 their selfishness level is finite. However, one can show that the selfishness level can be arbitrarily large (as $c_{\max}/c_{\min} \to \infty$) even for $n = 2$ and two facilities.

We next derive a bound for arbitrary fair cost sharing games with non-negative integer costs. Let L be the maximum number of facilities that any player can choose, i.e., $L := \max_{i\in N, s_i\in S_i} |s_i|$.

Proposition 3. *The selfishness level of a fair cost sharing game with non-negative integer costs is at most* $\max\{0, \frac{1}{2}Lc_{\max} - 1\}$. *Moreover, this bound is tight.*

Remark 1. We can bound the selfishness level of a fair cost sharing game with non-negative rational costs $c_e \in \mathbb{Q}_+$ for every facility $e \in E$ by using Proposition 3 and the following scaling argument: Simply scale all costs to integers, e.g., by multiplying them with the least common multiplier $q \in \mathbb{N}$ of the denominators. Note that this scaling does not change the selfishness level of the game by Proposition 1. However, it does change the maximum facility cost and thus q enters the bound.

Linear Congestion Games. In a congestion game $G := (N, E, \{S_i\}_{i\in N}, \{d_e\}_{e\in E})$ we are given a set of players $N = \{1, \dots, n\}$, a set of facilities E with a delay function $d_e : \mathbb{N} \to \mathbb{R}_+$ for every facility $e \in E$, and a strategy set $S_i \subseteq 2^E$ for every player $i \in N$. For a joint strategy $s \in S_1 \times \cdots \times S_n$, define $x_e(s)$ as the number of players using facility $e \in E$, i.e., $x_e(s) = |\{i \in N \mid e \in s_i\}|$. The goal of a player is to minimize his individual cost $c_i(s) = \sum_{e\in s_i} d_e(x_e(s))$. The social cost function is given by $SC(s) = \sum_{i\in N} c_i(s)$. Here we call a congestion game *symmetric* if there is some common strategy set $S \subseteq 2^E$ such that $S_i = S$ for all i. It is *singleton* if every strategy $s_i \in S_i$ is a singleton set, i.e., for every $i \in N$ and for every $s_i \in S_i$, $|s_i| = 1$. In a *linear* congestion game, the delay function of every facility $e \in E$ is of the form $d_e(x) = a_e x + b_e$, where $a_e, b_e \in \mathbb{R}_+$ are non-negative real numbers.

We first derive a bound on the selfishness level for symmetric singleton linear congestion games. As it turns out, a bound similar to the one for singleton cost sharing games does not extend to symmetric singleton linear congestion games. Instead, the crucial insight here is that the selfishness level depends on the *discrepancy* between any two facilities in a stable social optimum. We make this notion more precise.

Let s be a stable social optimum and let x_e refer to $x_e(s)$. Define the *discrepancy* between two facilities e and e' under s as

$$\lambda(x_e, x_{e'}) = \frac{2a_e x_e + b_e}{a_e + a_{e'}} - \frac{2a_{e'} x_{e'} + b_{e'}}{a_e + a_{e'}}. \tag{2}$$

It can be shown that $\lambda(x_e, x_e') \in (-1, 1)$. Let $\lambda_{\max}(s)$ be the maximum discrepancy between any two facilities under s. Further, let $\lambda_{\max}$ be the maximum discrepancy over all stable social optima, i.e., $\lambda_{\max} = \max_{s\in SSO} \lambda_{\max}(s)$.

Let $\Delta_{\max} := \max_{e\in E}(a_e + b_e)$ and $\Delta_{\min} := \min_{e\in E}(a_e + b_e)$. Further, let $a_{\min}$ be the minimum non-zero coefficient of a latency function, i.e., $a_{\min} = \min_{e\in E: a_e>0} a_e$.

Proposition 4. *The selfishness level of a symmetric singleton linear congestion game is at most* $\max\{0, \frac{1}{2}(\Delta_{\max} - \Delta_{\min})/((1 - \lambda_{\max})a_{\min}) - \frac{1}{2}\}$. *Moreover, this bound is tight.*

Observe that the selfishness level depends on the ratio $(\Delta_{\max} - \Delta_{\min})/a_{\min}$ and $1/(1 - \lambda_{\max})$. In particular, the selfishness level becomes arbitrarily large as $\lambda_{\max}$ approaches 1.

We next state a bound for the selfishness level of arbitrary congestion games with linear delay functions and non-negative integer coefficients, i.e., $d_e(x) = a_e x + b_e$ with $a_e, b_e \in \mathbb{N}$ for every $e \in E$. Let L be the maximum number of facilities that any player can choose, i.e., $L := \max_{i\in N, s_i\in S_i} |s_i|$.

Proposition 5. *The selfishness level of a linear congestion game with non-negative integer coefficients is at most* $\max\{0, \frac{1}{2}(L\Delta_{\max} - \Delta_{\min} - 1)\}$. *Moreover, this bound is tight.*

For linear congestion games, the price of anarchy is known to be $\frac{5}{2}$, see [18, 19]. In contrast, our bound shows that the selfishness level depends on the maximum number of facilities in a strategy set and the magnitude of the coefficients of the delay functions.

Remark 2. We can use Proposition 5 and the scaling argument outlined in Remark 1 to derive bounds on the selfishness level of congestion games with linear delay functions and non-negative rational coefficients.

Prisoner's Dilemma for n Players. We assume that each player $i \in N = \{1, \ldots, n\}$ has two strategies, 1 (cooperate) and 0 (defect). We put $p_i(s) := 1 - s_i + 2\sum_{j \neq i} s_j$.

Proposition 6. *The selfishness level of the n-players Prisoner's Dilemma game is $\frac{1}{2n-3}$.*

Intuitively, this means that when the number of players in the Prisoner's Dilemma game increases, a smaller share of the social welfare is needed to resolve the underlying conflict. That is, its 'acuteness' diminishes with the number of players. The formal reason is that the appeal factor of each unilateral deviation from the social optimum is inversely proportional to the number of players.

In particular, for $n = 2$ we get, as already argued in Example 1, that the selfishness level of the original Prisoner's Dilemma game is 1.

Public Goods. We consider the public goods game with n players. Every player $i \in N = \{1, \ldots, n\}$ chooses an amount $s_i \in [0, b]$ that he contributes to a public good, where $b \in \mathbb{R}_+$ is the budget. The game designer collects the individual contributions of all players, multiplies their sum by $c > 1$ and distributes the resulting amount evenly among all players. The payoff of player i is thus $p_i(s) := b - s_i + \frac{c}{n}\sum_{j \in N} s_j$.

Proposition 7. *The selfishness level of the n-players public goods game is* $\max\{0, \frac{1-\frac{c}{n}}{c-1}\}$.

In this game, every player has an incentive to "free ride" by contributing 0 to the public good (which is a dominant strategy). The above proposition reveals that for fixed c, in contrast to the Prisoner's Dilemma game, this temptation becomes stronger as the number of players increases. Also, for a fixed number of players this temptation becomes weaker as c increases.

Cournot Competition. We consider Cournot competition for n firms with a linear inverse demand function and constant returns to scale, see, e.g., [1, pages 174–175]. So we assume that each player $i \in N = \{1, \ldots, n\}$ has a strategy set $S_i = \mathbb{R}_+$ and payoff function $p_i(s) := s_i(a - b\sum_{j \in N} s_j) - cs_i$ for some given a, b, c, where $a > c \geq 0$ and $b > 0$.

The price of the product is represented by the expression $a - b\sum_{j \in N} s_j$ and the production cost corresponding to the production level s_i by cs_i. In what follows we rewrite the payoff function as $p_i(s) := s_i(d - b\sum_{j \in N} s_j)$, where $d := a - c$.

Proposition 8. *The selfishness level of the n-players Cournot competition game is ∞.*

Intuitively, this result means that in this game no matter how much we 'involve' the players in sharing the social welfare we cannot achieve that they will select a social optimum.

Tragedy of the Commons. Assume that each player $i \in N = \{1, \ldots, n\}$ has the real interval $[0,1]$ as its set of strategies. Each player's strategy is his chosen fraction of a common resource. Let (see [20, Exercise 63.1]): $p_i(s) := \max\{0, s_i(1 - \sum_{j \in N} s_j)\}$. This payoff function reflects the fact that player's enjoyment of the common resource depends positively from his chosen fraction of the resource and negatively from the total fraction of the common resource used by all players. Additionally, if the total fraction of the common resource by all players exceeds a feasible level, here 1, then player's enjoyment of the resource becomes zero.

Proposition 9. *The selfishness level of the n-players Tragedy of the Commons game is ∞.*

Bertrand Competition. Next, we consider Bertrand competition, a game concerned with a simultaneous selection of prices for the same product by two firms, see, e.g., [1, pages 175–177]. The product is then sold by the firm that chose a lower price. In the case of a tie the product is sold by both firms and the profits are split. We assume that each firm has identical marginal costs $c > 0$ and no fixed cost, and that each strategy set S_i equals $[c, \frac{a}{b})$, where $c < \frac{a}{b}$. The payoff function for player $i \in \{1, 2\}$ is given by

$$p_i(s_i, s_{3-i}) := \begin{cases} (s_i - c)(a - bs_i) & \text{if } c < s_i < s_{3-i} \\ \frac{1}{2}(s_i - c)(a - bs_i) & \text{if } c < s_i = s_{3-i} \\ 0 & \text{otherwise.} \end{cases}$$

Proposition 10. *The selfishness level of the Bertrand competition game is ∞.*

5 Concluding Remarks and Extensions

We introduced the selfishness level of a game as a new measure of discrepancy between the social welfare in a Nash equilibrium and in a social optimum. Our studies reveal that the selfishness level often provides more refined insights than other measures of inefficiency.

The definition of the selfishness level naturally extends to other solution concepts and other forms of games. For example, for mixed Nash equilibria we simply adapt our definitions by stipulating that a strategic game G is α-selfish if the social welfare of a mixed Nash equilibrium of $G(\alpha)$ is equal to the optimum social welfare of $G(\alpha)$. The selfishness level of G is then defined as before in (1). For example, with this notion the selfishness level of the Matching Pennies game is 0.

We can also consider subgame perfect equilibria and extensive games. We leave for future work the study of such alternatives.

Acknowledgements. We acknowledge initial discussions with Po-An Chen and thank anonymous reviewers for their valuable comments.

References

1. Jehle, G., Reny, P.: Advanced Microeconomic Theory, 3rd edn. Addison Wesley, New York (2011)
2. Koutsoupias, E., Papadimitriou, C.H.: Worst case equilibria. In: Annual IEEE Symposium on Theoretical Aspects of Computer Science, pp. 404–413 (1999)
3. Schulz, A.S., Moses, N.E.S.: On the performance of user equilibria in traffic networks. In: SODA, pp. 86–87 (2003)
4. Axelrod, R.: The Evolution of Cooperation. Basic Books (1984)
5. Ledyard, J.O.: 2. The Handbook of Experimental Economics. In: Public Goods: A Survey of Experimental Research, pp. 111–194. Princeton University Press (1995)
6. Marco, G.D., Morgan, J.: Slightly altruistic equilibria in normal form games. Working paper No 185, Center for Studies in Economics and Finance, University of Salerno, Italy (2007), `http://www.csef.it/WP/wp185.pdf`
7. Caragiannis, I., Kaklamanis, C., Kanellopoulos, P., Kyropoulou, M., Papaioannou, E.: The Impact of Altruism on the Efficiency of Atomic Congestion Games. In: Wirsing, M., Hofmann, M., Rauschmayer, A. (eds.) TGC 2010, LNCS, vol. 6084, pp. 172–188. Springer, Heidelberg (2010)
8. Chen, P.-A., de Keijzer, B., Kempe, D., Schäfer, G.: The Robust Price of Anarchy of Altruistic Games. In: Chen, N., Elkind, E., Koutsoupias, E. (eds.) WINE 2011. LNCS, vol. 7090, pp. 383–390. Springer, Heidelberg (2011)
9. Chen, P.A., Kempe, D.: Altruism, selfishness, and spite in traffic routing. In: Proc. 10th ACM Conference on Electronic Commerce, pp. 140–149 (2008)
10. Elias, J., Martignon, F., Avrachenkov, K., Neglia, G.: Socially-aware network design games. In: Proc. INFOCOM 2010, pp. 41–45 (2010)
11. Hoefer, M., Skopalik, A.: Altruism in Atomic Congestion Games. In: Fiat, A., Sanders, P. (eds.) ESA 2009. LNCS, vol. 5757, pp. 179–189. Springer, Heidelberg (2009)
12. Roughgarden, T.: Intrinsic robustness of the price of anarchy. In: Proc. 41st Annual ACM Symposium on Theory of Computing, pp. 513–522 (2009)
13. Beckmann, M., McGuire, B., Winsten, C.: Studies in the Economics of Transportation. Yale University Press, New Haven (1956)
14. Sharma, Y., Williamson, D.P.: Stackelberg thresholds in network routing games or the value of altruism. Games and Economic Behavior 67(1), 174–190 (2009)
15. Monderer, D., Shapley, L.S.: Potential games. Games and Economic Behaviour 14, 124–143 (1996)
16. Rosenthal, R.W.: A class of games possessing pure-strategy Nash equilibria. International Journal of Game Theory (2), 65–67 (1973)
17. Anshelevich, E., Dasgupta, A., Kleinberg, J., Tardos, E., Wexler, T., Roughgarden, T.: The price of stability for network design with fair cost allocation. In: Proc. 45th Annual IEEE Symposium on Foundations of Computer Science, pp. 295–304 (2004)
18. Christodoulou, G., Koutsoupias, E.: The price of anarchy of finite congestion games. In: Proc. 37th Annual ACM Symposium on Theory of Computing (2005)
19. Awerbuch, B., Azar, Y., Epstein, A.: Large the price of routing unsplittable flow. In: Proc. 37th Annual ACM Symposium on Theory of Computing, pp. 57–66 (2005)
20. Osborne, M.J.: An Introduction to Game Theory. Oxford University Press, Oxford (2005)

Mechanisms for Scheduling with Single-Bit Private Values*

Vincenzo Auletta[1], George Christodoulou[2,**], and Paolo Penna[1]

[1] Dipartimento di Informatica, Università di Salerno, Italy
{auletta,penna}@dia.unisa.it

[2] Computer Science Department, University of Liverpool, United Kingdom
gchristo@liv.ac.uk

Abstract. We consider randomized mechanisms for multi-dimensional scheduling. Following Lavi and Swamy [10], we study a setting with restrictions on the domain, while still preserving multi-dimensionality. In a sense, our setting is the simplest multi-dimensional setting, where each machine holds privately only a single-bit of information.

We prove a separation between truthful-in-expectation and universally truthful mechanisms for makespan minimization: We first show how to design an optimal truthful-in-expectation mechanism, and then prove lower bounds on the approximation guarantee of universally truthful mechanisms.

1 Introduction

Designing truthful mechanisms for scheduling problems was first suggested in the seminal paper by Nisan and Ronen [15], as a paradigm to demonstrate the applicability of Mechanism Design to an optimization problem. In its general form, where the machines are *unrelated*, there are n jobs to be assigned to m machines. The time needed by a machine i to process job k is described by a nonnegative real value t_{ik}. Given such an input matrix, a standard task from the algorithm designer's point of view, is to allocate the jobs in a way such that some global objective is optimized; a typical objective is to minimize the maximum completion time (i.e. the makespan). In a game-theoretic setting, it is assumed that each entry of this matrix is not known to the designer, but instead is a *private* value held by a selfish agent that controls the machine. Therefore this value might be misreported to the designer if this is advantageous to the agent. Mechanism design suggests using monetary compensation to incentivize agents to report truthfully. *Truthfulness* is desired, because it facilitates the prediction of the outcome and at the same time simplifies the agents' way of reasoning.

* Research partially supported by the PRIN 2008 research project COGENT (COmputational and GamE-theoretic aspects of uncoordinated NeTworks), funded by the Italian Ministry of University and Research.

** Part of this work was done while the author was at the Max-Planck Institute for Informatics, Saarbrücken, and while visiting University of Salerno.

M. Serna (Ed.): SAGT 2012, LNCS 7615, pp. 25–36, 2012.

The challenge is to design truthful mechanisms that optimize/approximate the makespan. When the entries of the matrix t are unrelated, the domain of input for each machine i is an n-valued vector t_i. For this multi-dimensional domain, the constraints imposed by truthfulness make the problem hard. Nisan and Ronen [15], showed that it is impossible to design a truthful mechanism with approximation factor better than 2, even for two machines. Later this bound was further improved to 2.41 [5] for 3 machines, and to 2.618 [9] for many machines. In [15], it was also shown that applying the VCG mechanism [17,3,8] achieves an approximation ratio of m, and it has been conjectured that this bound is tight. This conjecture still remains open, but it was further strengthened by Ashlagi et al. [2], who proved the conjecture for the intuitively very natural case of anonymous mechanisms (where roughly the allocation algorithm does not base its decisions on the machines' ids).

Randomization provably helps for this problem. There are two notions of truthfulness for randomized mechanisms. Roughly, a mechanism is *universally truthful* if it is defined by a probability distribution of truthful mechanisms, while it is *truthful in expectation*, if in expectation no player can benefit by lying. Already in [15], a universally truthful mechanism was suggested, for two machines. The mechanism was extended for the case of m machines by Mu'alem and Schapira [14] with an approximation guarantee of $0.875m$, and this was further improved in [12] to $0.837m$. Lu and Yu [13] showed a truthful-in-expectation mechanism with a guarantee of $(m+5)/2$. In [14] it was also shown a lower bound of $2-1/m$, for both randomized versions, while in [4] the lower bound was extended for fractional mechanisms, and an upper bound of $(m+1)/2$ was provided. Surprisingly, even for the special case of two machines a tight answer has not been given for randomized mechanisms. Currently the lower bound is 1.5 [14], while the best upper bound is 1.5963 due to [13].

Setting restrictions to the input domain can make the problem easier. The single-dimensional counterpart of the problem is the scheduling on *related* machines. In that case it is assumed that machine i has speed s_i, then $t_{ik} = w_k/s_i$, where the weights w_k of the jobs are known to the designer. Notice that the only missing information is the speed of the machines. In that case, the constraints imposed by truthfulness seem harmless; the optimal allocation is truthfully implementable [1], although it takes exponential running time, while the best possible approximation guarantee, a PTAS, can be achieved by polynomial time truthful mechanisms [7,6]. An immediate conclusion is that when one restricts the domain, then truthfulness becomes less and less stringent.

A prominent approach suggested by Lavi and Swamy [10], is to restrict the input domain, but still keep the multi-dimensional flavour. They assumed that each entry can take only two possible values L, H, that are publicly known to the designer. In this case, a very elegant deterministic mechanism achieves an approximation factor of 2, that is a great improvement comparing to the m upper bound that is the best known for the general problem. Surprisingly, even for this special case the lower bound is 11/10. Yu [18] extended this study, for more than two values, but where the inputs are restricted in balls around two values.

1.1 Our Contribution

The focus of this work is, following [10] to restrict even further the domain. We still allow only two possible values L, H, that are publicly known, but we even restrict the way these values are placed in a player i's input vector. We assume that for each machine some known partition of the tasks into two parts is given to the designer. The only missing information is in which part player assigns low values, and in which part it assigns the high ones. Therefore, the only missing information is a *single bit* for each player[1]. The lower bound given in [10] is still valid for our setting. It is important to emphasize that all the aforementioned lower bounds are due to truthfulness, and hold even for exponential running time algorithms. We explore the effects of truthfulness (both randomized and deterministic) in this restricted setting:

(1) Power of truthful-in-expectation mechanisms. There is a class of two-values scheduling problems for which every algorithm (thus including optimal ones) can be turned into a truthful-in-expectation mechanism with the same approximation guarantee (Theorem 3). On the contrary, randomized universally truthful mechanisms cannot achieve an approximation better than 31/30 (Theorem 17), and the 11/10 lower bound for deterministic mechanisms in [10] also applies.

Notice that such a separation was not known for the general problem since, although Lu [11], showed a lower bound higher than 1.5 for universally truthful mechanisms, the result holds only for scale-free mechanisms. This is arguably a very natural assumption, but it is still needed to be proven that it is without loss of generality.

(2) Two-values vs three-values domains. For two machines, three-values domains are as difficult as the general unrelated machines: the lower bound of 2 for deterministic mechanisms still hold (Theorem 18). We give a *partial* evidence of the fact that two-values domains are easier by giving a deterministic truthful 3/2-approximation mechanisms for the subcase of given partitions (the most general domain we consider in this work – see Section 1.2 – still a restriction of the two-values domains).

Due to space limitations some of the proofs are omitted. We refer the reader to the full version of this work.

1.2 Preliminaries

We have n jobs to be scheduled on m machines. Each job must be assigned to exactly one machine. In the unrelated-machines setting, each machine i has a vector of processing times or *type* $t_i = (t_{ih})_h$, where $t_{ih} \in \Re_{\geq 0}$ is i's processing time for job h. In the **two-values** domains by Lavi and Swamy [10], the time

[1] Notice that the information missing is just a single bit, much less than that of the related machines case, where the missing information is a positive real number. However, ours is not a single-dimensional domain. We refer the reader to Chapter 9 and 12 of [16] for the precise definition of a single-dimensional domain.

for executing job h on machine i is either L (low) or H (high), with $H > L$ (the case $L = H$ is trivial). We say that machine i is an L_S-*machine* (respectively, H_S-*machine*) if all jobs in S take time L (respectively, H), and all jobs not in S take time H (respectively, L). That is, the type t_i of an L_S-machine i is such that for any job h

$$t_{ih} = L_S^h := \begin{cases} L \text{ if } h \in S \\ H \text{ otherwise} \end{cases} \tag{1}$$

and similarly for H_S-machines.

In this work, we consider the following special case of the two-values domains [10]. In the case with **given partitions**, for each machine i we are given a (publicly known) subset S_i and the private information is whether i is an L_{S_i}-machine or an H_{S_i}-machine. Hence, the type t_i must belong to a simple domain of two elements only (i.e., $t_i \in \{L_{S_i}, H_{S_i}\}$). Intuitively, a type $t_i = L_{S_i}$ indicates that machine i is "good" for the jobs in S_i and "bad" for other jobs, while for $t_i = H_{S_i}$ it is the other way around (notice that $H_{S_i} = L_{\bar{S}_i}$ where $\bar{S}_i = [n] \setminus S_i$). We shall further distinguish between three restrictions (of increasing difficulty) of the given partitions domain: (i) **Identical partitions**, where all subsets S_i are identical; (ii) **Uniform partitions**, where all subsets S_i have size s for some $s \geq 0$; and (iii) **(Unrestricted) Given partitions**, which impose no restriction on the subsets S_i.

We say that job h is an L-job (respectively, H-job) for machine i if $t_{ih} = L$ (respectively, $t_{ih} = H$), with t_i being the type of machine i. We represent an allocation by a matrix $x = (x_{ih})$, where $x_{ih} \in \{0, 1\}$ and $x_{ih} = 1$ iff job h is assigned to machine i (since every job is assigned to exactly one machine, $\sum_i x_{ih} = 1$). Given an allocation x and machine types t, we define the *load* of machine i as the set of jobs allocated to i in x and denote by $C_i(x, t) := \sum_h x_{ih} t_{ih}$ the *cost* of machine i. The *makespan* of x with respect to t is $\max_i C_i(x, t)$. An *exact* or *optimal* allocation x is an allocation that, for the given input t, minimizes the makespan. A *c-approximation* is an allocation whose makespan is at most c times that of the optimal allocation. A deterministic algorithm A outputs an allocation $x = A(t)$. For randomized algorithms, $A(t)$ is a probability distribution over all possible allocations; we call $A(t)$ a randomized allocation.

In order to characterize truthful mechanisms, we consider the allocations that are given in output for two inputs which differ only in one machine's type. We let $(\hat{t}_i, t_{-i})$ denote the vector $(t_1, \ldots, t_{i-1}, \hat{t}_i, t_{i+1}, \ldots, t_m)$. Given types t and a job allocation x, we count the number of L-jobs and the number of H-jobs allocated to machine i in x: $n_L^i(x, t) := |\{h : t_{ih} = L \text{ and } x_{ih} = 1\}|$ and $n_H^i(x, t) := |\{h : t_{ih} = H \text{ and } x_{ih} = 1\}|$.

Definition 1 (monotone algorithm). *An algorithm A is monotone (in expectation) if, for any machine i and for any two inputs $t = (t_i, t_{-i})$ and $\hat{t} = (\hat{t}_i, t_{-i})$, the following inequality holds (in expectation):*

$$n_L^i - n_H^i + \hat{n}_L^i - \hat{n}_H^i \geq 0 \tag{2}$$

where $n_L^i = n_L^i(A(t), t)$, $n_H^i = n_H^i(A(t), t)$, $\hat{n}_L^i = n_L^i(A(\hat{t}), \hat{t})$, and $\hat{n}_L^i = n_L^i(A(\hat{t}), \hat{t})$.

By applying [10, Proposition 5.7] to our given partitions domains, we obtain that truthfulness is equivalent to the monotonicity condition above:

Theorem 2. *For the case of identical/uniform/unrestricted given partitions, there exist prices P such that the mechanism (A, P) is truthful (in expectation) iff A is monotone (in expectation).*

Throughout the paper, we refer to the quantity $n^i_L(x,t) - n^i_H(x,t)$ as the *unbalance* of machine i. We also refer to the quantity in (2) as the *overall unbalance* of machine i. For any instance t, for any two machines i and j, and for any $\alpha, \beta \in \{L, H\}$, we consider the subset of jobs whose execution time is α on machine i and β on machine j as $J^{ij}_{\alpha\beta}(t) := \{h : t_{ih} = \alpha \text{ and } t_{jh} = \beta\}$.

1.3 An Illustrative Example

We begin with an example and show how to use randomization. Consider the following instances along with their optimal allocation (gray box), and the quantities $n^i_L - n^i_H$ for each of the two machines (numbers outside the box):

-3	4	5	1
2 2 5 5 | 5 5 5	5 5 2 | 2 2 2 2	5 5 | 2 2 2 2 2	2 2 5 | 5 5 5 5
2 2 5 5 | 5 5 5	2 2 5 | 5 5 5 5	5 5 | 2 2 2 2 2	5 5 2 | 2 2 2 2
0	1	-2	4

(3)

Let machine 1 and machine 2 correspond to top and bottom machine, respectively. Observe that the monotonicity condition is violated for machine $i = 2$ by looking at the two middle instances (they differ in the machine connected by dotted line). Indeed the quantity in (2) is $1 - 2 = -1$. Alternatively, we can swap the allocation in the first and in the third input:

0	4	-2	1
2 2 5 5 | 5 5 5	5 5 2 | 2 2 2 2	5 5 | 2 2 2 2 2	2 2 5 | 5 5 5 5
2 2 5 5 | 5 5 5	2 2 5 | 5 5 5 5	5 5 | 2 2 2 2 2	5 5 2 | 2 2 2 2
-3	1	5	4

(4)

Now, however, the monotonicity condition is violated by machine $i = 1$ for the last two instances. These instances are used by Lavi and Swamy [10] to prove a lower bound for deterministic mechanisms. However, if we choose *randomly* between the allocation in (3) and the one in (4) with the same probability, the corresponding optimal algorithm satisfies monotonicity *in expectation* (for example, in the leftmost instance the unbalance becomes $-3/2$ for both machines, while in the second instance it remains unchanged).

Our simple example shows that for two machines with identical partitions as above, and values $L = 2$ and $H = 5$, there exists an exact truthful-in-expectation mechanism but no truthful mechanism can achieve an approximation factor better than 1.1 [10]. In the sequel we show that this positive result holds in general for some of our domains, and not only in the very special instance where deterministic mechanisms cannot be optimal.

2 Identical Partitions

In this section we consider the case of machines with identical partitions and give a general ("black box") method to convert scheduling algorithms into mechanisms that are truthful-in-expectation. The main result of this section is summarized by the following theorem.

Theorem 3. *Every deterministic algorithm A for scheduling jobs on machines with identical partitions can be turned into a randomized mechanism M which is truthful in expectation and such that the allocation returned by M has makespan not worse than the one returned by A.*

Let $(S, \bar{S})$ be a partition of the jobs, with $|S| = s$ and $|\bar{S}| = n - s$, such that for each machine i we have $S_i = S$. Without loss of generality we can reorder the jobs in such a way that $S = \{1, 2, \ldots, s\}$ and $\bar{S} = \{s+1, s+2, \ldots, n\}$. Since the partition of the jobs is public the only information that is private to each machine is which side of the partition contains its L-jobs. Thus, the type's domain of each machine contains only two elements: $L_S = (L \cdots LH \cdots\cdots H)$ and $H_S = (H \cdots HL \cdots\cdots L)$. For any instance t, we denote by $m_S(t)$ and $m_{\bar{S}}(t)$ the numbers of L_S-machines and H_S-machines in t, respectively. Clearly, $m_S(t) + m_{\bar{S}}(t) = m$. It is convenient to count, for each side of the partition, the number of jobs that x allocates as L-jobs:

$$\ell_S(x, t) = |h \in S : x_{ih} = 1 \text{ and } t_{ih} = L| \tag{5}$$

The quantity $\ell_{\bar{S}}(x, t)$ is defined similarly, and $\ell(x, t) := \ell_S(x, t) + \ell_{\bar{S}}(x, t)$ is the overall number of allocated L-jobs. Following the idea described in Section 1.3, we show now how to obtain a randomized allocation from a deterministic one by randomly "shuffling" machines of the same type:

Definition 4. *For any deterministic allocation x we denote by $x^{(rand)}$ the randomized allocation obtained as follows:*

- *Pick an integer $r \in \{0, \ldots, m_S - 1\}$ uniformly at random, and set $x_i^{(rand)} := x_{i+r \mod m_S}$ for each L_S-machine i;*
- *Pick an integer $\bar{r} \in \{0, \ldots, m_{\bar{S}} - 1\}$ uniformly at random, and set $x_i^{(rand)} := x_{i+\bar{r} \mod m_{\bar{S}}}$ for each H_S-machine i.*

For any deterministic algorithm A, we let $A^{(rand)}$ be the randomized algorithm that, on input t, returns the randomized allocation $x^{(rand)}$ where $x = A(t)$.

Notice that $\ell_S(x^{(rand)}, t) = \ell_S(x, t)$ and $\ell_{\bar{S}}(x^{(rand)}, t) = \ell_{\bar{S}}(x, t)$. In the following discussion we fix x and t and simply write ℓ_S and $\ell_{\bar{S}}$.

For any L_S-machine i, its expected load consists of $n_L^i = \ell_S / m_S$ L-jobs and $n_H^i = (n - s - \ell_{\bar{S}})/m_S$ H jobs. Thus the expected unbalance of an L_S-machine is $n_L^i - n_H^i = \frac{1}{m_S}[\ell_S - (n - s - \ell_{\bar{S}})] = \frac{1}{m_S}[\ell - (n - s)]$. Similarly, the expected load of an H_S-machine i consists of $n_L^i = \ell_{\bar{S}} / m_{\bar{S}}$ L-jobs and $n_H^j = (s - \ell_S)/m_{\bar{S}}$ H-jobs and its expected unbalance is equal to $n_L^i - n_H^i = \frac{1}{m_{\bar{S}}}[\ell_{\bar{S}} - (s - \ell_S)] = \frac{1}{m_{\bar{S}}}(\ell - s)$.

Lemma 5. *Algorithm $A^{(rand)}$ is monotone-in-expectation if the deterministic algorithm A satisfies the following condition. For any $t = (L_S, t_{-i})$ and $\hat{t} = (H_S, t_{-i})$, it holds that*

$$\frac{\ell - (n-s)}{m_S} + \frac{\hat{\ell} - s}{m_{\bar{S}} + 1} \geq 0 \tag{6}$$

where $\ell = \ell(A(t), t)$ and $\hat{\ell} = \ell(A(\hat{t}), \hat{t})$ denote the number of L-jobs allocated on input t and $\hat{t}$, respectively.

Proof. Consider two instances $t = (L_S, t_{-i})$ and $\hat{t} = (H_S, t_{-i})$. By the previous discussion we have that the total unbalance of machine i is $n_L^i - n_H^i + \hat{n}_L^i - \hat{n}_H^i = \frac{\ell-(n-s)}{m_S} + \frac{\hat{\ell}-s}{\hat{m}_{\bar{S}}}$, where $\hat{m}_{\bar{S}}$ is the number of H_S-machines in $\hat{t}$. Since $\hat{m}_{\bar{S}} = m_{\bar{S}}+1$, then (6) is equivalent to (2) and thus $A^{(rand)}$ is monotone-in-expectation.

2.1 Canonical Allocations

Lemma 5 says that in order to design an exact truthful-in-expectation mechanism it is enough to design a deterministic exact algorithm A which obeys the condition in (6). We show that this is always possible by transforming the allocation of the algorithm into a "canonical" allocation (specified below).

Definition 6. *Given an allocation x and an instance t, for any $\alpha, \beta \in \{L, H\}$ and for any two machines i and j, we let $n_{\alpha\beta}^{ij}(x,t)$ be the number of α-jobs that are allocated to machine i and that are β-jobs for machine j: $n_{\alpha\beta}^{ij}(x,t) := |\{k : x_{ik} = 1,\ t_{ik} = \alpha, \text{ and } t_{jk} = \beta\}|$.*

Notice that $n_{\alpha\beta}^{ij}$ is different from $n_{\beta\alpha}^{ji}$ because the first index denotes the machine that gets the jobs. We can now define our canonical allocations.

Definition 7 (canonical allocation). *A canonical allocation (for the instance t) is an allocation obtained by modifying a deterministic allocation x as follows:*

1. *Apply the following **Rule R1** until possible: Suppose jobs h and k are allocated to machines i and j, respectively ($x_{ih} = 1 = x_{jk}$). If $t_{ik} \leq t_{ih}$ and $t_{jh} < t_{jk}$ (no machine gets worse and at least one gets better if we swap the jobs), then move job h to machine j and job k to machine i (set $x_{ik} = x_{jh} = 1$ and $x_{ih} = x_{jk} = 0$).*
2. *Apply the following **Rule R2** until possible: If $n_{HL}^{ij}(x,t) > n_{LH}^{ji}(x,t)$ and j gets only jobs from $J_{LH}^{ji}(t)$, then move all n_{HL}^{ij} jobs in $J_{HL}^{ij}(t)$ from i to j, and move all jobs from $J_{LH}^{ji}(t)$ from j to i (see Figure 1.).*

Remark 8. Both Rule R1 and R2 decrease the overall number of H-jobs by at least one. Thus, given x and t, it is possible to compute in polynomial time a canonical allocation (following the two steps in Definition 7) whose cost is not larger than the cost of x.

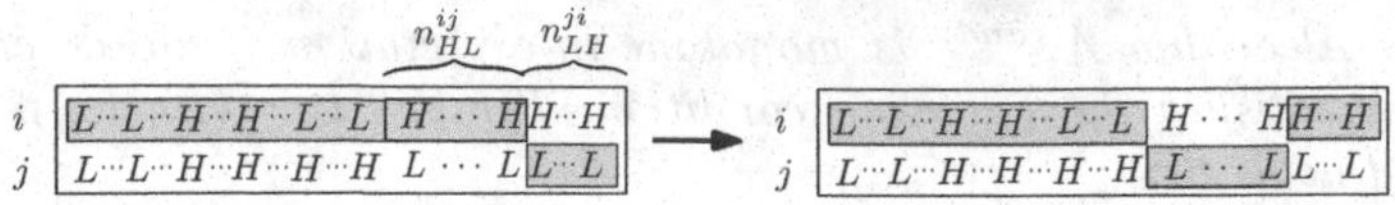

Fig. 1. The swapping Rule R2 in the definition of canonical allocation

2.2 A Black-Box Construction (Proof of Theorem 3)

Lemma 9. *Every deterministic algorithm A can be turned into a randomized algorithm $A^{(rand)}$ which is monotone in expectation and whose output allocation has makespan not worse than the one returned by A.*

Proof (Proof Idea). The proof is based on the following structural properties of canonical allocations. Either $\ell_S(x,t) = |S|$ or $\ell_{\bar{S}}(x,t) = |\bar{S}|$. Moreover these bounds also hold:

$$\ell_S(x,t) = |S| \text{ implies } \ell_{\bar{S}}(x,t) \geq \frac{m_{\bar{S}}}{m} \cdot |\bar{S}|$$

$$\ell_{\bar{S}}(x,t) = |\bar{S}| \text{ implies } \ell_S(x,t) \geq \frac{m_S}{m} \cdot |S|$$

To apply Lemma 5 we need to prove that, for any $t = (L_S, t_{-i})$ and $\hat{t} = (H_S, t_{-i})$

$$\frac{\ell - (n-s)}{m_S} + \frac{\hat{\ell} - s}{m - m_S + 1} \geq 0$$

We distinguish these four possible cases: (1) $\ell_S = |S|$ and $\hat{\ell}_{\bar{S}} = |\bar{S}|$, (2) $\ell_{\bar{S}} = |\bar{S}|$ and $\hat{\ell}_S = |S|$, (3) $\ell_S = |S|$ and $\hat{\ell}_S = |S|$, (4) $\ell_{\bar{S}} = |\bar{S}|$ and $\hat{\ell}_{\bar{S}} = |\bar{S}|$. Cases (1)-(2) use the fact that $\ell + \hat{\ell} \geq n$. Cases (3)-(4) use the bounds above and $\hat{m}_{\bar{S}} = m_{\bar{S}} + 1$.

3 Mechanisms for Two Machines

In this section we restrict our attention to the case of $m = 2$ machines. We give an exact truthful-in-expectation mechanism for the case of uniform partitions (Sect. 3.1) and a deterministic 3/2-approximation mechanism for the case of unrestricted given partitions (Sect. 3.2).

3.1 Exact Randomized Mechanisms for Uniform Partitions

We can always assume that any deterministic algorithm can produce canonical allocations (see Remark 8). We show that every exact algorithm can be turned into a monotone-in-expectation algorithm (thus an exact truthful-in-expectation mechanism). Unlike the case of identical partitions, we apply a randomization step only in some cases. Specifically, we will randomize if the two machines have types $t = (L_{S_1}, L_{S_2})$ or $t' = (H_{S_1}, H_{S_2})$, and give in output directly the allocation of the deterministic algorithm in the other two cases, inputs (L_{S_1}, H_{S_2}) and (H_{S_1}, L_{S_2}).

We start by characterizing exact canonical allocations.

Definition 10 (allocation classes). *An allocation x is classified with respect to the instance t as $J^{ij}_{LH}(t)$ if machine i gets only a proper subset of the jobs in $J^{ij}_{LH}(t)$; x is symmetric if each machine i gets all jobs in $J^{ij}_{LH}(t)$, no job in $J^{ij}_{HL}(t)$, and a (possibly empty) subset of the jobs in $J^{ij}_{LL}(t) \cup J^{ij}_{HH}(t)$.*

Lemma 11. *Every instance t admits an exact canonical allocation that is either symmetric or of class $J^{ij}_{LH}(t)$, with $i \in \{1,2\}$.*

We next show that *certain* symmetric allocations can be converted into randomized allocations having a "good" unbalance, without increasing their cost.

Definition 12 (randomizable allocation). *A symmetric allocation x is* randomizable *if the following holds. Let i be the machine such that $|J^{ij}_{LH}(t)| \geq |J^{ij}_{HL}(t)|$.*[2] *Then, one of the following two conditions holds: (1) $n^{ji}_{LL}(x,t) \geq |J^{ij}_{LH}(t)| - |J^{ij}_{HL}(t)|$, or (2) $n^{ji}_{HH}(x,t) \leq n^{ij}_{HH}(x,t)$.*

Lemma 13. *For every deterministic allocation x that is randomizable with respect to the instance t, there exists a randomized allocation $x^{(rand)}$ which gives an expected unbalance of $\frac{n}{2} - |J_{HH}(t)|$ to both machines and has the same makespan as the makespan of x.*

Proof (Sketch). Let x be an allocation that is randomizable with respect to t. We build a new allocation y as follows. Let i be a machine such that $|J^{ij}_{LH}(t)| \geq |J^{ij}_{HL}(t)|$ and let $\delta := |J^{ij}_{LH}(t)| - |J^{ij}_{HL}(t)|$. The new allocation is obtained by swapping all jobs in $J^{ij}_{HH}(t)$ and *some* of the jobs in $J^{ij}_{LL}(t)$ according to the following two cases:

1. ($n^{ji}_{LL}(x,t) \geq \delta$) In this case we move $n^{ji}_{LL}(x,t) - \delta$ jobs from machine j to machine i, and $n^{ij}_{LL}(x,t)$ jobs from machine i to machine j.
2. ($n^{ji}_{LL}(x,t) < \delta$) In this case we move $n^{ij}_{LL}(x,t) + \delta - n^{ji}_{LL}(x,t)$ jobs from machine i to machine j (this quantity is nonnegative by Definition 12).

Finally, build the randomized allocation $x^{(rand)}$ by picking at random x or y.

The above lemma and its randomization procedure guarantees that the monotonicity condition can be satisfied whatever are the canonical allocations that we *do not randomize*, inputs (L_{S_1}, H_{S_2}) and (H_{S_1}, L_{S_2}). Moreover, for the case of uniform partitions, it turns out that the exact allocations of instances $t = (L_{S_1}, L_{S_2})$ and $t' = (H_{S_1}, H_{S_2})$ are both randomizable. Thus the following holds:

Theorem 14. *There exists an exact truthful-in-expectation mechanism for the case of two machines with uniform partitions.*

3.2 A Deterministic 3/2-Approximation Mechanism

In this section we prove the following result:

Theorem 15. *There exists a 3/2-approximation deterministic truthful mechanism for two machines with (unrestricted) given partitions.*

In order to prove this result we exhibit a monotone 3/2-approximation algorithm.

[2] Recall that by definition $J^{ij}_{LH}(t) = J^{ji}_{HL}(t)$ and thus such machine must exist.

The algorithm. On input t, the algorithm partitions the jobs into three subsets: $J_{LL}(t)$, $J_{HH}(t)$ and $J_{LHHL}(t) := J^{12}_{LH}(t) \cup J^{12}_{HL}(t)$.

First, allocates jobs in $J_{LHHL}(t)$, and then completes the allocation by dividing "evenly" the other jobs in $J_{LL}(t)$ and in $J_{HH}(t)$. Some careful "tie breaking rule" must be used here to deal with the case in which some of these subsets of jobs have odd cardinality.

The algorithm consists of the following two steps (in the sequel we do not specify the input "t"):

1. **Step 1 (allocate jobs in J_{LHHL}).** We allocate these jobs depending on the class of the canonical exact allocation for *all* jobs:
   (a) **(class J^{12}_{LH} or J^{21}_{LH}).** Compute a canonical exact allocation for J_{LHHL}.
   (b) **(class symmetric).** Assign all jobs in J_{LHHL} as L-jobs.

   We denote by J^i_{LHHL} the set of jobs that are assigned to machine i in this first step, and $C_i^{(LHHL)}$ be the corresponding cost.
2. **Step 2 (allocate jobs in J_{LL} and J_{HH}).** For a set of jobs S and a nonnegative integer q, we denote by $\lfloor S/q \rfloor$ and $\lceil S/q \rceil$ an arbitrary subset of S of cardinality $\lfloor |S|/q \rfloor$ and $\lceil |S|/q \rceil$, respectively. We define the set J^i of all jobs that are assigned to machine i at the end of this step (which includes the jobs J^i_{LHHL} assigned in the previous step) as follows: For i and j satisfying $C_i^{(LHHL)} \geq C_j^{(LHHL)}$, we allocate to i the set J^i that is equal to $J^i_{LHHL} \cup \left\lfloor \frac{J_{HH}}{2} \right\rfloor \cup \left\lceil \frac{J_{LL}}{2} \right\rceil$, if both $|J_{LL}|$ and $|J_{HH}|$ are odd, and $J^i_{LHHL} \cup \left\lfloor \frac{J_{HH}}{2} \right\rfloor \cup \left\lfloor \frac{J_{LL}}{2} \right\rfloor$ otherwise. Machine j gets all the other jobs $J^j := \bar{J}^i = [n] \setminus J^i$.

Approximation guarantee. The proof of the approximation guarantee is based on the following lemma that proves that the two steps of the algorithm keep a small difference between the completion time of the two machines:

Lemma 16. *After each step, the difference between the two completion times is at most the optimum, that is* $\max_i C_i^{(LHHL)} \leq OPT$ *and* $\max_i C_i \leq \min_i C_i + OPT$.

To prove the 3/2-approximation we have to distinguish two cases, depending on which class of canonical allocation has been used in Step 1 to allocate jobs in J_{LHHL}. We observe that in the first case (class J^{12}_{LH} or J^{21}_{LH}) at least one of these jobs is allocated as an H-job; in the second case, instead, all jobs in J_{LHHL} are allocated as L-jobs and the remaining jobs are distributed among the two machines in order to minimize the makespan. In both the cases, using Lemma 16 we can prove that $APX \leq (3/2) \cdot OPT$.

It is easy to verify that the mechanism can indeed reach a 3/2 approximation guarantee.

Monotonicity. First observe that the algorithm assigns to machine i at least half (rounded up or down) of its L-jobs in t, and at most half (rounded up or down) of its H-jobs in t. In particular, $n_{LH}^{ij} \geq \left\lceil \frac{J_{LH}^{ij}}{2} \right\rceil$, $n_{LL}^{ij} \geq \left\lfloor \frac{J_{LL}^{ij}}{2} \right\rfloor$ and $n_{HL}^{ij} \leq \left\lfloor \frac{J_{HL}^{ij}}{2} \right\rfloor$, $n_{HH}^{ij} \leq \left\lceil \frac{J_{HH}^{ij}}{2} \right\rceil$. For the jobsets J_{LL} and J_{HH} this is immediate. Now, let us consider the jobset J_{LH}^{ij} (the other case is symmetric). If we are in the symmetric case, the algorithm assigns all these jobs to i. Otherwise, the algorithm computes the canonical optimum on J_{LHHL}. Suppose that $n_{LH}^{ij} < \lceil \frac{J_{LH}^{ij}}{2} \rceil$. Optimality implies that machine i gets at least one H job from the set J_{HL}^{ij}. But then the allocation is not canonical (nor optimal) since Rule 2 of Definition 7 can be applied. Therefore we obtain

$$n_L^i \geq \left\lceil \frac{J_{LH}^{ij}}{2} \right\rceil + \left\lfloor \frac{J_{LL}^{ij}}{2} \right\rfloor, \qquad n_H^i \leq \left\lceil \frac{J_{HH}^{ij}}{2} \right\rceil + \left\lfloor \frac{J_{HL}^{ij}}{2} \right\rfloor. \tag{7}$$

Second, observe that if the type of machine i flips from t_i to $\hat{t}_i$, then $|\hat{J}_{\bar{\alpha}\beta}^{ij}| = |J_{\alpha\beta}^{ij}|$ (here $\bar{L} = H$ and $\bar{H} = L$). Applying (7) for both $t_i, \hat{t}_i$ and using the last identity, we finally obtain (2).

4 Lower Bounds and Separation Results

The next theorem says that truthful-in-expectation mechanisms are provably *more powerful* than universally truthful mechanisms. Indeed, for this problem version, *exact* truthful-in-expectation mechanism exist for any number of machines (see Theorem 3)

Theorem 17. *No universally truthful mechanism can achieve an approximation factor better than* 31/30 *for scheduling on two machines, even for the case of identical partitions.*

The proof combines the idea of the lower bound by Lavi and Swamy [10] with the use of Yao's Min-Max Principle suggested by Mu'alem and Schapira [14]. The above bounds can be strengthen by considering three-values domains (which are no-longer "single-bit").

Theorem 18. *For two machines and the case in which the processing times can take three values, no (deterministic) truthful mechanism can achieve an approximation factor better than* 2. *Moreover, no randomized universally truthful mechanism can achieve and approximation factor better than* 9/8.

The first part is an alternative proof for the lower bound of 2 for two machines, first showed by Nisan and Ronen in [15]. The proof in [15] requires that the input domain consists of at least 4 different values. Here, we extend the proof in order to hold even when the domain consists of only 3 different values. The second part adapts this proof via Yao's Min-Max Principle.

5 Conclusion

This work leaves several open questions. First of all, we are able to derive mechanisms for an arbitrary number of machines only in one case (identical partitions). Second, exact truthful-in-expectation mechanisms for two machines are given only for uniform partitions. Is it possible to extend the result to more general cases, like (1) any number of machines with uniform partitions, or (2) two machines with unrestricted given partitions? Finally we note that the lower bound of 2 for three-values domains does not consider jobs' partitions (as our upper bounds do) and thus it would be natural/interesting to prove a lower bound for three-values domains with given partitions.

References

1. Archer, A., Tardos, E.: Truthful Mechanisms for One-Parameter Agents. In: Proc. 42nd FOCS, pp. 482–491 (2001)
2. Ashlagi, I., Dobzinski, S., Lavi, R.: An Optimal Lower Bound for Anonymous Scheduling Mechanisms. In: Proc. 10th EC, pp. 169–176 (2009)
3. Clarke, E.H.: Multipart pricing of public goods. Public Choice 11(1), 17–33 (1971)
4. Christodoulou, G., Koutsoupias, E., Kovács, A.: Mechanism design for fractional scheduling on unrelated machines. ACM Trans. on Algorithms 6(2) (2010)
5. Christodoulou, G., Koutsoupias, E., Vidali, A.: A Lower Bound for Scheduling Mechanisms. Algorithmica 55(4), 729–740 (2009)
6. Christodoulou, G., Kovács, A.: A deterministic truthful PTAS for scheduling related machines. In: Proc. of 21st SODA, pp. 1005–1016 (2010)
7. Dhangwatnotai, P., Dobzinski, S., Dughmi, S., Roughgarden, T.: Truthful Approximation Schemes for Single-Parameter Agents. SIAM J. on Computing 40(3), 915–933 (2011)
8. Groves, T.: Incentives in teams. Econometrica 41(4), 617–631 (1973)
9. Koutsoupias, E., Vidali, A.: A Lower Bound of $1 + \phi$ for Truthful Scheduling Mechanisms. In: Kučera, L., Kučera, A. (eds.) MFCS 2007. LNCS, vol. 4708, pp. 454–464. Springer, Heidelberg (2007)
10. Lavi, R., Swamy, C.: Truthful mechanism design for multidimensional scheduling via cycle monotonicity. Games and Economic Behavior 67(1), 99–124 (2009)
11. Lu, P.: On 2-Player Randomized Mechanisms for Scheduling. In: Leonardi, S. (ed.) WINE 2009. LNCS, vol. 5929, pp. 30–41. Springer, Heidelberg (2009)
12. Lu, P., Yu, C.: An Improved Randomized Truthful Mechanism for Scheduling Unrelated Machines. In: Proc. of 25th STACS. LIPIcs, vol. 1, pp. 527–538 (2008)
13. Lu, P., Yu, C.: Randomized Truthful Mechanisms for Scheduling Unrelated Machines. In: Papadimitriou, C., Zhang, S. (eds.) WINE 2008. LNCS, vol. 5385, pp. 402–413. Springer, Heidelberg (2008)
14. Mu'alem, A., Schapira, M.: Setting Lower Bounds on Truthfulness. In: Proc. of 18th SODA, pp. 1143–1152 (2007)
15. Nisan, N., Ronen, A.: Algorithmic Mechanism Design. Games and Economic Behavior 35, 166–196 (2001)
16. Nisan, N., Roughgarden, T., Tardos, E., Vazirani, V.: Algorithmic Game Theory. Cambridge University Press (2007)
17. Vickrey, V.: Counterspeculations, auctions and competitive sealed tenders. Journal of Finance 16, 8–37 (1961)
18. Yu, C.: Truthful mechanisms for two-range-values variant of unrelated scheduling. Theoretical Computer Science 410(21-23), 2196–2206 (2009)

The Complexity of Decision Problems about Nash Equilibria in Win-Lose Games*

Vittorio Bilò[1] and Marios Mavronicolas[2]

[1] Department of Mathematics and Physics "Ennio De Giorgi", University of Salento, Provinciale Lecce-Arnesano, P.O. Box 193, 73100 Lecce, Italy
vittorio.bilo@unisalento.it

[2] Department of Computer Science, University of Cyprus, Nicosia CY-1678, Cyprus
mavronic@ucy.ac.cy

Abstract. We revisit the complexity of deciding, given a (finite) *strategic game*, whether *Nash equilibria* with certain natural properties exist; such *decision problems* are well-known to be $\mathcal{NP}$-complete [2, 6, 10]. We show that this complexity remains unchanged when all *utilities* are restricted to be 0 or 1; thus, *win-lose games* are as complex as general games with respect to such decision problems.

1 Introduction

Among the most fundamental problems in *Algorithmic Game Theory* are those concerning the *Nash equilibria* [13, 14] of a *strategic game*: states where no *player* could unilaterally deviate to improve her *utility*. Such algorithmic problems, including their *decision*, *search* and *approximation* variants, have been studied extensively in the last few years — see, e.g., [1–8, 11]. The fundamental theorem of Nash [13, 14] that Nash equilibria are guaranteed to exist makes the search problem for Nash equilibria *total*, which implies that the search problem is *not* $\mathcal{NP}$-complete unless $\mathcal{NP} = co\text{-}\mathcal{NP}$ [12, Theorem 2.1].

In a wake of breakthrough results on *search problems* about Nash equilibria, it has been shown [3, 8] that computing or approximating (additively) a Nash equilibrium even for two-player games with rational utilities is complete for $\mathcal{PPAD}$ [15], a complexity class capturing the computation of discrete fixed points; ditto for multiplicative approximation [7].

Abbott *et al.* [1] present a polynomial time reduction, specifically a *Nash-homomorphism*, mapping a two-player game with rational utilities to a two-player ***win-lose game*** where all utilities are 0 or 1; the Nash homomorphism comes with a polynomial time map which guarantees *(i)* to return a Nash equilibrium for the two-player general game when presented with a Nash equilibrium for the two-player win-lose game, so that *(ii)* every Nash equilibrium for the two-player general game is returned for *some* Nash equilibrium of the two-player

* This work was partially supported by "Progetto 5 per mille per la ricerca": "Collisioni fra vortici puntiformi e fra filamenti di vorticità: singolarità, trasporto e caos" at the University of Salento and by research funds at the University of Cyprus.

M. Serna (Ed.): SAGT 2012, LNCS 7615, pp. 37–48, 2012.

win-lose game (*surjectivity* property). Hence, the search problem for win-lose games remains $\mathcal{PPAD}$-complete.

Decision problems about Nash equilibria result naturally by twisting the search problem in one of several simple ways that deprive it from its existence guarantees. Here is a (non-exhaustive) list of (informally stated here, see Section 3 for formal statements) decision problems about Nash equilibria: Given a strategic game, does it have: *(i)* A Nash equilibrium where each player has utility at least a given number? [10],*(ii)* A Nash equilibrium where each player has utility at most a given number?, *(iii)* At least two Nash equilibria? [10], *(iv)* A Nash equilibrium whose support contains a set of strategies? [10], *(v)* A Nash equilibrium whose support is contained in a set of strategies? [10], *(vi)* A Nash equilibrium whose support has size greater than a given number? [10], *(vii)* A Nash equilibrium whose support has size smaller than a given number? [10], *(viii)* A Nash equilibrium in which the total utility of players is at least a given number? [6], *(ix)* A Nash equilibrium in which the total utility of players is at most a given number?, *(x)* A *rational* Nash equilibrium (i.e., one with all probabilities rational)? [2].

Some of these decision problems are $\mathcal{NP}$-complete for *symmetric* two-player games; this was originally shown by Gilboa and Zemel [10] and later by Conitzer and Sandholm [6] via a unifying reduction from the *satisfiability* problem (which covered some additional decision problems over those considered in [10]). The last problem in the list is $\mathcal{NP}$-complete even for three-player games [2] — recall that all Nash equilibria of a two-player game are rational, so that the problem is trivial for two-player games.

We emphasize that the polynomial time reduction (Nash-homomorphism) from two-player games with rational utilities to two-player win-lose games [1] does not imply that the *decision problems* about Nash equilibria for two-player win-lose games have the same complexity as for two-player general games. This is because the Nash-homomorphism provides no guarantee that any property of the Nash equilibrium of the two-player win-lose game is preserved in the returned (by the map) Nash equilibrium of the general two-player game.

It follows that a polynomial time reduction from an $\mathcal{NP}$-hard problem to the decision problems about Nash equilibria for two-player games (such as the ones given in [6, 10]) **composed** with the Nash-homomorphism from two-player games to two-player win-lose games in [1] does *not* suffice to provide a polynomial time reduction from the same $\mathcal{NP}$-hard problem to the decision problems about Nash equilibria for two-player win-lose games. Thus, the complexity of decision problems about Nash equilibria for two-player win-lose games has remained open.

In this work, we settle the complexity of the natural decision problems about Nash equilibria previously considered in [6, 10] (or introduced here) for win-lose games. Specifically, we show, as our main result, that these decision problems are $\mathcal{NP}$-complete for two-player win-lose games (Theorem 2). In a similar vein, the decision problem asking whether a given game has a rational Nash equilibrium [2] is shown $\mathcal{NP}$-complete for three-player win-lose games (Theorem 3). Thus, these decision problems about Nash equilibria have the same complexity for win-lose games as for general games.

To show our results (Theorems 2 and 3), we first prove a significant milestone, which we describe. Say that a game has the ***positive utility property*** if each player has always a response to the choices of the other players that makes her utility greater than zero. Note that for two-player win-lose games, the positive utility property implies that the utility matrix of the row player (resp., column player) cannot have a column (resp., row) containing only zeros. We revisit the decision problem from [2] asking whether a given game has the same set of Nash equilibria with a ***gadget game***, and additionally we assume that the gadget game has the positive utility property; we show that, *when restricted to win-lose games*, this problem is co-$\mathcal{NP}$-hard for any choice of a win-lose gadget game (with the positive utility property) (Theorem 1).

Theorem 1 is the backbone technical result in the paper; its proof utilizes a reduction from the satisfiability problem, which establishes that the *unsatisfiability* of a given formula is equivalent with the fact that the constructed win-lose game does not have a Nash equilibrium with properties opposite to those possessed by the Nash equilibria of the gadget game (Proposition 1), while the *satisfiability* of the formula always guarantees the existence of at least one Nash equilibrium with some particular properties which are independent from both the formula and the gadget game (i.e., they hold *a priori* as a feature of the reduction) (Proposition 2). This implies, in particular, that deciding whether the two win-lose games have the same set of Nash equilibria is co-$\mathcal{NP}$-hard (Theorem 1), improving [2, Theorem 1] which applies to general two-player games.

The reduction used for the proof of Theorem 1 constitutes a major improvement over previous reductions from the satisfiability problem in [2, 6] to **yield a win-lose game** (rather than a general game) **while preserving** the relation between its Nash equilibria and the satisfiability of the formula.

Moreover, by suitable choices of the gadget game so that the properties possessed by its Nash equilibria dismatch the properties of the Nash equilibria induced when the formula is satisfiable, particular $\mathcal{NP}$-hardness results follow (Theorems 2 and 3). These results extend the corresponding results from [6, 10] which apply to two-player symmetric games with rational utilities.

For example, choosing the gadget game as a two-player win-lose game where each player has a single strategy and all utilities are 1 implies that a handful of properties are $\mathcal{NP}$-hard to decide for two-player, win-lose games (Theorem 2). Choosing the gadget game as a three-player win-lose game with a single irrational Nash equilibrium implies that deciding the existence of a rational Nash equilibrium is $\mathcal{NP}$-hard for three-player win-lose games (Theorem 3).

2 Framework

A ***(strategic) game*** is a triple $\mathsf{SG} = \langle [r], \{\Sigma_i\}_{i\in[r]}, \{\mathsf{U}_i\}_{i\in[r]}\rangle$, where: *(i)* $[r] = \{1, \ldots, r\}$ is a finite set of ***players*** with $r \geq 2$, and *(ii)* for each player $i \in [r]$, Σ_i is the set of ***strategies*** for player i, and U_i is the ***utility function***

$\mathsf{U}_i : \times_{k \in [r]} \Sigma_k \to \mathbb{R}$ for player i. A ***win-lose game*** is a game SG such that for each player $i \in [r]$, $\mathsf{U}_i : \times_{k \in [r]} \Sigma_k \to \{0, 1\}$. For each player $i \in [r]$, denote $\Sigma_{-i} = \times_{k \in [r] \setminus \{i\}} \Sigma_k$; denote $\Sigma = \times_{k \in [r]} \Sigma_k$. For each integer $r \geq 2$, denote as r-$\mathcal{SG}$ the set of *r-player games*; so, $\mathcal{SG} = \bigcup_{r \geq 2} r$-$\mathcal{SG}$ is the set of all strategic games. A ***profile*** is a tuple $\mathbf{s}$ of r strategies, one for each player. For a profile $\mathbf{s}$, the vector $\mathsf{U}(\mathbf{s}) = \langle \mathsf{U}_1(\mathbf{s}), \ldots, \mathsf{U}_r(\mathbf{s}) \rangle$ is called the ***utility vector***. A ***partial profile*** $\mathbf{s}_{-i}$ is a tuple of $r - 1$ strategies, one for each player other than i; so $\mathbf{s}_{-i} \in \Sigma_{-i}$. For a profile $\mathbf{s}$ and a strategy $t_i \in \Sigma_i$ of player i, denoted as $\mathbf{s}_{-i} \diamond t_i$ the profile obtained by substituting t_i for s_i in $\mathbf{s}$.

A ***mixed strategy*** for player $i \in [r]$ is a probability distribution σ_i on her strategy set Σ_i: a function $\sigma_i : \Sigma_i \to [0, 1]$ such that $\sum_{s \in \Sigma_i} \sigma_i(s) = 1$. Denote as $\mathsf{Supp}(\sigma_i)$ the set of strategies $s \in \Sigma_i$ such that $\sigma_i(s) > 0$. The mixed strategy $\sigma_i : \Sigma_i \to [0, 1]$ is ***rational*** if all values of σ_i are rational numbers. A ***mixed profile*** $\boldsymbol{\sigma} = (\sigma_i)_{i \in [r]}$ is a tuple of mixed strategies, one for each player. A ***partial mixed profile*** $\boldsymbol{\sigma}_{-i}$ is a tuple of $r - 1$ mixed strategies, one for each player other than i. For a mixed profile $\boldsymbol{\sigma}$ and a mixed strategy τ_i of player $i \in [r]$, denote as $\boldsymbol{\sigma}_{-i} \diamond \tau_i$ the mixed profile obtained by substituting τ_i for σ_i in $\boldsymbol{\sigma}$. A mixed profile is ***rational*** if all of its mixed strategies are rational. So, a profile is the degenerate case of a mixed profile where all probabilities are either 0 or 1. A mixed profile $\boldsymbol{\sigma}$ induces a probability measure $\mathbb{P}_{\boldsymbol{\sigma}}$ on the set of profiles in the natural way. Say that the profile $\mathbf{s}$ is ***enabled***, and write $\mathbf{s} \sim \sigma$, in the mixed profile $\boldsymbol{\sigma}$ if $\mathbb{P}_{\boldsymbol{\sigma}}(\mathbf{s}) > 0$; note that for a profile $\mathbf{s}$, $\mathbb{P}_{\boldsymbol{\sigma}}(\mathbf{s}) = \prod_{k \in [r]} \sigma_k(s_k)$. Under the mixed profile $\boldsymbol{\sigma}$, the utility of each player becomes a random variable. So, associated with the mixed profile $\boldsymbol{\sigma}$ is the ***expected utility*** for each player $i \in [r]$, denoted as $\mathsf{U}_i(\boldsymbol{\sigma})$ and defined as the expectation according to $\mathbb{P}_{\boldsymbol{\sigma}}$ of her utility for a profile $\mathbf{s}$ enabled in the mixed profile $\boldsymbol{\sigma}$; so, $\mathsf{U}_i(\boldsymbol{\sigma}) = \mathbb{E}_{\mathbf{s} \sim \boldsymbol{\sigma}} (\mathsf{U}_i(\mathbf{s})) = \sum_{\mathbf{s} \in \Sigma(\mathsf{SG})} \mathbb{P}_{\boldsymbol{\sigma}}(\mathbf{s}) \cdot \mathsf{U}_i(\mathbf{s}) = \sum_{\mathbf{s} \in \Sigma(\mathsf{SG})} \left(\prod_{k \in [r]} \sigma_k(s_r) \right) \cdot \mathsf{U}_i(\mathbf{s})$.

A ***pure Nash equilibrium***, is a profile $\mathbf{s} \in \Sigma$ such that for each player $i \in [r]$ and for each strategy $t_i \in \Sigma_i$, $\mathsf{U}_i(\mathbf{s}) \geq \mathsf{U}_i(\mathbf{s}_{-i} \diamond t_i)$. A ***best-response*** for player i in $\boldsymbol{\sigma}$ is a pure strategy $t \in \Sigma_i$ such that $\mathsf{U}_i(\boldsymbol{\sigma}_{-i} \diamond t) \geq \mathsf{U}_i(\boldsymbol{\sigma}_{-i} \diamond s)$ for each $s \in \Sigma_i$. A ***mixed Nash equilibrium***, or ***Nash equilibrium*** for short, is a mixed profile $\boldsymbol{\sigma}$ such that for each player $i \in [r]$ and for each mixed strategy τ_i, $\mathsf{U}_i(\boldsymbol{\sigma}) \geq \mathsf{U}_i(\boldsymbol{\sigma}_{-i} \diamond \tau_i)$. Denote as $\mathcal{NE}(\mathsf{SG})$ the set of Nash equilibria for a strategic game SG. We shall later use the following basic property:

Lemma 1. *A mixed profile $\boldsymbol{\sigma}$ is a Nash equilibrium if and only if for each player $i \in [r]$, (i) for each pair of strategies $s_i, t_i \in \mathsf{Supp}(\sigma_i)$, $\mathsf{U}_i(\boldsymbol{\sigma}_{-i} \diamond s_i) = \mathsf{U}_i(\boldsymbol{\sigma}_{-i} \diamond t_i)$, (ii) for each strategy $t_i \in \mathsf{Supp}(\sigma_i)$, $\mathsf{U}_i(\boldsymbol{\sigma}) = \mathsf{U}_i(\boldsymbol{\sigma}_{-i} \diamond t_i)$, and (iii) for each pair of strategies $s_i \in \mathsf{Supp}(\sigma_i)$ and $t_i \notin \mathsf{Supp}(\sigma_i)$, $\mathsf{U}_i(\boldsymbol{\sigma}_{-i} \diamond s_i) \geq \mathsf{U}_i(\boldsymbol{\sigma}_{-i} \diamond t_i)$.*

Given two mixed profiles $\boldsymbol{\sigma}$ and $\widehat{\boldsymbol{\sigma}}$, define $\mathsf{Diff}(\boldsymbol{\sigma}, \widehat{\boldsymbol{\sigma}}) := \{i \in [r] : \sigma_i \neq \widehat{\sigma}_i\}$ as the set of players playing different mixed strategies in $\boldsymbol{\sigma}$ and $\widehat{\boldsymbol{\sigma}}$. A Nash equilibrium $\boldsymbol{\sigma}$ is ***Strong*** if for each mixed profile $\widehat{\boldsymbol{\sigma}}$ such that $\mathsf{U}_i(\widehat{\boldsymbol{\sigma}}) > \mathsf{U}_i(\boldsymbol{\sigma})$ for some $i \in [r]$, it holds $\mathsf{U}_j(\widehat{\boldsymbol{\sigma}}) \leq \mathsf{U}_j(\boldsymbol{\sigma})$ for some $j \in \mathsf{Diff}(\boldsymbol{\sigma}, \widehat{\boldsymbol{\sigma}})$. A Nash equilibrium $\boldsymbol{\sigma}$ is ***Pareto-Optimal*** if for each mixed profile $\widehat{\boldsymbol{\sigma}}$ such that $\mathsf{U}_i(\widehat{\boldsymbol{\sigma}}) > \mathsf{U}_i(\boldsymbol{\sigma})$ for some $i \in [r]$, it holds $\mathsf{U}_j(\widehat{\boldsymbol{\sigma}}) < \mathsf{U}_j(\boldsymbol{\sigma})$ for some $j \in [r]$.

3 Statement of Results and Related Work

This section collects together the formal statements of *(i)* the various decision problems we shall consider, all in the style of Garey and Johnson [9] where I. and Q. stand for INSTANCE and QUESTION, respectively, and *(ii)* the main results (Theorems 1, 2 and 3). It concludes with a summary of known results.

MAXIMUM UTILITY
I.: A game SG and a number u.
Q.: Is there a Nash equilibrium $\boldsymbol{\sigma}$ s.t. for each player $i \in [r]$, $\mathsf{U}_i(\boldsymbol{\sigma}) \geq u$?

MINIMUM UTILITY
I.: A game SG and a number u.
Q.: Is there a Nash equilibrium $\boldsymbol{\sigma}$ s.t. for each player $i \in [r]$, $\mathsf{U}_i(\boldsymbol{\sigma}) \leq u$?

∃ SECOND NASH
I.: A game SG.
Q.: Does SG have at least two Nash equilibria?

NASH IN A SUBSET
I.: A game SG, and a subset of strategies $\mathsf{T}_i \subseteq \Sigma_i$ for each player $i \in [r]$.
Q.: Is there a Nash equilibrium $\boldsymbol{\sigma}$ s.t. for each player $i \in [r]$, $\mathsf{Supp}(\sigma_i) \subseteq \mathsf{T}_i$?

A SUBSET IN NASH
I.: A game SG, and a subset of strategies $\mathsf{T}_i \subseteq \Sigma_i$ for each player $i \in [r]$.
Q.: Is there a Nash equilibrium $\boldsymbol{\sigma}$ s.t. for each player $i \in [r]$, $\mathsf{T}_i \subseteq \mathsf{Supp}(\sigma_i)$?

NASH MAXIMUM SUPPORT
I.: A game SG and an integer $k \geq 1$.
Q.: Is there a Nash equilibrium $\boldsymbol{\sigma}$ s.t. for each player $i \in [r]$, $|\mathsf{Supp}(\sigma_i)| \geq k$?

NASH MINIMUM SUPPORT
I.: A game SG and an integer $k \geq 1$.
Q.: Is there a Nash equilibrium $\boldsymbol{\sigma}$ s.t. for each player $i \in [r]$, $|\mathsf{Supp}(\sigma_i)| \leq k$?

MAXIMUM TOTAL UTILITY
I.: A game SG and a number u.
Q.: Is there a Nash equilibrium $\boldsymbol{\sigma}$ s.t. $\sum_{i\in[r]} \mathsf{U}_i(\boldsymbol{\sigma}) \geq u$?

MINIMUM TOTAL UTILITY
I.: A game SG and a number u.
Q.: Is there a Nash equilibrium $\boldsymbol{\sigma}$ s.t. $\sum_{i\in[r]} \mathsf{U}_i(\boldsymbol{\sigma}) \leq u$?

∃ PARETO OPTIMAL NASH
I.: A game SG.
Q.: Is there a Pareto Optimal Nash equilibrium?

∃ NON-PARETO OPTIMAL NASH
I.: A game SG.
Q.: Is there a non-Pareto Optimal Nash equilibrium?

∃ STRONG NASH
I.: A game SG.
Q.: Is there a Strong Nash equilibrium?

∃ NON-STRONG NASH

I.: A game SG.

Q.: Is there a non-Strong Nash equilibrium?

Clearly, restricted to two-player games with rational utilities, all these decision problems belong to $\mathcal{NP}$. We shall start with two milestone decision problems related to the notion of ***Nash-equivalence***: Two strategic games $\widehat{\mathsf{SG}}$ and SG are ***Nash-equivalent*** [2] if $\mathcal{NE}(\widehat{\mathsf{SG}}) = \mathcal{NE}(\mathsf{SG})$: they have the same set of Nash equilibria. This leads to the following decision problem [2]:

NASH-EQUIVALENCE

I.: Two games $\widehat{\mathsf{SG}}$ and SG from r-$\mathcal{SG}$, for some integer $r \geq 2$.

Q.: Are $\widehat{\mathsf{SG}}$ and SG Nash-equivalent?

For a fixed game $\widehat{\mathsf{SG}}$, called the ***gadget game***, a parameterized restriction of NASH-EQUIVALENCE with a single input (the game SG) results to the following decision problem [2]:

NASH-EQUIVALENCE($\widehat{\mathsf{SG}}$)

I.: A game SG from r-$\mathcal{SG}$ (where $\widehat{\mathsf{SG}}$ is from r-$\mathcal{SG}$).

Q.: Are $\widehat{\mathsf{SG}}$ and SG Nash-equivalent?

So, NASH-EQUIVALENCE($\widehat{\mathsf{SG}}$)$\leq_{\mathrm{P}}$NASH-EQUIVALENCE. It has been shown that NASH-EQUIVALENCE($\widehat{\mathsf{SG}}$) is co-$\mathcal{NP}$-hard [2, Theorem 1]; hence, so is NASH-EQUIVALENCE. A strategic game SG satisfies the **positive utility property** if for each player $i \in [r]$ and each partial pure profile $\mathbf{s}_{-i} \in \Sigma_{-i}$, there is a strategy $t(\mathbf{s}_{-i}) \in \Sigma_i$ for which $\mathsf{U}_i(\mathbf{s}_{-i} \diamond t(\mathbf{s}_{-i})) > 0$.
The main results follow:

Theorem 1. *Fix a win-lose game* $\widehat{\mathsf{SG}}$ *with the positive utility property. Then, restricted to win-lose games,* NASH-EQUIVALENCE($\widehat{\mathsf{SG}}$) *is co-*$\mathcal{NP}$*-hard.*

By suitable choices for the gadget game $\widehat{\mathsf{SG}}$, the Nash-equivalence of the given game SG to the gadget game $\widehat{\mathsf{SG}}$ becomes equivalent to the fact that SG does not have a Nash equilibrium with certain properties. Since deciding the Nash-equivalence is co-$\mathcal{NP}$-hard, we obtain:

Theorem 2. *Restricted to two-player win-lose games, the following decision problems are* $\mathcal{NP}$*-complete:*

Group I	*Group II*
MINIMUM UTILITY	MAXIMUM UTILITY
NASH MAXIMUM SUPPORT	NASH MINIMUM SUPPORT
MINIMUM TOTAL UTILITY	MAXIMUM TOTAL UTILITY
∃ SECOND NASH	
NASH IN A SUBSET	
A SUBSET IN NASH	
∃ NON-PARETO OPTIMAL NASH	
∃ NON-STRONG NASH	

Problem	$\mathcal{NP}$-complete for Two-player		
	Games	Symmetric Games	Win-lose Games
NASH-EQUIVALENCE($\widehat{\mathsf{SG}}$)	[2]	?	Theorem 1
NASH-EQUIVALENCE	[2]	?	Theorem 1
MAXIMUM UTILITY	$\Leftarrow$	[6, 10]	Theorem 2
MAXIMUM TOTAL UTILITY	$\Leftarrow$	[6]	Theorem 2
MINIMUM UTILITY	$\Leftarrow$	?	Theorem 2
MINIMUM TOTAL UTILITY	$\Leftarrow$	?	Theorem 2
$\exists$ SECOND NASH	$\Leftarrow$	[6, 10]	Theorem 2 & [5]
NASH IN A SUBSET	$\Leftarrow$	[6, 10]	Theorem 2
A SUBSET IN NASH	$\Leftarrow$	[6, 10]	Theorem 2
NASH MAXIMUM SUPPORT	$\Leftarrow$	[6, 10]	Theorem 2
NASH MINIMUM SUPPORT	$\Leftarrow$	[10]	Theorem 2
$\exists$ STRONG NASH	$\Leftarrow$	[6]	?
$\exists$ NON-STRONG NASH	$\Leftarrow$	?	Theorem 2
$\exists$ PARETO-OPTIMAL NASH	$\Leftarrow$	[6]	?
$\exists$ NON-PARETO-OPTIMAL NASH	$\Leftarrow$	?	Theorem 2
NASH-REDUCTION($\widehat{\mathsf{SG}}$)	[2]	?	?
NASH-REDUCTION	[2]	?	?

Problem	$\mathcal{NP}$-complete for Three-player		
	Games	Symmetric Games	Win-lose Games
$\exists$ RATIONAL NASH	[2]	?	Theorem 3
$\exists$ IRRATIONAL NASH*	[2]	?	?

Fig. 1. Summary of results and comparison to previous work. The symbol "?" indicates that the $\mathcal{NP}$-hardness of the corresponding decision problem remains open. (*: The problem $\exists$ IRRATIONAL NASH was only shown to be $\mathcal{NP}$-hard [2].) An arrow $\Leftarrow$ in the column for two-player (general) games indicated that the result follows immediately from the corresponding result in some of the other two columns.

Group I and *Group II* include decision problems whose proof of $\mathcal{NP}$-hardness will use the two-player win-lose gadget games $\widehat{\mathsf{SG}}_1$ and $\widehat{\mathsf{SG}}_n$, respectively.

Theorem 3. *Restricted to three-player win-lose games,* $\exists$ RATIONAL NASH *is* $\mathcal{NP}$*-complete.*

Figure 1 provides a tabular summary of our results in direct comparison to previous results. Roughly speaking, our results extend almost all previous $\mathcal{NP}$-completeness results for decision problems about Nash equilibria in two-player symmetric games [6, 10] (resp., three-player general games) to two-player win-lose games (resp., three-player win-lose games).

4 Proofs

Theorem 1:

Given an instance ϕ of 3-SAT with $n = |\mathsf{V}(\phi)| \geq 5$ and an r-player win-lose gadget game $\widehat{\mathsf{SG}}$ with the positive utility property, construct (in polynomial time) the r-player win-lose game $\mathsf{SG}(\phi) = \left\langle [r], \{\Sigma_i\}_{i\in[r]}, \{\mathsf{U}_i\}_{i\in[r]} \right\rangle$ as follows:

For each player $i \in [2]$, $\Sigma_i := \widehat{\Sigma}_i \cup \mathsf{L}(\phi) \cup \mathcal{C}(\phi) \cup \mathsf{V}$; for each player $i \in r \setminus [2]$, $\Sigma_i := \widehat{\Sigma}_i \cup \{\delta\}$, where each strategy in $\widehat{\Sigma}_i$ with $i \in [r]$ is inherited from the gadget game $\widehat{\mathsf{SG}}$, $\mathsf{L}(\phi)$ and $\mathcal{C}(\phi)$ are strategies coming from the formula ϕ, while $\mathsf{V} = \{\mathsf{v}_{i,j} : 0 \leq i, j < n \text{ and } i \neq j\}$ and δ are special strategies needed to force each Nash equilibrium $\boldsymbol{\sigma} \in \mathcal{NE}(\mathsf{SG}(\phi)) \setminus \mathcal{NE}(\widehat{\mathsf{SG}})$ to have some desired properties. Players 1 and 2 are *special*; they are the only players whose sets of strategies are influenced by the formula ϕ.

Fix a profile $\mathbf{s} = \langle s_1, \ldots, s_r \rangle$ from $\Sigma = \Sigma_1 \times \ldots \times \Sigma_r$. Use $\mathbf{s}$ to partition $[r]$ into $\widehat{\mathcal{P}}(\mathbf{s}) = \{i \in [r] \mid s_i \in \widehat{\Sigma}_i\}$ and $\mathcal{P}(\mathbf{s}) = \{i \in [r] \mid s_i \notin \widehat{\Sigma}_i\}$; loosely speaking, $\widehat{\mathcal{P}}(\mathbf{s})$ and $\mathcal{P}(\mathbf{s})$ are the sets of players choosing and not choosing strategies inherited from $\widehat{\mathsf{SG}}$, respectively. We use c to denote a generic clause belonging to $\mathcal{C}(\phi)$ and either ℓ or $\overline{\ell}$, eventually indexed with constants or variables in the set $I_n := \{0, 1, \ldots, n-1\}$, to denote a generic literal in $\mathsf{L}(\phi)$. Moreover, all the arithmetic operations defined on I_n are taken modulo n. Finally, we denote as π a permutation on $[r]$. The utility vector $\mathsf{U}(\mathbf{s})$ is depicted in Figure 2. For brevity,

Case	Condition on the profile $\mathbf{s} = \langle s_1, s_2, \ldots, s_r \rangle$	Utility vector $\mathsf{U}(\mathbf{s})$
(1)	$\mathbf{s} = \langle \ell, \overline{\ell}, \delta, \ldots, \delta \rangle$	$\langle 0, 0, 1, \ldots, 1 \rangle$
(2)	$\mathbf{s} = \langle \ell_i, \ell_j, \delta, \ldots, \delta \rangle$ with $(j = i \wedge \ell_i \neq \overline{\ell}_j) \vee j = i + 1$	$\langle 1, 0, 1, \ldots, 1 \rangle$
(3)	$\mathbf{s} = \langle \ell_i, \ell_j, \delta, \ldots, \delta \rangle$ with $j = i + 2 \vee j = i + 3$	$\langle 0, 1, 1, \ldots, 1 \rangle$
(4)	$\mathbf{s} = \langle \ell_i, \ell_j, \delta, \ldots, \delta \rangle$ with $j \notin \{i, i+1, i+2, i+3\}$	$\langle 0, 0, 1, \ldots, 1 \rangle$
(5)	$\mathbf{s} = \langle \mathsf{v}_{ij}, \ell_k, \delta, \ldots, \delta \rangle$ with $k \in \{i, j\}$	$\langle 1, 0, 1, \ldots, 1 \rangle$
(6)	$\mathbf{s} = \langle \mathsf{c}, \ell, \delta, \ldots, \delta \rangle$ with $\overline{\ell} \in \mathsf{c}$	$\langle 1, 0, 1, \ldots, 1 \rangle$
(7)	For each $i \in [r]$, $s_i \in \widehat{\Sigma}_i$	$\widehat{\mathsf{U}}(\langle s_1, \ldots, s_r \rangle)$
(8)	$\lvert\widehat{\mathcal{P}}(\mathbf{s})\rvert > 1$ and $\lvert\mathcal{P}(\mathbf{s})\rvert > 0$	$\mathsf{U}_i(\mathbf{s}) = 1$ if $i \in \widehat{\mathcal{P}}(\mathbf{s})$ $\mathsf{U}_i(\mathbf{s}) = 0$ if $i \in \mathcal{P}(\mathbf{s})$
(9)	$\widehat{\mathcal{P}}(\mathbf{s}) = \{i\}$ with $i \in [r] \setminus [2]$	$\mathsf{U}_i(\mathbf{s}) = 1$ $\mathsf{U}_{j \neq i}(\mathbf{s}) = 0$
(10)	$\widehat{\mathcal{P}}(\mathbf{s}) = \{i\}$ with $i \in [2]$ and $s_{[2] \setminus \{i\}} \in \{\ell_0, \overline{\ell}_0, \ell_1, \overline{\ell}_1\} \cup \mathcal{C}(\phi) \cup \mathsf{V}$	$\mathsf{U}_i(\mathbf{s}) = 1$ $\mathsf{U}_{j \neq i}(\mathbf{s}) = 0$
(11)	$\mathbf{s} = \pi(\mathbf{t})$, where $\mathbf{t}$ falls in one of the Cases (1) through (10)	$\pi(\mathsf{U}(\mathbf{t}))$
(12)	None of the above	$\langle 0, \ldots, 0 \rangle$

Fig. 2. Utility functions for the game $\mathsf{SG}(\phi)$

we shall also write SG, L and $\mathcal{C}$ for $\mathsf{SG}(\phi)$, $\mathsf{L}(\phi)$ and $\mathcal{C}(\phi)$. We first prove two significant properties of the game SG which come from the definition of the utility functions and the positive utility property of the gadget game $\widehat{\mathsf{SG}}$.

Lemma 2. *Let $\boldsymbol{\sigma}$ be a mixed strategy profile of SG such that $\mathsf{Supp}(\sigma_i) \subseteq \widehat{\Sigma}_i$ for some $i \in [r]$. Then, for any $[r] \ni j \neq i$, the set of best-responses for player j in $\boldsymbol{\sigma}$ is contained in $\widehat{\Sigma}_j$.*

Lemma 3. *Let $\boldsymbol{\sigma}$ be a mixed strategy profile of* SG *such that* $\mathsf{Supp}(\sigma_i) \subseteq \widehat{\Sigma}_i \cup \mathcal{C} \cup \mathsf{V}$ *for some $i \in [2]$. Then, the set of best-responses for player $j = [2] \setminus \{i\}$ in $\boldsymbol{\sigma}$ is contained in $\widehat{\Sigma}_j$.*

We first prove that the only Nash equilibria for SG in which one of the players only plays strategies inherited from the gadget game are those belonging to $\mathcal{NE}(\widehat{\mathsf{SG}})$; the proof uses Lemma 2.

Lemma 4. $\mathcal{NE}(\widehat{\mathsf{SG}}) \subseteq \mathcal{NE}(\mathsf{SG})$ *and there is no Nash equilibrium $\boldsymbol{\sigma} \in \mathcal{NE}(\mathsf{SG}) \setminus \mathcal{NE}(\widehat{\mathsf{SG}})$ such that $\mathsf{Supp}(\sigma_i) \subseteq \widehat{\Sigma}_i$ for some player $i \in [r]$.*

For a given mixed profile $\boldsymbol{\sigma}$, we denote as $\sigma_i(\mathsf{L})$, $\sigma_i(\mathcal{C} \cup \mathsf{V})$ and $\sigma_i(\widehat{\Sigma}_i)$, the total probability that player i puts on strategies belonging to L, $\mathcal{C} \cup \mathsf{V}$ and $\widehat{\Sigma}_i$, respectively. Using Lemma 3, we prove that in a Nash equilibrium $\boldsymbol{\sigma} \in \mathcal{NE}(\mathsf{SG}) \setminus \mathcal{NE}(\widehat{\mathsf{SG}})$, both special players play some literal with positive probability.

Lemma 5. *In a Nash equilibrium $\boldsymbol{\sigma} \in \mathcal{NE}(\mathsf{SG}) \setminus \mathcal{NE}(\widehat{\mathsf{SG}})$, $\sigma_1(\mathsf{L}) \cdot \sigma_2(\mathsf{L}) > 0$.*

Next, we prove that in a Nash equilibrium $\boldsymbol{\sigma} \in \mathcal{NE}(\mathsf{SG}) \setminus \mathcal{NE}(\widehat{\mathsf{SG}})$, the two special players play each pair of literals ℓ and $\overline{\ell}$ with the same positive probability.

Lemma 6. *In a Nash equilibrium $\boldsymbol{\sigma} \in \mathcal{NE}(\mathsf{SG}) \setminus \mathcal{NE}(\widehat{\mathsf{SG}})$, it holds that $\sigma_i(\ell) + \sigma_i(\overline{\ell}) = \sigma_i(\mathsf{L})/n > 0$ for any $i \in [2]$ and for any $\ell \in \mathsf{L}$.*

Next, we prove that in a Nash equilibrium $\boldsymbol{\sigma} \in \mathcal{NE}(\mathsf{SG}) \setminus \mathcal{NE}(\widehat{\mathsf{SG}})$, all the non-special players adopt δ as a pure strategy.

Lemma 7. *In a Nash equilibrium $\boldsymbol{\sigma} \in \mathcal{NE}(\mathsf{SG}) \setminus \mathcal{NE}(\widehat{\mathsf{SG}})$, it holds that $\sigma_i(\delta) = 1$ for each player $i \in [r] \setminus [2]$.*

As a consequence of Lemma 7, the utility of the two special players in either $\boldsymbol{\sigma}$ and $\boldsymbol{\sigma}_{-i} \diamond s$, for any $s \in \Sigma_i$ and $i \in [2]$, are entirely determined by their chosen strategies. So, henceforth, we shall only focus on the strategies adopted by the two special players when considering Nash equilibria in the set $\mathcal{NE}(\mathsf{SG}) \setminus \mathcal{NE}(\widehat{\mathsf{SG}})$.

We continue to refine the conditions satisfied by a Nash equilibrium by showing that, in a Nash equilibrium other than the ones inherited from the gadget game, no special player puts positive probability on strategies from $\mathcal{C} \cup \mathsf{V}$.

Lemma 8. *In a Nash equilibrium $\boldsymbol{\sigma} \in \mathcal{NE}(\mathsf{SG}) \setminus \mathcal{NE}(\widehat{\mathsf{SG}})$, it holds that $\sigma_i(\mathcal{C} \cup \mathsf{V}) = 0$, for each special player $i \in [2]$.*

Next step is to prove that in each Nash equilibrium other than those inherited from the gadget game, both special players only play literals.

Lemma 9. *In a Nash equilibrium $\boldsymbol{\sigma} \in \mathcal{NE}(\mathsf{SG}) \setminus \mathcal{NE}(\widehat{\mathsf{SG}})$, it holds that $\sigma_i(\mathsf{L}) = 1$, for each special player $i \in [2]$.*

The next claim, combined with Lemma 6, shows that a Nash equilibrium other than those inherited from the gadget game induces an assignment for ϕ.

Lemma 10. *In a Nash equilibrium $\boldsymbol{\sigma} \in \mathcal{NE}(\mathsf{SG}) \setminus \mathcal{NE}(\widehat{\mathsf{SG}})$, it holds that $\sigma_1(\ell) \cdot \sigma_2(\overline{\ell}) = 0$, for any $\ell \in \mathsf{L}$.*

Corollary 1. *In a Nash equilibrium $\boldsymbol{\sigma} \in \mathcal{NE}(\mathsf{SG}) \setminus \mathcal{NE}(\widehat{\mathsf{SG}})$, it holds that $\mathsf{U}_1(\boldsymbol{\sigma}) = \mathsf{U}_2(\boldsymbol{\sigma}) = 2/n$.*

As a consequence of Lemmas 6 and 10, we have the following corollary.

Corollary 2. *In a Nash equilibrium $\boldsymbol{\sigma} \in \mathcal{NE}(\mathsf{SG}) \setminus \mathcal{NE}(\widehat{\mathsf{SG}})$, both players uniformly randomizes over the same assignment for ϕ.*

Proposition 1. *Fix a win-lose game $\widehat{\mathsf{SG}}$ with the positive utility property and an instance ϕ of 3-SAT. If ϕ is unsatisfiable, then it holds $\mathcal{NE}(\mathsf{SG}) = \mathcal{NE}(\widehat{\mathsf{SG}})$.*

Proposition 2. *Fix a win-lose game $\widehat{\mathsf{SG}}$ with the positive utility property and an instance ϕ of 3-SAT with $n = |\mathsf{V}(\phi)|$. If ϕ is satisfiable, then SG has a Nash equilibrium $\boldsymbol{\sigma} \in \mathcal{NE}(\mathsf{SG}) \setminus \mathcal{NE}(\widehat{\mathsf{SG}})$ such that for each $i \in [2]$, it holds (i) $|\mathsf{Supp}(\sigma_i)| = n$, (ii) $\mathsf{U}_i(\boldsymbol{\sigma}) = 2/n$, (iii) $\mathsf{Supp}(\sigma_i) \cap \widehat{\Sigma}_i = \emptyset$, (iv) $\sigma_i(s) = 1/n$ for each $s \in \mathsf{Supp}(\sigma_i)$. Moreover if ϕ is satisfiable with v_j set to true for some $j \in I_n$, SG has a Nash equilibrium $\boldsymbol{\sigma} \in \mathcal{NE}(\mathsf{SG}) \setminus \mathcal{NE}(\widehat{\mathsf{SG}})$ such that for each $i \in [2]$, (v) it holds $\ell_j \in \mathsf{Supp}(\sigma_i)$.*

Since $\widehat{\mathsf{SG}}$ satisfies the positive utility property, then, by Propositions 1 and 2, it holds $\mathcal{NE}(\mathsf{SG}) \neq \mathcal{NE}(\widehat{\mathsf{SG}})$ if and only if ϕ is satisfiable. So, SG and $\widehat{\mathsf{SG}}$ are Nash-equivalent if and only if ϕ is unsatisfiable. Hence, $\overline{\text{3-SAT}} \leq_{\mathrm{P}}$ NASH-EQUIVALENCE($\widehat{\mathsf{SG}}$) and NASH-EQUIVALENCE($\widehat{\mathsf{SG}}$) is co-$\mathcal{NP}$-hard.

Theorem 2:

Define $\widehat{\mathsf{SG}}_1$ as the two-player win-lose game in which both players have exactly one strategy and all utilities are 1. Since $\widehat{\mathsf{SG}}_1$ trivially satisfies the positive utility property, it follows by Proposition 1 that if ϕ is unsatisfiable then SG has a unique Nash equilibrium $\widehat{\boldsymbol{\sigma}}$ which is pure, Pareto-Optimal, Strong and has $\mathsf{U}_i(\widehat{\boldsymbol{\sigma}}) = 1$ for each $i \in [2]$.

By Propositions 1 and 2, there is a Nash equilibrium $\boldsymbol{\sigma} \in \mathcal{NE}(\mathsf{SG}) \setminus \mathcal{NE}(\widehat{\mathsf{SG}}_1)$ if and only if ϕ is satisfiable, thus showing $\mathcal{NP}$-hardness of $\exists$ SECOND NASH.

By property (i) of Proposition 2, it follows that if ϕ is satisfiable, then for any $1 \leq k \leq n$, SG has a Nash equilibrium $\boldsymbol{\sigma}$ such that $|\mathsf{Supp}(\sigma_i)| \geq k$ for each player $i \in [2]$. Hence, for any $2 \leq k \leq n$, SG has a Nash equilibrium $\boldsymbol{\sigma}$ such that $|\mathsf{Supp}(\sigma_i)| \geq k$ for each player $i \in [2]$ if and only if ϕ is satisfiable, thus showing $\mathcal{NP}$-hardness of NASH MAXIMUM SUPPORT.

The arguments for MINIMUM UTILITY, MINIMUM TOTAL UTILITY, NASH IN A SUBSET, A SUBSET IN NASH, $\exists$ NON-PARETO-OPTIMAL NASH and $\exists$ NON-STRONG NASH are deferred to the Appendix.

Given a positive integer n, define $\widehat{\mathsf{SG}}_n$ as the two-player win-lose game in which both players have the same set of strategies $\{s_0, \ldots, s_{n-1}\}$ and $\mathsf{U}_1(\mathbf{s}) = 1$ if and only if $\mathbf{s} = \langle s_i, s_i \rangle$, while $\mathsf{U}_2(\mathbf{s}) = 1$ if and only if $\mathbf{s} = \langle s_i, s_{i+1} \rangle$, where addition is taken modulo n; so player 1 wins if the two players concur while player 2 wins if the two players choose successive strategies (with player 2 following).

Proposition 3. $\widehat{\mathsf{SG}}_n$ *satisfies the positive utility property and has a unique Nash equilibrium in which both players play all the strategies with probability* $\frac{1}{n}$ *and achieve expected utility equal to* $\frac{1}{n}$.

For any instance ϕ of 3-SAT with $|\mathsf{V}(\phi)| \geq 5$, consider the gadget game $\widehat{\mathsf{SG}}_h$ for some $h > n$. Since $\widehat{\mathsf{SG}}_h$ satisfies the positive utility property, it follows by Propositions 1 and 3 that if ϕ is unsatisfiable, then SG has a unique Nash equilibrium $\widehat{\boldsymbol{\sigma}}$ such that for each $i \in [2]$ it holds $\mathsf{U}_i(\widehat{\boldsymbol{\sigma}}) = \frac{1}{h}$ and $|\mathsf{Supp}(\sigma_i)| = h$.

By property (i) of Proposition 2, it follows that if ϕ is satisfiable, then for any $n \leq k \leq h$, SG has a Nash equilibrium $\boldsymbol{\sigma}$ such that $|\mathsf{Supp}(\sigma_i)| \leq k$ for each player $i \in [2]$. Hence, for any $n \leq k < h$, SG has a Nash equilibrium $\boldsymbol{\sigma}$ such that $|\mathsf{Supp}(\sigma_i)| \leq k$ for each player $i \in [2]$ if and only if ϕ is satisfiable, thus showing $\mathcal{NP}$-hardness of NASH MINIMUM SUPPORT. The arguments for MAXIMUM UTILITY and MAXIMUM TOTAL UTILITY are deferred to the Appendix.

<u>**Theorem 3:**</u>
Define $\widehat{\mathsf{SG}}_2$ as the gadget game in which there are three players, the set of strategies for players 1 and 2 is $\{0, 1\}$, the set of strategies for player 3 is $\{0, 1, 2\}$ and the utility functions are defined as follows:

Profile s	Utility vector $\widehat{\mathsf{U}}(\mathsf{s})$	Profile s	Utility vector $\widehat{\mathsf{U}}(\mathsf{s})$
$\langle 0,0,0\rangle$	$\langle 1,0,1\rangle$	$\langle 1,0,0\rangle$	$\langle 0,0,1\rangle$
$\langle 0,0,1\rangle$	$\langle 1,1,0\rangle$	$\langle 1,0,1\rangle$	$\langle 0,1,1\rangle$
$\langle 0,0,2\rangle$	$\langle 0,1,0\rangle$	$\langle 1,0,2\rangle$	$\langle 1,0,0\rangle$
$\langle 0,1,0\rangle$	$\langle 0,1,0\rangle$	$\langle 1,1,0\rangle$	$\langle 1,1,0\rangle$
$\langle 0,1,1\rangle$	$\langle 1,0,0\rangle$	$\langle 1,1,1\rangle$	$\langle 0,0,1\rangle$
$\langle 0,1,2\rangle$	$\langle 0,0,1\rangle$	$\langle 1,1,2\rangle$	$\langle 1,1,0\rangle$

Proposition 4. $\widehat{\mathsf{SG}}_2$ *satisfies the positive utility property and has only one Nash equilibrium which is irrational.*

Since $\widehat{\mathsf{SG}}_2$ satisfies the positive utility property and has only irrational Nash equilibria, it follows from Proposition 1 that if ϕ is unsatisfiable, then SG only admits irrational Nash equilibria. Conversely, from property (iv) of Proposition 2 it follows that if ϕ is satisfiable, then SG admits a rational Nash equilibrium. Hence, SG admits a rational Nash equilibrium if and only if ϕ is satisfiable.

5 Epilogue

We have shown that win-lose games are **as hard** as general games with respect to (all but four of) the decision problems about Nash equilibria that were previously considered in [2, 6, 10]. Our result complements the result of Abbott *et al.* [1] that win-lose games are as hard as general games with respect to the search problem for a Nash equilibrium.

Our work leaves open several interesting problems. The most obvious open problem is whether the decision problems ∃ PARETO-OPTIMAL NASH, ∃

STRONG NASH and NASH-REDUCTION (resp., $\exists$ IRRATIONAL NASH) are still $\mathcal{NP}$-hard for two-player (resp., three-player) win-lose games.

Win-lose (two-player) games and symmetric (two-player) games are individually hard, as shown in this paper and in [6], respectively. Perhaps the most interesting problem left open is whether symmetric, win-lose (two-player) games are still as hard as general games with respect to the decision problems.

References

1. Abbott, T., Kane, D., Valiant, P.: On the Complexity of Two-Player Win-Lose Games. In: Proceedings of the 46th Annual IEEE Symposium on Foundations of Computer Sciences, pp. 113–122 (October 2005)
2. Bilò, V., Mavronicolas, M.: Complexity of Rational and Irrational Nash Equilibria. In: Persiano, G. (ed.) SAGT 2011. LNCS, vol. 6982, pp. 200–211. Springer, Heidelberg (2011)
3. Chen, X., Deng, X., Teng, S.-H.: Settling the Complexity of Computing Two-Player Nash Equilibria. Journal of the ACM 56(3) (2009)
4. Chen, X., Teng, S.-H., Valiant, P.: The Approximation Complexity of Win-Lose Games. In: Proceedings of the 18th Annual ACM-SIAM Symposium on Discrete Algorithms, pp. 159–168 (January 2007)
5. Codenotti, B., Stefanovic, D.: On the Computational Complexity of Nash Equilibria for (0, 1) Bimatrix Games. Information Processing Letters 94(3), 145–150 (2005)
6. Conitzer, V., Sandholm, T.: New Complexity Results about Nash Equilibria. Games and Economic Behavior 63(2), 621–641 (2008)
7. Daskalakis, C.: On the Complexity of Approximating a Nash Equilibrium. In: Proceedings of the 22nd Annual ACM-SIAM Symposium on Discrete Algorithms, pp. 1498–1517 (January 2011)
8. Daskalakis, C., Goldberg, P.W., Papadimitriou, C.H.: The Complexity of Computing a Nash Equilibrium. SIAM Journal on Computing 39(1), 195–259 (2009)
9. Garey, M.R., Johnson, D.S.: Computers and Intractability — A Guide to the Theory of $\mathcal{NP}$-Completeness. W. H. Freeman (1979)
10. Gilboa, I., Zemel, E.: Nash and Correlated Equilibria: Some Complexity Considerations. Games and Economic Behavior 1(1), 80–93 (1989)
11. McLennan, A., Tourky, R.: Simple Complexity from Imitation Games. Games and Economic Behavior 68(2), 683–688 (2010)
12. Megiddo, N., Papadimitriou, C.H.: On Total Functions, Existence Theorems and Computational Complexity. Theoretical Computer Science 81, 317–324 (1991)
13. Nash, J.F.: Equilibrium Points in n-Person Games. Proceedings of the National Academy of Sciences of the United States of America 36, 48–49 (1950)
14. Nash, J.F.: Non-Cooperative Games. Annals of Mathematics 54(2), 286–295 (1951)
15. Papadimitriou, C.H.: On the Complexity of the Parity Argument and Other Inefficient Proofs of Existence. Journal of Computer and System Sciences 48(3), 498–532 (1994)

An Optimal Bound to Access the Core in TU-Games*

Sylvain Béal[1], Eric Rémila[2], and Philippe Solal[3]

[1] Université de Franche-Comté,
CRESE (EA 3190)
30 Avenue de l'Observatoire, 25009 Besançon, France
sylvain.beal@univ-fcomte.fr
[2] Université de Lyon,
GATE LSE (UMR CNRS 5824)
Université de Saint-Etienne,
6 rue basse des rives, 42023 Saint-Etienne, France
eric.remila@univ-st-etienne.fr
[3] Université de Saint-Etienne,
GATE LSE (UMR CNRS 5824)
6 rue basse des rives, 42023 Saint-Etienne, France
philippe.solal@univ-st-etienne.fr

Abstract. For any transferable utility game in coalitional form with a nonempty core, we show that that the number of blocks required to switch from an imputation out of the core to an imputation in the core is at most $n-1$, where n is the number of players. This bound exploits the geometry of the core and is optimal. It considerably improves the upper bounds found so far by Kóczy [7], Yang [13, 14] and a previous result by ourselves [2] in which the bound was $n(n-1)/2$.

1 Introduction

1.1 Preliminaries

TU Games. We consider cooperative games with transferable utility (TU-games for short). Formally, a *TU-game* is a pair (N, v) where

- $N = \{1, \ldots, n\}$ is a nonempty finite *player set*;
- $v : 2^N \longrightarrow \mathbb{R}$ is a real-valued function such that $v(\emptyset) = 0$.

A nonempty subset S of N is called a *coalition* and s stands for its cardinality. The real number $v(S)$ is interpreted as the *worth* of coalition S, i.e. the value generated by players of S when they cooperate without the help of players in $N \backslash S$. The set N is called the *grand coalition*.

* Financial support by the National Agency for Research (ANR) — research programs "Models of Influence and Network Theory" (MINT) ANR.09.BLANC-0321.03 and "Mathmatiques de la dcision pour l'ingnierie physique et sociale" (MODMAD) — and by IXXI (Complex System Institute, Lyon) is gratefully acknowledged.

M. Serna (Ed.): SAGT 2012, LNCS 7615, pp. 49–60, 2012.

An *allocation* $x \in \mathbb{R}^n$ on N is an n-dimensional vector giving a payoff $x_i \in \mathbb{R}$ to each player $i \in N$. We state: $x(S) = \sum_{i \in S} x_i$.

An allocation $x \in \mathbb{R}^n$ is *efficient* if $x(N) = v(N)$, *individually rational* if $x_i \geq v(\{i\})$ for each $i \in N$. The two properties above can be seen as the minimal requested for an acceptable allocation. The set of efficient allocations is denoted by $E(N, v)$. An individually rational and efficient allocation is referred to as an *imputation.* The (possibly empty) set of imputations is denoted by $I(N, v) \subseteq E(N, v)$.

The Core. Let Γ be the class of all finite TU-games. A *solution* on Γ is a function F which assigns to each $(N, v) \in \Gamma$ a set of allocations $F(N, v)$. The most famous solution for TU-games is the core introduced by Gillies [5]. The *core* is the solution C on Γ that assigns to each TU-game $(N, v) \in \Gamma$ the possibly empty set $C(N, v)$ of all efficient and *stable* allocations, i.e.

$$C(N, v) = \{x \in E(N, v) : \forall S \in 2^N, x(S) \geq v(S)\}.$$

Note that the core is a subset of the set of imputations. We denote by Γ^c the class of all TU-games with a nonempty core. In the rest of the article we only consider the class Γ^c. The core can be also defined using the notions of block and dominance.

Given an allocation $x \in \mathbb{R}^n$ and a coalition S, x_S denotes the restriction of x to S. For two allocations $x, y \in \mathbb{R}^n$, we write $x_S < y_S$ if $x_i < y_i$ for each $i \in S$ and $x_S \leq y_S$ if $x_i \leq y_i$ for each $i \in S$ but $x_S \neq y_S$.

Definition 1. *Assume that there exists a non empty coalition $S \in 2^N$ and two efficient allocations x and y of $E(N, v)$ such that both $x(S) < y(S) \leq v(S)$ and $x_S \leq y_S$ (resp. $x_S < y_S$). In such a case, we say that S weakly (resp. strongly) blocks x, and that y weakly dominates x (resp. y strongly dominates x) via coalition S, and we denote this relation by $x \preceq_S y$ (resp. $x \prec_S y$). We write $x \preceq y$ (resp. $x \prec y$) if there exists a coalition S such that $x \preceq_S y$ (resp. $x \prec_S y$) and say that y weakly (resp. strongly) dominates x.*

The strong dominance relation indicates that it is in the interest of all players in S to switch from x to y, while the weak dominance relation only imposes that the payoff of no player in S is reduced when moving from x to y and at least one of them is strictly better off. In the rest of the article, and with the notable exception of section 3, we will use the weak dominance relation.

Let x be an efficient allocation that lies out of the core. There necessarily exists an efficient allocation y such that $x(S) < y(S) \leq v(S)$. Thus, coalition S can propose to replace x by y. For instance, y can be any efficient allocation such that $y_i = x_i + (v(S) - x(S))/s$ for each $i \in S$, which makes every member of S strictly better off than in x. Thus, the players of S can threaten to split from the grand coalition since $x \prec_S y$. In a sense x fails to ensure the stability of the grand coalition. Such a situation cannot arise if x is a core allocation. Hence, the core can also be defined as the set of efficient allocations which are not (strongly or weakly) dominated i.e.

$$C(N,v) = \{x \in E(N,v) : \forall y \in E(N,v), \neg(x \preceq y)\}.$$

In other words, the maximal elements of the dominance relations over $E(N,v)$ coincide with the core allocations. As such, the core satisfies the internal stability property: elements of the core are not comparable under the weak or strong dominance relation. Nevertheless, the core is often criticized on two aspects. Firstly, it does not account for every imputation it excludes. More specifically, the core does not necessarily satisfy the external stability property: an imputation out of the core is not always dominated by an imputation of the core. Shapley [11] has proved the external stability for the class of convex TU-games.[1] Secondly, Harsanyi [6] and Chwe [3] consider that this solution concept is too myopic because it neglects the effect of successive blocks. Harsanyi [6] introduces a new indirect dominance relation, which consists of a chain of blocks, in order to cope with these lacks.

A weak *weak (resp. strong) chain of blocks* is a finite sequence $(x_0, x_1, \ldots, x_m)$ of efficient allocations such that, for each $k = 0, \ldots, m-1$, it holds that $x^k \preceq x^{k+1}$ (resp. $x^k \prec x^{k+1}$). The number m of allocations in the chain is called its *length*. An allocation y indirectly weakly (resp. strongly) dominates an allocation x if there exists a weak (resp. strong) chain of blocks starting at x and ending at y. Harsanyi originally applies this indirect dominance relation to study the von Neumann-Morgenstern stable sets (Von Neumann and Morgenstern [12]). Sengupta and Sengupta [10] employ it to show that the core is indirectly externally stable for the weak dominance relation: starting from any imputation that stands outside the core, there always exists a weak chain of blocks which terminates in the core. In other words, the core can be considered as a von Neumann-Morgenstern stable set under the indirect weak dominance relation.

1.2 The Results

This last result has initiated the literature on the accessibility of the core, on which the reader is referred to Bal et al. [2] and the references therein. The central question that has appeared is to determine a upper bound on the length of the chain of blocks needed to access the core. Several recent articles try to answer this question and this article improves on the existing answers. We only mention the two most recent approaches. In our previous article [2] we show that the core of any TU-game with n players can be accessed with a weak chain of blocks of length at most $n(n-1)/2$ blocks such that each element of this chain is an imputation. Yang [14] obtains the linear bound $2n-1$ but this result has two drawbacks. Firstly, this bound only holds for the class of cohesive TU-games i.e. the class of games in which no partition of the player set generates a larger cumulated worth than the grand coalition. Secondly, even if the starting allocation is an imputation, the other allocations in the weak chain of blocks need not be efficient, which is rather far from the spirit of the original idea of Harsanyi [6].

[1] A TU-game is convex if for each pair of coalitions $S, T \subseteq N$, it holds that $v(S \cup T) - v(S \cap T) \geq v(S) + v(T)$.

In the present article we also obtain a linear bound: the core can be accessed in at most $n-1$ blocks. More importantly, for the first time in the literature, we are able to prove that this bound is optimal. These results can be stated as follows.

Theorem 1. *Let $(N,v) \in \Gamma^c$ be a n-player TU-game with a nonempty core with $n \geq 3$. For each imputation $x \in I(N,v)$, there exists a core element $c \in C(N,v)$ and a weak chain of blocks from x to c with length at most $n-1$ and such that each allocation of the chain is an imputation.*

This bound is optimal: for each integer $n \geq 3$, there exists a n-player TU-game $(N,v) \in \Gamma^c$ with a nonempty core and an imputation $x \in I(N,v)$ such that each weak chain of blocks from x to any core allocation in $C(N,v)$ has length at least $n-1$.

Note that in the cases $n=1$ and $n=2$, the accessibility of the core is trivial, since either the set of imputations and the core coincide or the core is empty. The proof of Theorem 1 relies on a procedure which is similar to the one introduced by Sengupta and Sengupta [10]. In particular, both procedures share the idea of using a core element as a reference point. Nevertheless, we introduce three major differences detailed below.

The first difference is the choice, at each step of the chain of blocks, of the blocking coalition. While Sengupta and Sengupta [10] choose a coalition among the most unsatisfied coalitions with respect to the current imputation, we select a coalition S such that the hyperplane defined by $x(S) = v(S)$ for $x \in \mathbb{R}^n$ has a nonempty intersection with the core. As a consequence, the geometry of the core plays an important role in our analysis. This choice enables to simplify the procedure introduced in Sengupta and Sengupta [10].

The second difference is that the target core allocation can vary from one block to the next block in the weak chain of blocks while it is unique in Sengupta and Sengupta [10]. However, all target core allocations that are used along the weak chain of blocks have at least one common coordinate.

Thirdly, the result crucially relies on the use of the Davis-Maschler reduced-games ([4]). The Davis-Maschler reduced-games describe situations in which all the players agree that the left players get their core reference payoffs but continue to cooperate with the remaining players, subject to the foregoing agreement. The Davis-Maschler reduced-games are well known for being the basis of the so-called reduced-game property, which states that if an allocation is prescribed by some solution concept in a TU-game, then the restriction of this allocation to any coalition of players is also prescribed by the solution concept in the reduced-game associated with these coalition and allocation. Our previous article describes connections between a game and its Davis-Maschler reduced-games. In the current article, we explore more deeply these connections.

In [2], the first two aspects were not used while the third one was not essential to prove the results.

Core accessibility only using strong dominances can also be investigated. To the best of our knowledge, there does not exist any such study in the literature so far. This fact can be explained by our last result: the accessibility of the core is not always possible under the strong dominance relation.

Theorem 2. *For each $n \geq 3$, there exists a n-player TU-game $(N, v) \in \Gamma^c$ with a nonempty core and an imputation $x \in I(N, v)$ from which the core cannot be accessed by a strong chain of blocks.*

This article proves that the optimal number of blocks required to access the core of any n-player TU-game is $n - 1$. Our result provides a definitive answer to the question of the accessibility of the core. A challenging issue is to investigate whether this result still holds for the related concept of coalition structure core. From Yang [14] and Bal et al. [1] we know that the number of blocks required to access the coalition structure core is at most quadratic in the number of players.

2 Optimality under the Weak Dominance Relation

In this section, we start by stating some connections between a TU-game and its Davis-Maschler reduced-games. These intermediary results will be used later on to prove Theorem 1. Although the proofs are similar than in Bal et al. [2], we give them for completeness. Secondly, we describe the procedure. Thirdly, we prove the first part of Theorem 1, i.e. that the core of an n-player TU-game is accessible in at most $n - 1$ blocks. This part is obtained as a corollary of a more general results in which we consider efficient allocations instead of imputations. Fourthly, we show that our bound is optimal.

2.1 Reduced-Games Equivalences

Let $S \subset N$ be any coalition different from N and $x \in E(N, v)$ any efficient allocation. Davis and Maschler [4] introduce the *reduced-game* with respect to S and x, denoted by $(S, v_{S,x})$ and defined, for each $T \subseteq S$, by :

$$v_{S,x}(T) = \begin{cases} 0 & \text{if } T = \emptyset, \\ v(N) - x(N \backslash S) & \text{if } T = S, \\ \max_{R \in 2^{N \backslash S}} \big(v(T \cup R) - x(R)\big) & \text{otherwise.} \end{cases}$$

The weak dominance relation $\preceq$ will be used in the Davis-Maschler reduced-games as well. In such a case, we will specify in which game the dominance relation is applied in order to avoid any risk of confusion.

A solution F on Γ satisfies the *reduced-game property* if for each $(N, v) \in \Gamma$, each nonempty coalition $S \subset N$ and each $x \in F(N, v)$, it holds that $x_S \in F(S, v_{S,x})$. It is well-known that the core satisfies the reduced-game property. In fact, the reduced-game property is one of the axioms used by Peleg [8] in order to characterize the core.

For the class Γ^c of TU-games with a nonempty core, we will construct Davis-Maschler reduced-games with respect to core allocations only. This section establishes two interesting properties of such reduced-games.

Let $(N, v) \in \Gamma^c$ be any TU-game with a nonempty core, $S \subset N$ be any nonempty coalition and $c \in C(N, v)$ be any core allocation of (N, v). Observe that

$$v_{S,c}(S) = v(N) - c(N \backslash S) = c(N) - c(N \backslash S) = c(S).$$

The first lemma establishes connections between the sets of efficient allocations and the cores of a TU-game and of its Davis-Maschler reduced games.

Lemma 1. *Consider any $(N, v) \in \Gamma^c$, any nonempty coalition $S \subset N$ and any $c \in C(N, v)$. Pick any allocation $x \in \mathbb{R}^n$ such that $x_{N \backslash S} = c_{N \backslash S}$. Then*

- *$x \in E(N, v)$ if and only if $x_S \in E(S, v_{S,c})$,*
- *$x \in C(N, v)$ if and only if $x_S \in C(S, v_{S,c})$.*

Proof. Firstly, suppose that $x \in E(N, v)$. It holds that

$$x_S(S) = x(N) - x(N \backslash S) = v(N) - c(N \backslash S) = c(N) - (c(N) - c(S)) = c(S) = v_{S,c}(S),$$

so that $x_S \in E(S, v_{S,c})$. Conversely, suppose that $x_S \in E(S, v_{S,c})$. Since $x_{N \backslash S} = c_{N \backslash S}$ and $x_S(S) = v_{S,c}(S) = c(S)$, we get

$$x(N) = x(S) + x(N \backslash S) = c(S) + c(N \backslash S) = c(N) = v(N),$$

proving that $x \in E(N, v)$.

Secondly, suppose that $x \in C(N, v)$. Since $x \in E(N, v)$, we have that $x_S \in E(S, v_{S,c})$. Now, choose any coalition $T \in 2^S$. It holds that:

$$v_{S,c}(T) = v(T \cup \overline{T}) - c(\overline{T}) \leq x(T \cup \overline{T}) - c(\overline{T}) = x(T) = x_S(T),$$

which means that x_S is also a stable allocation in $(S, v_{S,c})$. We conclude that $x_S \in C(S, v_{S,c})$.

Conversely, suppose that $x_S \in C(S, v_{S,c})$. Since $x_S \in E(S, v_{S,c})$, we have $x \in E(N, v)$. Next, choose any coalition $T \in 2^N$. The definition of $v_{S,c}$ and $x_S \in C(S, v_{S,c})$ imply that

$$\begin{aligned} v(T) &= v((T \cap S) \cup (T \backslash S)) \\ &= v((T \cap S) \cup (T \backslash S)) - c(T \backslash S) + c(T \backslash S) \\ &\leq v_{S,c}(T \cap S) + c(T \backslash S) \\ &\leq x_S(T \cap S) + c(T \backslash S) \\ &= x(T), \end{aligned}$$

from which we obtain $x \in C(N, v)$.

The second lemma describes the connections between the weak dominance relations in a TU-game and in its Davis-Maschler reduced games.

Lemma 2. *Consider any $(N, v) \in \Gamma^c$, any non-empty coalition $S \subset N$ and any $c \in C(N, v)$. Pick any two allocations $x, y \in E(N, v)$ such that $x_{N \backslash S} = y_{N \backslash S} = c_{N \backslash S}$ and a nonempty coalition $T \subset S$. Then:*

$$x \preceq_{T \cup \overline{T}} y \text{ in } E(N, v) \text{ if and only if } x_S \preceq_T y_S \text{ in } E(S, v_{S,c}).$$

Proof. Firstly, assume that $x \preceq_{T\cup\overline{T}} y$ in $E(N,v)$. Then $x(T\cup\overline{T}) < y(T\cup\overline{T})$ and $x_{T\cup\overline{T}} \leq y_{T\cup\overline{T}}$. In addition, $\overline{T} \in 2^{N\setminus S}$ implies $x_{\overline{T}} = y_{\overline{T}}$. It follows that $x(T) < y(T)$.

Next, $x \preceq_{T\cup\overline{T}} y$ in $E(N,v)$ also means that $y(T\cup\overline{T}) \leq v(T\cup\overline{T})$. Therefore, by definition of y, we get $v_{S,c}(T) = v(T\cup\overline{T}) - c(\overline{T}) \geq y(T\cup\overline{T}) - c(\overline{T}) = y(T)$. We conclude that $x_S \preceq_T y_S$ in $E(S, v_{S,c})$.

Secondly, assume that $x_S \preceq_T y_S$ in $(S, v_{S,c})$. Then $x(T) < y(T)$ and $x_T \leq y_T$. Since $x_{\overline{T}} = y_{\overline{T}}$ by assumption, this implies that $x_{T\cup\overline{T}} \leq y_{T\cup\overline{T}}$.

Furthermore, $x_S \preceq_T y_S$ in $(S, v_{S,c})$ also means that $y(T) \leq v_{S,c}(T) = v(T\cup\overline{T}) - c(\overline{T})$. Thus, $v(T\cup\overline{T}) \geq y(T) + c(\overline{T}) = y(T\cup\overline{T})$. Therefore $x \preceq_{T\cup\overline{T}} y$ in $E(N,v)$.

One directly obtains the following corollary.

Corollary 1. *Consider any $(N,v) \in \Gamma^c$, any $S \subset N$ and any $c \in C(N,v)$. Pick any two allocations $x, y \in E(N,v)$ such that $x_{N\setminus S} = y_{N\setminus S} = c_{N\setminus S}$. If $x_S \preceq y_S$ in $E(S, v_{S,c})$, then $x \preceq x$ in $E(N,v)$.*

A key-point is that this corollary does not hold if the weak dominance relation is replaced by the strong dominance relation. This explains the differences in term of accessibility of the core with respect to the weak and strong dominance relations. This aspect is formally investigated in section 3.

The intermediary results in this section also raise a difficulty. Equivalent results cannot be stated if the considered allocations are imputations: it may happen that, for an imputation $x \in I(N,v)$, the allocation x_S is not in $I(S, v_{S,c})$. We also discuss this point latter on.

2.2 The Procedure

Consider a TU-game $(N,v) \in \Gamma^c$. From now on, we fix an efficient allocation $x \in E(N,v)$. In order to exploit the results of the previous section, we will exhibit an allocation which satisfies several properties. It will be used to define the allocations along our weak chain of blocks. The construction of this allocation relies on the following result about the geometry of $C(N,v)$.

Let $B(N,v)$ denote a minimal collection, with respect to set inclusion, of coalitions in $2^N\setminus\{N\}$ such that

$$C(N,v) = \{z \in E(N,v) : \forall S \in B(N,v), z(S) \geq v(S)\}.$$

Lemma 3. *For each $S \in B(N,v)$, there exists a core allocation $c \in C(N,v)$ such that $c(S) = v(S)$.*

Proof. First note that such a minimal collection $B(N,v)$ trivially exists. By way of contradiction, assume that there exists T in $B(N,v)$ such that for each $z \in C(N,v)$ it holds that $v(T) < z(T)$. Consider the set

$$D(N,v) = \{z \in E(N,v) : \forall S \in B(N,v)\setminus\{T\}, z(S) \geq v(S)\}.$$

Since $B(N,v)$ is a minimal set with respect to inclusion, we have $C(N,v) \subset D(N,v)$. Hence, we can choose $x \in D(N,v) \backslash C(N,v)$. Pick also any $y \in C(N,v)$. By definition of x and y and by the initial assumption on T, we have $x(T) < v(T) < y(T)$. Now, define

$$\alpha = \frac{v(T) - y(T)}{x(T) - y(T)},$$

and observe that $\alpha \in (0,1)$. Next, construct the allocation $c = \alpha x + (1-\alpha)y$. In particular, it holds that $c(T) = v(T)$. Moreover, c belongs to $D(N,v)$ because it is a convex combination of two elements of the convex set C'. In addition, since both $x(S) \geq v(S)$ and $y(S) \geq v(S)$ for each $S \in B(N,v) \backslash \{T\}$, we conclude that $c \in C(N,v)$. This contradicts the initial assumption.

Because the core is a polytope, the statement of Lemma 3 can be strengthen if the core is full-dimensional, i.e. has a nonempty interior. More precisely, it is known that for each full-dimensional core, there exists a unique (up to a multiplication by positive scalars) minimal collection of constraint inequalities that determines it. Moreover, for each distinct pair of constraint inequalities, there is an element is the core that saturates the first constraint and lies strictly above the second constraint inequality. On this point we refer the reader to Schrijver [9].

Lemma 4. *Let $x \in E(N,v)$ be an efficient allocation such that $x \notin C(N,v)$. There exists an allocation y satisfying the following four properties.*

1. *$y \in E(N,v)$;*
2. *$x \preceq y$;*
3. *there exists $c \in C(N,v)$ and $i \in N$ such that $y_i = c_i$;*
4. *for each player $i \in N$, we have $y_i \geq \min(x_i, v(\{i\}))$.*

Proof. Consider any $x \in E(N,v) \backslash C(N,v)$. There is some $S \in B(N,v)$ such that $x(S) < v(S)$. Moreover, by Lemma 3, there isa core allocation $c \in C(N,v)$ such that $c(S) = v(S)$. We define the allocation y as follows:

$$y_i = \begin{cases} x_i + \dfrac{v(S) - x(S)}{s} & \text{if } i \in S, \\ c_i & \text{if } i \in N \backslash S. \end{cases}$$

Now we prove that y satisfies the claimed properties. Actually, we do not use the particular structure of y for players in S. The important coordinates of y are on the players in $N \backslash S$.

For the first property, the equality $v(S) = c(S)$ implies that

$$y(N) = x(S) + s\frac{v(S) - x(S)}{s} + c(N \backslash S) = v(S) + c(N \backslash S) = c(S) + c(N \backslash S) = v(N),$$

which proves that $y \in E(N,v)$.

For the second property, the claim $x \preceq y$ is obviously satisfied, since $v(S) - x(S) > 0$ and $y(S) = v(S)$ imply $x \preceq_S y$.

For the third property, it is enough to show that $N\backslash S \neq \emptyset$, which is ensured by $S \in B(N,v)$ since $N \notin B(N,v)$. For the fourth property, the inequality $x(S) < v(S)$ implies that $y_S > x_S$. Finally, for each $i \in N\backslash S$, it holds that $y_i = c_i \geq v(\{i\})$.

The selection of the unsatisfied coalition S such that $x \preceq_S y$ is the cornerstone of our construction. It is a main difference with Sengupta and Sengupta [10] and Béal et al. [2]. We choose $S \in B(N,v)$ whereas Sengupta and Sengupta [10] choose S such that the positive excess $v(S) - x(S)$ is maximal, and Béal et al. [2] select a coalition among the smallest coalitions with positive excess. Selecting S in $B(N,v)$ is necessary to construct an allocation y in a simpler way than in Sengupta and Sengupta [10] and Béal et al. [2]. Another difference with Sengupta and Sengupta [10] and Béal et al. [2] is related to the use of the core allocations as a target. Both use a unique core allocation along their weak chain of blocks. Here, we take in account the whole geometry of the core since it is the chosen coalition $S \in B(N,v)$ that determines which core allocation can be used to construct the current block.

2.3 The Upper Bound

We now have the material to prove that the core of any n-player TU-game can be accessed in at most $n-1$ blocks.

Proposition 1. *Let $(N,v) \in \Gamma^c$ be a n-player TU-game with a nonempty core. For each efficient allocation $x \in E(N,v)$, there exists a core element $c \in C(N,v)$ and a weak chain of blocks from x to c such that*

- *the length of this weak chain of blocks is at most $n-1$;*
- *each blocking efficient allocation z in this weak chain of blocks satisfies the condition that, for each $j \in N$, $z_j \geq \min(x_j, v(\{j\}))$.*

Proof. Consider any arbitrary TU-game $(N,v) \in \Gamma^c$. The proof is done by induction on the number n of players in (N,v).
INITIALIZATION: For $n = 1$, the unique efficient allocation is also the unique core allocation so that the result trivially holds.
INDUCTION HYPOTHESIS: Assume that the statement of the Proposition 1 holds for any k-player TU-game, $k \in \{1,\ldots,n-1\}$.
INDUCTION STEP: Consider any n-player TU-game $(N,v) \in \Gamma^c$ and any $x \in E(N,v)$. If $x \in C(N,v)$, then we are done. Otherwise, using the procedure described in Lemma 4, we construct an imputation y satisfying the four properties stated in that Lemma.

In particular, there exists a player $i \in N$ and an core allocation $c \in C(N,v)$ such that $y_i = c_i$. Consider the Davis-Maschler reduced-game $(N\backslash\{i\}, v_{N\backslash\{i\},c})$. This TU-game is an $(n-1)$-player TU-game with a nonempty core since Lemma 1 states that $c_{N\backslash\{i\}} \in C(N\backslash\{i\}, v_{N\backslash\{i\},c})$. The induction hypothesis can be used: there exists a core element $d \in C(N\backslash\{i\}, v_{N\backslash\{i\},c})$ and a weak chain of blocks $(z^0, z^1, \ldots, z^m)$ such that $z^0 = y_{N\backslash\{i\}}$ and $z^m = d$ with $m \leq n-2$. For each

$k \in \{0, \ldots, m\}$, it also holds that $z^k \in E(N\backslash\{i\}, v_{N\backslash\{i\},c})$. Furthermore, for each $j \in N\backslash\{i\}$, we have $z_j^0 \geq \min(y_j, v_{N\backslash\{i\},c}(\{j\}))$ and, for each $k \in \{1, \ldots, m\}$, $z_j^k \geq \min(z_j^{k-1}, v_{N\backslash\{i\},c}(\{j\}))$. Hence, for each $k \in \{1, \ldots, m\}$ and each $j \in N\backslash\{i\}$, it holds that $z_j^k \geq \min(y_j, v_{N\backslash\{i\},c}(\{j\}))$.

For each $k \in \{0, \ldots, m\}$, define the allocation $z'^k \in \mathbb{R}^n$ as $z'^k_{N\backslash\{i\}} = z^k$ and $z'^k_i = c_i$. Lemma 1 yields that $z'^m \in C(N, v)$ since $z'^m_{N\backslash\{i\}} = z^m = d$ and $d \in C(N\backslash\{i\}, v_{N\backslash\{i\},c})$. Furthermore, from Lemma 2, the sequence of allocations $(z'^0, z'^1, \ldots, z'^m)$ is a weak chain of blocks. This implies that $(x, z'^0, z'^1, \ldots, z'^m)$ is a weak chain of blocks of length $m + 1$. Since $m \leq n - 2$, the length of the weak chain of blocks $(x, z'^0, z'^1, \ldots, z'^m)$ is bounded by $n - 1$, which proves the first part of the Proposition 1.

Regarding the second part of the Proposition 1, consider any player $j \in N\backslash\{i\}$. We have already proved that $z_j^k \geq \min(y_j, v_{N\backslash\{i\},c}(\{j\}))$. We also have $y_j \geq \min(x_j, v(\{j\}))$. By definition of $v_{N\backslash\{i\},c}$, it holds that $v_{N\backslash\{i\},c}(\{j\})) \geq v(\{j\})$. Altogether, this implies that

$$z_j^k \geq \min(x_j, v(\{j\})),$$

or equivalently, that $z'^k_j \geq \min(x_j, v(\{j\}))$. Lastly, the inequality $z'^k_i = c_i \geq v(\{i\}) \geq \min(x_i, v(\{i\}))$ completes the proof.

The *first part of Theorem 1* is a corollary of Proposition 1. In fact, if the set of efficient allocations is replaced by the set of imputations in the statement of Proposition 1, then the length of the weak chain of blocks required to access the core is still at most $n - 1$. Moreover, for each player $j \in N$, the condition $z_j \geq \min(x_j, v(\{j\}))$ for each blocking allocation z reduces to $z_j \geq v(\{j\}$ since $x_j \geq v(\{j\})$ whenever x is an imputation. This ensures that the weak chain of blocks only contains imputations. It is nevertheless important to state Proposition 1. The reason is that, in the induction step, a blocking imputation can be lead to an allocation which is not individually rational in the associated Davis-Maschler reduced-game.

2.4 Optimality

This section is devoted to the proof of the second part of Theorem 1, i.e. it is impossible to improve upon the bound $n - 1$. More specifically, for each $n \geq 3$, we construct an n-player TU-game with a nonempty core and an imputation $x \in I(N, v)$ such that the length of each chain of blocks from x to any core allocation of $C(N, v)$ is at least $n - 1$.

Proof (Theorem 1 — second part). Let $n \geq 3$ and (N, v) be the n-player TU-game such that $v(S) = s$ if $s \geq n - 1$ and $v(S) = 0$ otherwise. If c belongs to $C(N, v)$, then, for each player $i \in N$, it holds that

$$c_i = c(N) - c(N\backslash\{i\}) \leq v(N) - v(N\backslash\{i\}) = 1.$$

Combined with the efficiency of c, we obtain that $c_i = 1$ for each $i \in N$. Since the allocation $(1, \ldots, 1)$ belongs to $C(N, v)$, we conclude this allocation is the unique core allocation.

Now, pick any $x \in I(N,v)\backslash C(N,v)$. The imputation x is acceptable for all coalitions of size at most $n-2$ and it is not acceptable for a coalition $N\backslash\{i\}$ if and only if $x_i > 1$. Then, the number $\Delta(x)$ of coalitions for which x is not acceptable is the number of players $i \in N$ such that $x_i > 1$.

Consider any imputation $y \in I(N,v)$ such that $x \preceq y$. Note that $x \preceq_S y$ is only possible if $S = N\backslash\{i\}$ for some $i \in N$ such that $x_i > 1$. Consider any such block. By definition of a block, we get that $y(N\backslash\{i\}) \leq v(N\backslash\{i\})$. Now, consider any other coalition S for which x is not acceptable, which means that $S = N\backslash\{j\}$ for some $j \in N\backslash\{i\}$. Since $y_j \geq x_j$ and $y, x \in E(N,v)$, it holds that

$$y(N\backslash\{j\}) \leq x(N\backslash\{j\}) < v(N\backslash\{j\}).$$

In other words, if x is not acceptable for a coalition other than $N\backslash\{i\}$, then y is also not acceptable for this coalition. still unsatisfied with respect to y. Thus $\Delta(y) \geq \Delta(x) - 1$. It follows that any weak chain of blocks of length p and starting from x terminates in an allocation z such that $\Delta(z) \geq \Delta(z) - p$.

Now suppose that the starting imputation x is given $x_1 = 0$ and, for each $i \in N\backslash\{1\}$, $x_i = 1 + 1/(n-1)$. It holds that $\Delta(x) = n - 1$. This implies that each weak chain of blocks of length at most $n-2$ and starting at x terminates in an allocation z, such that $\Delta(z) \geq n - 1 - (n-2) = 1$. In other words, each weak chain of blocks of length at most $n-2$ and starting at x cannot access the core $C(N,v)$. We conclude that a weak chain of blocks of length at most $n-1$ is necessary to access the core if x is the initial imputation.

3 An Impossibility Result under the Strong Dominance Relation

In this section, we prove Theorem 2 stating that the core is not always accessible if the blocks are constructed from the strong dominance relation. For each $n \geq 3$, the proof relies on the n-player TU-game depicted in section 2.4, which is used to show that there exists an imputation from which the core is not accessible through a strong chain of blocks.

Proof (Theorem 2). For each $n \geq 3$, consider the n-player TU-game (N,v) introduced in section 2.4. Consider the set $A(N,v)$ of imputations of (N,v) such that $x \in A(N,v)$ if the two following conditions are satisfied:

- for each $i \in N\backslash\{1\}$, $x_i \geq 1$;
- there exists at most one $i \in N\backslash\{1\}$ such that $x_i = 1$.

Note that if $x \in A(N,v)$, these two conditions imply that $0 \leq x_1 < 1$. The set $A(N,v)$ is clearly nonempty since it contains the allocation x constructed at the end of the proof of the second part of Theorem 1 in the section 2.4. Now consider any imputation $x \in A(N, v)$. Pick any $i \in N$ and any efficient allocation $y \in E(N, v)$ such that $x \prec_{N\setminus\{i\}} y$. Recall that no coalition of size at most $n-2$ can block x. Furthermore, observe that $i \neq 1$ since $x(N\setminus\{1\}) > n-1 = v(N\setminus\{1\})$. By definition, for each player $j \in N\setminus\{1, i\}$, it holds that $y_j > x_j \geq 1$. Moreover, we have $y_i = v(N) - y(N\setminus\{i\}) \geq v(N) - v(N\setminus\{i\}) = 1$. This proves that y belongs to $A(N, v)$.

As a consequence, any strong block starting in $A(N, v)$ also ends in $A(N, v)$. It follows that any strong chain of blocks starting in $A(N, v)$ also terminates in $A(N, v)$. Since the unique core allocation $(1, \ldots, 1)$ does not belong to $A(N, v)$, we can conclude that the core of the TU-game (N, v) is not accessible from $A(N, v)$ by means of the strong dominance relation.

References

[1] Béal, S., Rémila, E., Solal, P.: On the Number of Blocks Required to Access the Coalition-Structure Core. MPRA Paper No. 29755 (2011)
[2] Béal, S., Rémila, E., Solal, P.: On the number of blocks required to access the core. Discrete Applied Mathematics 160(7-8), 925–932 (2012)
[3] Chwe, M.S.-Y.: Farsighted Coalitional Stability. Journal of Economic Theory 63, 299–325 (1994)
[4] Davis, M., Maschler, M.: The Kernel of a Cooperative Game. Naval Research Logistics Quarterly 12, 223–259 (1965)
[5] Gillies, D.B.: Some Theorems on n-Person Games. Ph.D. Dissertation. Princeton University, Department of Mathematics (1953)
[6] Harsanyi, J.C.: An Equilibrium Point Interpretation of Stable Sets and a Proposed Alternative Definition. Management Science 20, 1472–1495 (1974)
[7] Kóczy, L.Á.: The Core can be Accessed with a Bounded Number of Blocks. Journal of Mathematical Economics 43, 56–64 (2006)
[8] Peleg, B.: On the Reduced Game Property and its Converse. International Journal of Game Theory 15, 187–200 (1986)
[9] Schrijver, A.: Polyhedral combinatorics. In: Graham, Grotschel, Lovasz (eds.) Handbook of Combinatorics. Elsevier Science B.V. (1995)
[10] Sengupta, A., Sengupta, K.: A Property of the Core. Games and Economic Behavior 12, 266–273 (1996)
[11] Shapley, L.S.: Cores of Convex Games. International Journal of Game Theory 1, 11–26 (1971)
[12] von Neunamm, J., Morgenstern, O.: The Theory of Games and Economic Behavior. Princeton University Press, Princeton (1953)
[13] Yang, Y.-Y.: On the Accessibility of the Core. Games and Economic Behavior 69, 194–199 (2010)
[14] Yang, Y.-Y.: Accessible Outcomes versus Absorbing Outcomes. Mathematical Social Sciences 62, 65–70 (2011)

Convergence of Ordered Improvement Paths in Generalized Congestion Games

K. Ruben Brokkelkamp and Mees J. de Vries

Korteweg–de Vries Institute,
University of Amsterdam

Abstract. We consider generalized congestion games, a class of games in which players share a set of strategies and the payoff functions depend only on the chosen strategy and the number of players playing the same strategy, in such a way that fewer such players results in greater payoff. In these games we consider improvement paths. As shown by Milchtaich [2] such paths may be infinite. We consider paths in which the players deviate in a specific order, and prove that ordered best response improvement paths are finite, while ordered better response improvement paths may still be infinite.

1 Introduction

Congestion games, first introduced by Rosenthal [5], are a class of games characterized by a finite set of resources, from which players choose a subset as a strategy. Each resource then gives each of its users a cost, which depends only on the number of players using it, and a player's cost is the sum of her resources' costs. Rosenthal showed that congestion games always admit a potential function, which implies not only that congestion games have Nash equilibria in pure strategies, but also that they have the finite improvement property: starting at any joint strategy, when players change their strategies one by one in a way that improves their individual payoffs, a Nash equilibrium is eventually reached. Later, Monderer and Shapley [4] showed the converse: every finite potential game is isomorphic to a congestion game.

Expanding on the idea of modeling congestion situations with games, Milchtaich [2] introduced a class of games we will call generalized congestion games. These games are similar to congestion games, dropping the assumption that the payoff for using a strategy is not player-dependent, but introducing the limitations that a strategy consists of a single resource, and a resource's payoff depends nonincreasingly on the number of players using it. Generalized congestion games do not, in general, admit a potential function. In fact, they do not generally have the finite improvement property or even the weaker finite best response property, which states that players shifting to a best response move towards an equilibrium. Milchtaich did however show that all generalized congestion games are weakly acyclic. Weakly acyclic games are games in which from every joint strategy there is a way for players to consecutively play best response that reaches a Nash equilibrium.

M. Serna (Ed.): SAGT 2012, LNCS 7615, pp. 61–71, 2012.

Meanwhile, motivated by a different class of weakly acyclic games, Apt and Simon [1] introduced scheduler functions. These functions are a way to formalize the idea of an arbiter who determines which player is allowed to deviate to a new strategy. This allows one to consider whether the paths generated by such an arbiter converge to an equilibrium.

With this in mind we take a second look at generalized congestion games. We consider ordered paths, better and best response paths in which players move only when players which are ordered higher are playing best response. Although we avoid using schedulers because in this case the notation does not add anything, these paths correspond to the paths generated by Apt and Simon's local scheduler functions. These ordered paths are important because they model common situations, for example ones in which some players respond faster than others, or a situation with an arbiter with a very simple method of selection: a fixed ordering of players in which the higher player has higher priority.

We prove that best response ordered improvement paths are necessarily finite in generalized congestion games, improving on Milchtaich's result, which did not classify which kinds of best response improvement paths must be finite. Better response ordered improvement paths on the other hand may still be infinite.

In [3] a somewhat related result is proved in the context of extensive games. Given a strategic game, a sequential-move version of the game is an extensive game of perfect information in which the players are put in a fixed order and whose actions are the strategies of the original game. Milchtaich proved that for every ordering of the players the resulting sequential-move version has a subgame perfect equilibrium such that the actions selected by the players in this equilibrium form a Nash equilibrium of the original game.

2 Notation

Definition 1. A *generalized congestion game* is a noncooperative game wherein *players* $N = \{1, \ldots, n\}$ share a set of *strategies* $R = \{1, \ldots, r\}$, also called resources. Given a joint strategy $s \in R^n$, we write s^i for the strategy played by player i and s^{-i} for the strategies played by the other players, so that $s = (s^i, s^{-i})$. For a joint strategy s, we refer to the number of players playing strategy j as the *congestion* $c_j(s)$ at that strategy, and call the vector of congestions for all strategies $c(s) = (c_1(s), \ldots, c_r(s))$ the *congestion vector*. A player's payoff depends only on her chosen strategy and the congestion at that strategy. This allows us to write $p_i(j, c)$ for the payoff of player i for playing strategy j with congestion c. For a joint strategy $s \in R^n$ the payoff to player i is then written as $p_i(s^i, c_{s^i}(s))$. A player's payoff does not increase as the congestion at her strategy increases: for any $i \in N$, $j \in R$ we have

$$c_1 \geq c_2 \implies p_i(j, c_1) \leq p_i(j, c_2).$$

We now formalize what better and best response mean in this context.

Definition 2. Given a joint strategy $s \in R^n$, a strategy $j \in R$ is called a *better response* for player i when

$$p_i(j, c_j(s) + 1) > p_i(s^i, c_{s^i}(s))$$

and a *best response* for player i when

$$j \in \arg\max_{\tilde{\jmath} \in R} p_i(\tilde{\jmath}, c_{\tilde{\jmath}}((\tilde{\jmath}, s^{-i}))).$$

Next, we introduce our object of study.

Definition 3. A path is a finite or infinite sequence of joint strategies $(s_0, s_1, \ldots)$, such that every joint strategy can be reached by a single player changing strategy from the previous joint strategy. Formally, for any s_k in the sequence with $k > 0$, there are $i \in N$ and $j \in R$ such that $j \neq s^i_{k-1}$ and

$$s_k = (j, s^{-i}_{k-1}).$$

Player i is called the deviator at this step.

A path is called a *better response improvement path* if at every step the deviator changes her strategy to a better response strategy. A path is called a *best response improvement path* if at every step the deviator is not playing best response, but changes her strategy to a best response strategy.

A game is said to have the *finite improvement property* (FIP) if all its better response improvement paths are finite, and the *finite best response property* (FBRP) if all its best response improvement paths are finite. A game is called *weakly acyclic* (WA) if for every joint strategy there exists a best response improvement path that starts in that joint strategy and ends in a Nash equilibrium.

Milchtaich [2] has characterized which of these properties generalized congestion games have (as detailed in the next section). To make his classification finer, we introduce the notion of ordered improvement paths.

Definition 4. An *ordered (best or better response) improvement path* is an improvement path for which the players can be ordered in such a way, that whenever a player deviates, all players higher in the order than her are already playing best response.

We say a game has the *finite ordered improvement property* (FOIP) if all its ordered better response improvement paths are finite; it has the *finite ordered best response property* (FOBRP) if all its ordered best response improvement paths are finite.

3 Weak Acyclicity and Finite Path Properties

Milchtaich [2] showed that generalized congestion games do not generally have the FIP or even the FBRP. We reproduce his counterexample here, to give a simple example of a generalized congestion game, and also to introduce some notation that will be used later at a more complex example.

Milchtaich's example is a three-player three-strategy game, in which the players' payoffs satisfy the following inequalities.

- $p_1(3,1) > p_1(2,1) \geq p_1(2,2) > p_1(3,2)$.
- $p_2(1,1) > p_2(3,1) \geq p_2(3,2) > p_2(1,2)$.
- $p_3(2,1) > p_3(1,1) \geq p_3(1,2) > p_3(2,2)$.

Further, any payoff not mentioned in the above list is minimal. In this game, the path $((\underline{2},1,1),(3,\underline{1},1),(3,3,\underline{1}),(\underline{3},3,2),(2,\underline{3},2),(2,1,\underline{2}),(2,1,1))$ is a best response improvement path. This path is represented by the following diagram.

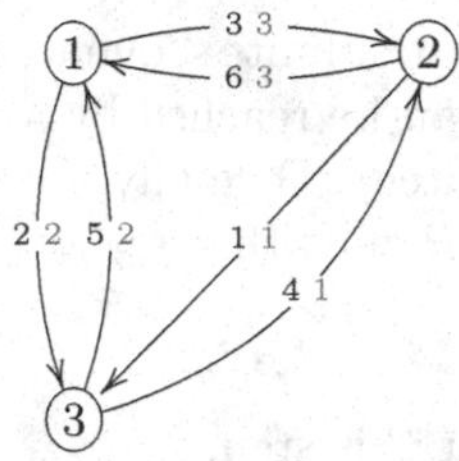

Fig. 1. Diagram representing the deviations in the infinite best response improvement path

We visualize the strategies as nodes, and the deviations as arrows. Every arrow is labeled with two numbers. The first, black number represents the deviation (which are numbered in order). The second, red number represents the deviator.

This best response improvement path is in particular a cycle, so it can be repeated arbitrarily, showing that generalized congestion games do not generally have the FBRP. However, the joint strategies $(3,1,2)$ and $(2,3,1)$ are Nash equilibria, and it is not difficult to see that from any joint strategy there exists a best response improvement path to one of these: the game is weakly acyclic. More generally, Milchtaich [2] showed that all generalized congestion games are weakly acyclic.

4 Convergence of Ordered Paths

4.1 Ordered Best Response Paths

If generalized congestion games do not possess the FBRP, we can ask what restrictions we can put on improvement paths to ensure that the paths do always end in an equilibrium. Here we consider ordered paths, inspired by Apt and Simon [1]. They formalized the notion of an arbiter who determines which player is the next deviator in a path by introducing *scheduler functions.* Our theorem states, in the language of those functions, that every generalized congestion game respects every best response local scheduler.

Theorem 1. *All generalized congestion games have the finite ordered best response property.*

Proof. Consider an arbitrary n player generalized congestion game and an ordered best response improvement path $(s_0, s_1, \ldots)$. Renumber the players so that player one is first in the order, then player two, and so on.

We proceed the proof by induction on the number of players. For one player the theorem trivially holds. Assume the theorem holds for $n-1$ players.

We need to show that the improvement path $(s_0, s_1, \ldots)$ is necessarily finite. If the nth player deviates only finitely many times in this path, it must be finite. Indeed, given finitely many deviations of player n, from some point on during the path, player n's strategy remains the same; then the game can be viewed as an $n-1$ player generalized congestion game, and by the induction hypothesis the path can only be finitely much longer. It thus suffices to show that the nth player deviates only finitely many times. We will show that from the nth player's *first* deviation, her payoff never decreases, and since it must increase with each of her deviations, she can deviate only finitely many times.

Let the kth step of the improvement path be the step immediately before any of the nth player's deviations. Since player n is the next deviator, in s_k players 1 through $n-1$ are playing best response. The congestion vector $c(s_k) = (c_1(s_k), \ldots, c_r(s_k))$ plays a special role in the proof: we want to consider every congestion vector that follows it relative to $c(s_k)$. For any strategy $s_l \in R^n$ in our path we define the *relative congestion vector*

$$\tilde{c}(s_l) = c(s_l) - c(s_k)$$

and call the sum of the positive entries of $\tilde{c}(s_l)$ the *congestion difference* between s_l and s_k.

For instance, say that the nth player is playing strategy j_0 at step k, and that in changing her strategy to a best response to s_k, she deviates to strategy j_1. This gives (assuming without loss of generality that $j_0 < j_1$)

$$\begin{aligned} c(s_{k+1}) &= (c_1(s_{k+1}), \ldots, c_{j_0}(s_{k+1}), \ldots, c_{j_1}(s_{k+1}), \ldots, c_r(s_{k+1})) \\ &= (c_1(s_k), \ldots, c_{j_0}(s_k) - 1, \ldots, c_{j_1}(s_k) + 1, \ldots, c_r(s_k)), \end{aligned}$$

and

$$\tilde{c}(s_{k+1}) = (0, \ldots, -1, \ldots, 1, \ldots, 0).$$

So s_{k+1} has 1 congestion difference from s_k.

The nth player's deviation may have upset the equilibrium of the first $n-1$ players, and the path will continue until they have reached an equilibrium again. If during this process the number of players playing strategy j_1 does not increase from $c_{j_1}(s_k) + 1$, the payoff of the nth player does not decrease during the other players' moves. In order to have a joint strategy in which the number of players playing strategy j_1 is at least $c_{j_1}(s_k) + 2$, that joint strategy must have a congestion difference from s_k of at least 2. We claim that no joint strategies with congestion difference greater than one from s_k occur.

Suppose to the contrary that one does. Then there must be a *first* deviation changing a joint strategy with congestion difference 1 from s_k to one with 2. We denote this latter strategy s_u, and call the player who caused this increase in congestion difference player i. For player i we consider the sequence of strategies $l_0, \ldots, l_m$ she plays, where l_0 is her strategy at the kth step, l_1 is her strategy after her first deviation, and so on, and the deviation from l_{m-1} to l_m results in s_u. (Note that the steps where these deviations occur are not consecutive; other players deviate while player i stays at her strategy.) Let $s_{b_1}, \ldots, s_{b_m}$ be the joint strategies that occur immediately before each of the ith player's deviations. More precisely, the shift from s_{b_p} to s_{b_p+1} is caused by player i deviating from l_{p-1} to l_p, for $1 \leq p \leq m$.

To arrive at a contradiction, we look at each of the ith player's deviations in reverse, and for each deduce what the relative congestion vector immediately before the deviation must be. We visually represent the argument by showing each of the ith player's deviations separately, with what we know about the relative congestion vector of the joint strategy before the deviation written below it. In this visualisation, the strategies $l_0, \ldots, l_m$ are shown distinct. Of course, we can not a priori say anything about the distinctness of these strategies, but while considering the relative congestion vectors of $s_{b_1}, \ldots, s_{b_m}$ we will see that the strategy l_m appears only once in the sequence, which is what is necessary for the proof.

First, consider the deviation from l_{m-1} to l_m. By assumption, the sum of the positive entries of $\tilde{c}(s_{b_m})$ is 1, while the sum of the positive entries of $\tilde{c}(s_{b_m+1})$ is 2. This means that at s_{b_m}, the relative congestion at l_{m-1} is either -1 or 0, and the relative congestion at l_m is either 0 or 1. Figure 2 shows this in a diagram:

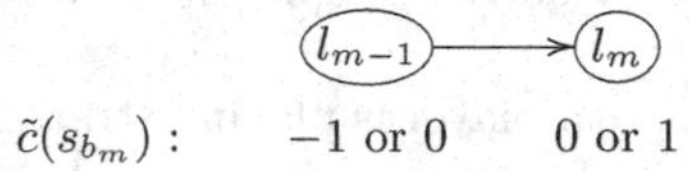

	l_{m-1} $\longrightarrow$	l_m
$\tilde{c}(s_{b_m})$:	-1 or 0	0 or 1

Fig. 2. Diagram with possible values of the relative congestion vector when considering player i's last deviation

This shows in particular that $m > 1$, because if this were the only deviation of the ith player, she would also have wanted to deviate from $l_0 = l_{m-1}$ to l_m at s_k, which contradicts her playing best response at s_k.

This justifies considering the ith player's deviation from l_{m-2} to l_{m-1}. At $s_{b_{m-1}}$, the relative congestion at l_{m-1} is at least -1, because the congestion difference from s_k is at most 1. Therefore, after the deviation to l_{m-1}, the payoff of player i is at most $p_i(l_{m-1}, c_{l_{m-1}}(s_k))$. But from the diagram above, the relative congestion at l_{m-1} was at most zero when player i deviated away from it, improving her payoff. So the payoff to player i has risen between $s_{b_{m-1}}$ and s_u. It follows that $l_{m-2} \neq l_m$, because if they were equal, the congestion at l_m must have been higher during $s_{b_{m-1}}$ than during s_u. This is impossible, because at s_u there is a relative congestion at l_m of at least 1, and at $s_{b_{m-1}}$ there is by assumption a relative congestion at any strategy of at most 1. Then, since the deviation to l_{m-1} was by assumption a *best* response, we must have that $p_i(l_m, c_{l_m}(s_{b_{m-1}})) < p_i(l_m, c_{l_m}(s_{b_m}))$ (otherwise the deviation would have been to l_m). From the definition of generalized congestion games it follows that $c_{l_m}(s_{b_{m-1}}) > c_{l_m}(s_{b_m})$, so in particular $\tilde{c}_{l_m}(s_{b_{m-1}}) = 1$ and $\tilde{c}_{l_m}(s_{b_m}) = 0$. This accounts for the sole positive entry of $\tilde{c}(s_{b_{m-1}})$ and $\tilde{c}(s_{b_{m-1}+1})$. This means that we must have $\tilde{c}_{l_{m-1}}(s_{b_{m-1}}) = -1$, which is sufficient information to add to our diagram, shown in Fig. 3.

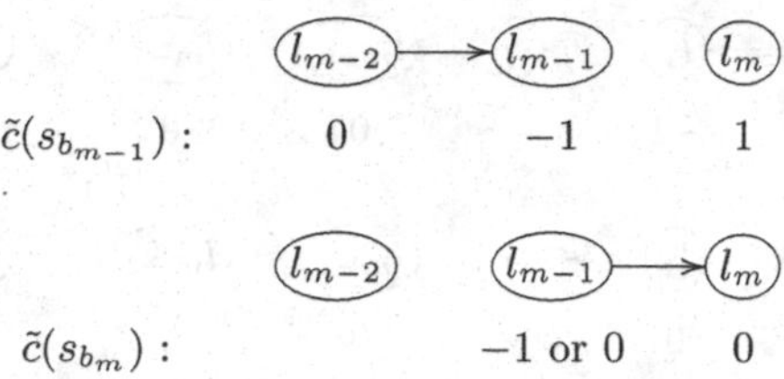

Fig. 3. Diagram with possible values of the relative congestion vector when considering player i's last two deviations

The relative congestion vector $\tilde{c}(s_{b_{m-1}})$ takes a very specific form: we have $\tilde{c}_{l_m}(s_{b_{m-1}}) = 1$, $\tilde{c}_{l_{m-1}}(s_{b_{m-1}}) = -1$, and all its other entries equal 0. We will show that all relative congestion vectors associated with $s_{b_1}, \ldots, s_{b_{m-1}}$ are of this form. We forgo the formal induction step to avoid even more cumbersome notation, and instead show the next step, trusting that this will illustrate the general process.

So consider the ith player's deviation from l_{m-3} to l_{m-2}. Since $s_{b_{m-2}}$ has congestion difference at most one from s_k, $\tilde{c}_{l_{m-2}}(s_{b_{m-2}})$ is at least -1. This means that after the deviation, the payoff to player i is at most $p_i(l_{m-2}, c_{l_{m-2}}(s_k))$. But as can be easily seen in the diagram above, this payoff is strictly smaller than $p_i(l_m, c_{l_m}(s_k) + 1)$, because the difference is two payoff increasing deviations; so again, $l_{m-3} \neq l_m$, since that would imply a relative congestion greater than 1 at l_m during $s_{b_{m-3}}$, and it follows that at steps $s_{b_{m-2}}$ and $s_{b_{m-2}+1}$, the relative congestion at l_m must be 1. This means that at $s_{b_{m-2}+1}$ the relative congestion at l_{m-2} may be at most 0, which means it must precisely be -1 at $s_{b_{m-2}}$. Again, we can fill out our diagram (Fig. 4):

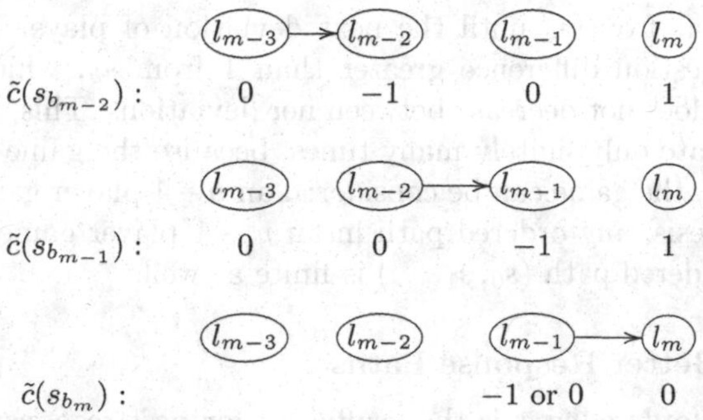

Fig. 4. Diagram with possible values of the relative congestion vector when considering player i's last three deviations

For every possible previous deviation, the same argument holds. That is, before any deviation, there is a relative congestion of 1 at l_m, a relative congestion of -1 at the strategy to which the ith player deviates, and a relative congestion of 0 elsewhere. This yields the complete diagram in Fig. 5:

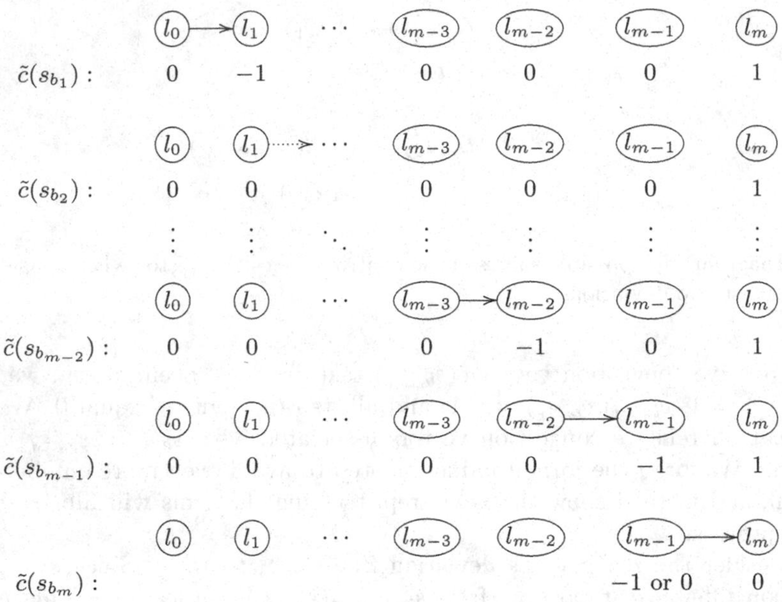

Fig. 5. Diagram with possible values of the relative congestion vector when considering all of player i's deviations

In this figure a contradiction clearly appears. Namely, when player i deviates from l_0, the strategy she played at s_k, the relative congestion there is 0; between each of her deviations, her payoff stays the same or goes up, and she ends up at strategy l_m, with relative congestion 1. This contradicts the fact that in s_k she plays a best response, because she would then, too, have gotten a better payoff at l_m.

This means that from s_k until the next deviation of player n, no strategy occurs with congestion difference greater than 1 from s_k, which means that player n's payoff does not decrease between her deviations. This, in turn, means that she can deviate only finitely many times, because the game is finite. After her last deviation, the game can be considered an $n-1$ player game, and by the induction hypothesis, any ordered path in an $n-1$ player game is finite. This shows that the ordered path $(s_0, s_1, \ldots)$ is finite as well. □

4.2 Ordered Better Response Paths

The condition of 'orderedness' is then sufficient for best response improvement paths to be finite. Next, we consider ordered better response improvement paths.

The following game is an example of a generalized congestion game in which an ordered better response improvement path is not finite. This demonstrates that generalized congestion games do not, generally, have the FOIP.

Example 1. An example of a generalized congestion game without the finite ordered improvement property is a seven-player eight-strategy game, in which the following equations hold:

- $p_1(2,1) = p_1(3,1) > p_1(3,2) > p_1(2,2)$,
- $p_2(3,1) > p_2(1,c) > p_2(3,2)$,
- $p_3(4,1) = p_3(5,1) > p_3(5,2) > p_3(4,2)$,
- $p_4(5,1) > p_4(1,c) > p_4(5,2)$,
- $p_5(1,1) > p_5(6,1) > p_5(4,c) > p_5(2,c) > p_5(1,2) > p_5(6,2)$,
- $p_6(8,1) > p_6(6,c) > p_6(5,c) > p_6(3,c) > p_6(8,2)$,
- $p_7(1,1) \geq p_7(1,2) > p_7(7,1) > p_7(8,1) \geq p_7(8,2) > p_7(1,3)$,

where c represents any possible congestion. All payoffs not mentioned above are assumed to be lower than the ones listed above. In such a game, the following path is an ordered better response improvement path:

1. $(2,3,4,5,1,8,\underline{7})$
2. $(2,3,4,5,\underline{1},8,1)$
3. $(\underline{2},3,4,5,2,8,1)$
4. $(3,\underline{3},4,5,2,8,1)$
5. $(3,1,4,5,\underline{2},8,1)$
6. $(3,1,\underline{4},5,4,8,1)$
7. $(3,1,5,\underline{5},4,8,1)$
8. $(3,1,5,1,\underline{4},8,1)$
9. $(3,1,5,1,6,8,\underline{1})$
10. $(3,1,5,1,6,\underline{8},8)$
11. $(\underline{3},1,5,1,6,3,8)$
12. $(2,1,5,1,6,\underline{3},8)$
13. $(2,\underline{1},5,1,6,5,8)$
14. $(2,3,\underline{5},1,6,5,8)$
15. $(2,3,4,1,6,\underline{5},8)$
16. $(2,3,4,\underline{1},6,6,8)$
17. $(2,3,4,5,\underline{6},6,8)$
18. $(2,3,4,5,1,6,\underline{8})$
19. $(2,3,4,5,1,\underline{6},7)$
20. $(2,3,4,5,1,8,7)$

This path is represented by the diagram in figure 6. The visualisation is the same as for the example in section 3.

Because the proof of theorem 1 revolves around the fact that the last player's payoff can not decrease, this example revolves around decreasing the seventh player's payoff. The reason this can occur is that, unlike in a best response ordered path, a player who starts moving can move almost arbitrarily between strategies, disturbing other players (who will then start moving) before reaching a best response.

The path falls into roughly two parts. The first part starts as player 7 deviates from strategy 7 to strategy 1. This causes player 5, who was playing strategy one, to become dissatisfied: she starts moving, and goes past strategies 2 and 4 to strategy 6. As she passes strategy 2, player 1, who has higher priority, is disturbed: player 1 moves to strategy 3, where she disturbs player 2, who in turn deviates to strategy 1. (This double movement, player 1 to strategy 3 and then player 2 to strategy 1 is necessary, because if player 1 had gone straight to strategy 1, she would deviate back as soon as player 5 left strategy 2). Similarly,

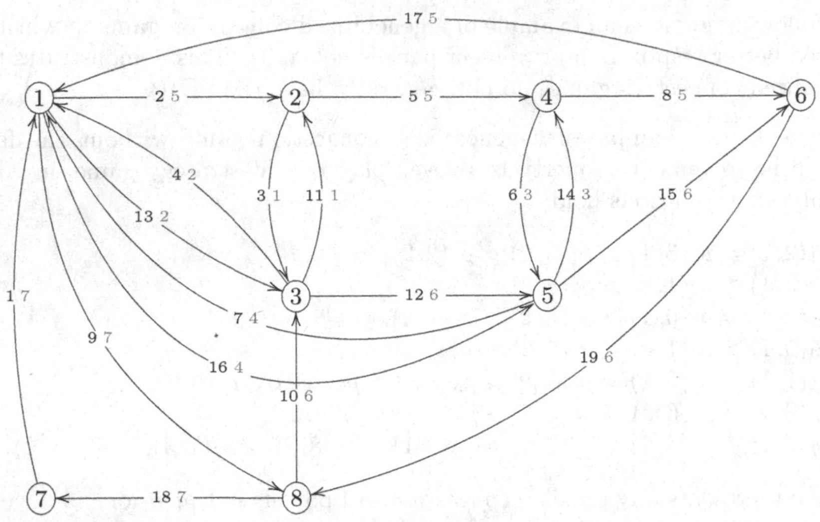

Fig. 6. Diagram representing deviations in the infinite ordered improvement path

as player 5 passes strategy 4, she disturbs player 3, who moves to strategy 5 where she disturbs player 4, who moves to strategy 1. At this point, one player has left strategy 1 since the arrival of player 7, and two extra players have arrived at it. This means that the payoff of player 7 has decreased, which is what we wanted to achieve.

The second part is about moving players back to their original positions. Player 7 deviates to strategy 8, where player 6 has been waiting. Player 6 will move between important strategies, disturbing players, causing them to move back to their original position. First, she moves to strategy 3, disturbing player 1, who deviates back to strategy 2. As player 6 departs from strategy 3, the strategy is empty again, so player 2 deviates back to it. Next, player 6 moves to strategy 5, similarly disturbing player 3, who deviates back to strategy 4. Again, when player 6 departs to strategy 6, strategy 5 is left empty, allowing player 4 to return to it. This in turn frees up strategy 1, which incentivizes player 5 to deviate back to it. Then player 6 has reached best response, and almost all players are in their original positions. Finally, player 7 completes her cycle, moving to strategy 7, freeing up strategy 8 for player 6 to return to, which restores the initial joint strategy.

5 Open Problems

Our original aim was to determine whether for every generalized congestion game there exists some order such that every improvement path conforming to that order is finite. By Theorem 1, this is true for best response improvement paths. For better response improvement paths the problem is still open. We believe that every generalized congestion game has an order for which all better response improvement paths are finite. The smallest counterexample (explained above) we have found is quite large and very fragile: for most orders, all paths with those orders are finite.

Milchtaich [2] showed in the proof of the weak acyclicity of generalized congestion games, that there always exists a best response improvement path which is no longer than $r\binom{n+1}{2}$. We have not provided a proof of an upper bound on the lengths of our paths. Numerical experiments suggest an upper bound of at most n^2 deviations in a path. In particular, we suspect that this bound does not depend on the number of strategies. So far, we have only been able to prove this for two players.

Acknowledgements. We would like to thank Krzysztof Apt for suggesting the problem and helpful discussions, and Neil Walton for reviewing our article at very short notice.

References

1. Apt, K.R., Simon, S.: A Classification of Weakly Acyclic Games. In: Serna, M. (ed.) SAGT 2012. LNCS, vol. 7615, pp. 1–12. Springer, Heidelberg (2012)
2. Milchtaich, I.: Congestion games with player-specific payoff functions. Games and Economic Behaviour 13, 111–124 (1996)
3. Milchtaich, I.: Crowding games are sequentially solvable. International Journal of Game Theory 27, 501–509 (1998)
4. Monderer, D., Shapley, L.S.: Potential games. Games and Economic Behaviour 14, 124–143 (1996)
5. Rosenthal, R.W.: A class of games possessing pure-strategy Nash equilibria. International Journal of Game Theory 2, 65–67 (1973)

Basic Network Creation Games with Communication Interests*

Andreas Cord-Landwehr, Martina Hüllmann, Peter Kling, and Alexander Setzer

Heinz Nixdorf Institute & Department of Computer Science
University of Paderborn, Germany

Abstract. Network creation games model the creation and usage costs of networks formed by a set of selfish peers. Each peer has the ability to change the network in a limited way, e.g., by creating or deleting incident links. In doing so, a peer can reduce its individual communication cost. Typically, these costs are modeled by the maximum or average distance in the network. We introduce a generalized version of the *basic network creation game* (BNCG). In the BNCG (by Alon et al., SPAA 2010), each peer may replace one of its incident links by a link to an arbitrary peer. This is done in a selfish way in order to minimize either the maximum or average distance to all other peers. That is, each peer works towards a network structure that allows himself to communicate efficiently with *all* other peers. However, participants of large networks are seldom interested in all peers. Rather, they want to communicate efficiently with a small subset only. Our model incorporates these (communication) *interests* explicitly.

Given peers with interests and a communication network forming a tree, we prove several results on the structure and quality of equilibria in our model. We focus on the MAX-version, i.e., each node tries to minimize the maximum distance to nodes it is interested in, and give an upper bound of $\mathcal{O}(\sqrt{n})$ for the private costs in an equilibrium of n peers. Moreover, we give an equilibrium for a circular interest graph where a node has private cost $\Omega(\sqrt{n})$, showing that our bound is tight. This example can be extended such that we get a tight bound of $\Theta(\sqrt{n})$ for the price of anarchy. For the case of general networks we show the price of anarchy to be $\Theta(n)$. Additionally, we prove an interesting connection between a maximum independent set in the interest graph and the private costs of the peers.

1 Introduction

In a network creation game (NCG), several selfish players create a network by egoistic modifications of its edges. One of the most famous NCG models is due to Fabrikant et al. [7]. Their model intends to capture the dynamics in large communication and computer networks built by the individual participants (peers, players) in a selfish way: participants try to ensure a network structure supporting their own communication needs whilst limiting their individual investment into the network. Since the players do not (necessarily) cooperate, the resulting network structure may be suboptimal from

* This work was partially supported by the German Research Foundation (DFG) within the Collaborative Research Centre "On-The-Fly Computing" (SFB 901) and by the Graduate School on Applied Network Science (GSANS).

M. Serna (Ed.): SAGT 2012, LNCS 7615, pp. 72–83, 2012.

a global point of view. The analysis of the resulting structure and its comparison to a (socially) optimal structure is a central aspect in the analysis of network creation games.

In the original model by Fabrikant et al., players may buy (or create) a single edge for a certain (fixed) cost of $\alpha > 0$. Their goal is to improve the network structure with respect to their individual communication needs. There are typically two ways to formalize the corresponding communication cost of a single peer: the maximum distance or the average distance to all other peers in the network. We refer to the different variants by MAX-version and AVG-version. Alon et al. [2] introduce a slightly simpler model, called *basic network creation games* (BNCG), that drops the cost parameter α. Instead, they limit the possible ways in which peers may change the network by restricting them to edge swaps: a peer may only replace one of its incident edges with a new edge to an arbitrary node in the network. Since peers are assumed to be selfish, only edge swaps (including simultaneous swapping of several edges at once) that improve the private communication cost of the corresponding peer are considered. In a *swap equilibrium*, no player can decrease its communication cost by an edge swap. This simpler variant of network creation games has the advantage of polynomially computable best responses of the players. Moreover, it still captures the inherent dynamic character and difficulty of communication networks formed by selfish participants, while avoiding the quite intricate dependence on the parameter α (see related work).

Our work generalizes the BNCG model of Alon et al. by introducing the concept of *interests*. In real communication networks, participants are typically only interested in a small subset of peers rather than the complete network. Thus, instead of trying to minimize the maximal or average distance to *all* other nodes, the individual players consider only the distances to nodes they are interested in. The main part of our analysis focuses on tree networks. Especially, we show that tree networks perform much better than general networks with respect to the price of anarchy. To avoid networks to become disconnected (note that in a BNCG peers want to communicate with all other peers and hence never disconnect the network), we restrict the peers to swaps that preserve connectivity. This restriction is valid from a practical point of view, where a lost network connectivity is to be avoided, since re-connecting a network causes high or even unpredictable costs. Moreover, if you consider that interests of the peers may change over time, it is also important for each single selfish peer to sustain connectivity.

Model and Notions. An instance of the *basic network creation game with interests (I-BNCG)* is given by a set of n players (peers, nodes) $V = \{v_1, v_2, \ldots, v_n\}$, an initial *connection graph* $G = (V,E)$, and an *interest graph* $G_I = (V,I)$. We use $I(v) := \{u \in V \mid \{v,u\} \in I\}$ to refer to the neighborhood of a player v in the interest graph and denote them as the *interests of* v. Both the connection graph and the interest graph are undirected. Thus, interests are always mutual. The connection graph represents the current communication network and can change during the course of the game. We consider only instances where the (initial) connection graph is a tree, whereas the interest graph G_I may be an arbitrary and not necessarily connected graph. Each player is assumed to have at least one interest. We study two different ways to formalize the private communication costs of nodes: the *MAX-version* and the *AVG-version*. In the first, the private cost $c(v) := \max\{d(v,u) \mid u \in I(v)\}$ of a node $v \in V$ is defined as the maximum distance from v to its interests. In the second, we define $c(v) := \sum_{u \in I(v)} \frac{d(v,u)}{|I(v)|}$

as the average distance to its interests. Here, $d(v,u)$ denotes the (shortest path) distance between u and v in the connection graph.

To improve its private cost, a player u may perform *edge swaps* in the connection graph: replace an incident edge $\{u,v\}$ with a new edge $\{u,w\}$ to an arbitrary player $w \in V$, written as $u : [v \to w]$. We refer to a single as well as to a series of simultaneously executed edge swaps of a player u as an *improving step* if u's private cost decreases. A player is only allowed to perform an improving step if the connection graph stays connected. If no player can perform an improving step, we say the connection graph is in a *MAX-equilibrium* or *AVG-equilibrium*, respectively. See Figure 1 for an example.

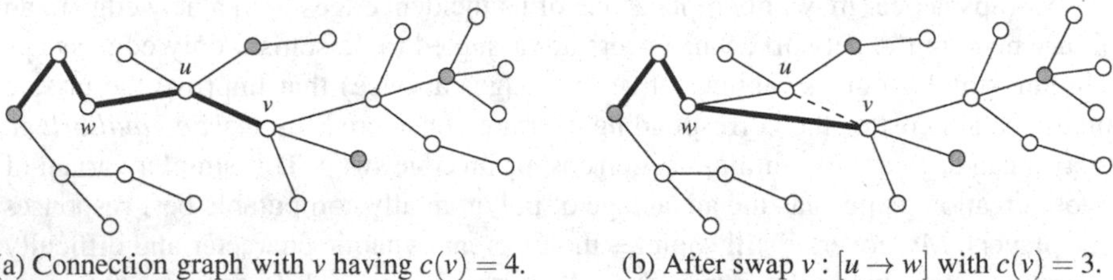

(a) Connection graph with v having $c(v) = 4$. (b) After swap $v : [u \to w]$ with $c(v) = 3$.

Fig. 1. MAX-version example of an improving swap performed by v. The gray nodes denote $I(v)$, the thick lines indicate the largest distance to a node in $I(v)$.

The quality of a connection graph G is measured by the *social cost* $c(G) = \sum_{v \in V} c(v)$ as the sum over all private costs. Our goal is to analyze the structure and social cost of worst case swap equilibria and compare them with a general optimal solution. As usual in algorithmic game theory, we use the ratio of these two values (*price of anarchy*, see Section 2.3) for this comparison [9]. Note that if the interest graph is the complete graph, I-BNCG coincides with the BNCG by Alon et al. [2].

Related Work. Network creation games combine two crucial aspects of modern communication networks: network design and routing. Many such networks consist of autonomous peers and have a highly dynamic character. Thus, it seems natural to use a game theoretic approach to study their evolution and behavior. Given the possibility to change the network structure (buy bandwidth, create new links, etc.), peers typically try to improve their individual communication experience. The question whether this selfish behavior results in an overall good network structure constitutes the central question of the study of network creation games as introduced by Fabrikant et al. [7]. In their model, the authors use a fixed cost parameter $\alpha > 0$ representing the cost of buying a single edge. The players (nodes) in such a game can buy edges to decrease their local communication cost (the average distance to all other nodes in the network). Each player's objective is to minimize the sum of its individual communication cost and the money spent on buying edges. In their seminal work, the authors proved (among other things) an upper bound of $\mathcal{O}\left(\sqrt{\alpha}\right)$ on the price of anarchy (PoA) in the case of $\alpha < n^2$. Albers et al. [1] proved a constant PoA for $\alpha \in \mathcal{O}\left(\sqrt{n}\right)$ and the first sublinear worst case bound of $\mathcal{O}\left(n^{1/3}\right)$ for general α. Demaine et al. [6] were the first to prove an $\mathcal{O}\left(n^{\varepsilon}\right)$ bound for α in the range of $\Omega\left(n\right)$ and $o\left(n \lg n\right)$. Furthermore, Demaine et al. introduced a new cost measure for the private cost, causing the individual nodes to consider their maximum distance to all remaining nodes instead of the average distance. For this variant they showed that the PoA is at most 2 for $\alpha \geq n$, $\mathcal{O}\left(\min\{4^{\sqrt{\lg n}}, (n/\alpha)^{1/3}\}\right)$ for $2\sqrt{\lg n} \leq \alpha \leq n$, and $\mathcal{O}\left(n^{2/\alpha}\right)$ for $\alpha < 2\sqrt{\lg n}$. Recently, Mihalák and Schlegel [11] could prove that for $\alpha > 273 \cdot n$ all equilibria in the AVG-version are trees (and thus the PoA is constant). The same result applies to the MAX-version if $\alpha > 129$.

While network creation games, as defined by Fabrikant et al., and their variants seem to capture the dynamics and evolution caused by the selfish behavior of peers in an accurate way, there is a major drawback of these models: most of them compute the private communication cost of the peers over the *complete* network. Given the immense size of such communication networks, this seems rather unrealistic. Typically, participants want to communicate only in small groups, with a small subset of participants they know. To the best of our knowledge, the only other work taking this into account is due to Halevi and Mansour [8]. They introduce a concept similar to our interests (see model description). For the objective of minimizing the average distance of a peer to its interests, they proved the existence of pure nash equilibria for $\alpha \leq 1$ and $\alpha \geq 2$ and upper bounded the PoA by $\mathcal{O}\left(\sqrt{n}\right)$ for general α. In the case of constant α or d (where d denotes the average degree in the interest graph) or $\alpha \in \mathcal{O}\left(nd\right)$, Halevi and Mansour upper bounded the PoA by a constant. Furthermore, the authors provided a family of problem instances for which the PoA is lower bounded by $\Omega\left(\log n / \log\log n\right)$.

Note that all these results largely depend on the cost parameter α. Moreover, as has been stated in [7], computing a player's best response for these models is NP-hard. This observation leads to a new, simplified formalization by Alon et al. [2], trying to capture the crux of the problem without the burden of this additional parameter. They introduce basic network creation games (BNCG), where players no longer have to pay for edges. Instead, possible actions are limited to *improving edge swaps*: replacing a single, incident edge by an edge to some arbitrary node which improves the node's private cost. Other than that, the general problem stays mostly untouched, especially the private cost function (average distance or maximum distance to all other nodes). Best responses in this game turn out to be polynomially computable. Restricting the initial network to trees, they show that the only equilibrium in the AVG-version is a star graph. Without restrictions, all swap equilibria are proven to have a diameter of $2^{\mathcal{O}\left(\sqrt{\lg n}\right)}$. For the MAX-version, the authors prove a maximum diameter of three if the resulting equilibrium is a tree. Furthermore, the authors construct an equilibrium of diameter $\Theta\left(\sqrt{n}\right)$. Our model is a direct generalization of these BNCGs, introducing the concept of interests. Up to now, the only other work on BNCGs we are aware of is due to Lenzner [10]. He studies the dynamics of the AVG-version of BNCGs and proves for the case of tree connection graphs a convergence to pure equilibria. Moreover, he proves that any sequence of improving edge swaps converges in at most $\mathcal{O}\left(n^3\right)$ steps to a star equilibrium.

Our Contribution. We introduce a generalized class of the BNCG by taking the different interests of individual peers into account. We analyze the structure and quality for the case that the initial connection graph is a tree. For the MAX-version, we derive a worst case upper bound of $\mathcal{O}\left(\sqrt{n}\right)$ for the private costs of the individual players in an equilibrium. Thereto, we introduce and apply a novel combinatorial technique that captures the structural properties of our equilibria (see MAX-arrangement, Definition 1). Furthermore, for interest graphs with a maximum independent set of size $M \leq \sqrt{n}$ (e.g., the clique graph with $M = 1$), we can improve the private cost upper bound to $\mathcal{O}(M)$. Using a circular interest graph, we construct an equilibrium with a player having private cost $\Omega(\sqrt{n})$, showing that our bound is tight. By extending this construction, we are able to prove a tight bound of $\Theta(\sqrt{n})$ on the price of anarchy (ratio between the social cost of a worst case equilibrium and an optimum [9]). Using a star-like connection graph, we show the existence of a MAX-equilibrium with small social cost, yielding a *price of stability* (ratio between the social cost of a best case equilibrium and an optimum [3,4]) of at most two for an I-BNCG. For the case of an I-BNCG featuring a general connection graph (instead of a tree), we show that the price of anarchy is $\Theta(n)$.

2 Quality of Equilibria in I-BNCGs

In this section we show a tight worst case private cost upper bound of $\Theta(\sqrt{n})$ for every MAX-equilibrium on trees as well as the same bound for the price of anarchy. The price of stability we can limit to be at most two. For general connection graphs we provide an instance with social cost $\Omega(n^2)$, yielding a price of anarchy of $\Theta(n)$.

2.1 Private Cost Upper Bound

In the following we prove the private cost upper bound as stated below:

Theorem 1. *Let I be a set of interests and $G = (V,E)$ a corresponding tree in a MAX-equilibrium, $n := |V|$. Then, for all $v \in V$ we have $c(v) \in \mathcal{O}(\sqrt{n})$.*

Outline of the proof: We consider a tree network in a MAX-equilibrium and take any node with maximal private cost. Starting with this node, we define a special node sequence, called MAX-arrangement, that will contribute the following properties: each two successive nodes are interested in each other and every node is "far away" from all previous nodes of the sequence. We will prove that such a sequence necessarily exists and that its length is proportional to the private cost of the starting node.

In detail, we prove with Lemma 3 and Lemma 4 that a shortest path traversal of a MAX-arrangement in the connection graph uses each edge at most twice and by this limits its length. Lemma 5 constructively shows that given a node with maximal private cost, there always exists a MAX-arrangement starting with this node and ending with a node with a private cost of 3. Lemma 2 gives us that the number of nodes in this MAX-arrangement is proportional to the maximal private cost of the first node. Comparing the maximum private cost of a node with the length of a shortest path traversal of any corresponding MAX-arrangement gives us the upper bound.

Remark 1. Note that in a MAX-equilibrium, each node v with $|I(v)| = 1$ has $c(v) = 1$. Hence, for a node v' with $c(v') > 1$, it holds $|I(v')| > 1$.

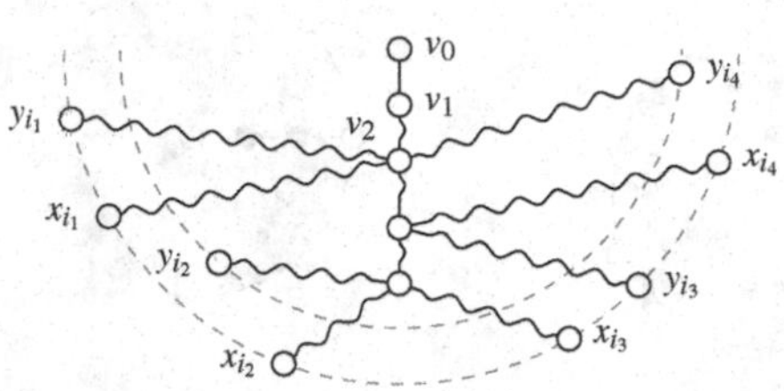

Fig. 2. Visualization for Lemma 1. v_0 can perform improving swap $v_0 : [v_1 \to v_2]$.

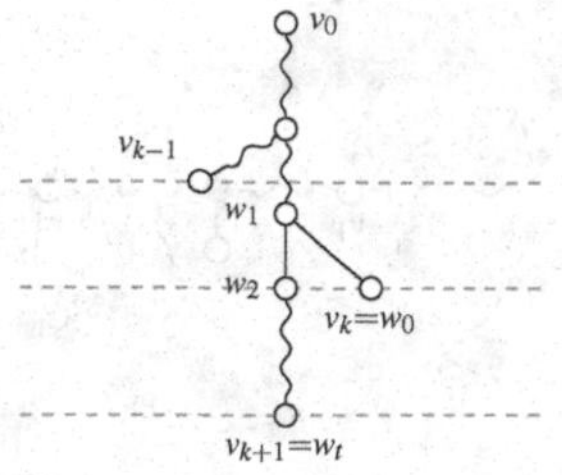

Fig. 3. Visualization for proof of Lemma 4. Edge $\{w_0, w_1\}$ is used twice.

Lemma 1 (T-configuration). *Let I be a set of interests, $G = (V, E)$ a corresponding tree in a MAX-equilibrium and $v \in V$ with $|I(v)| \geq 2$. Then there exist nodes $x, y \in I(v)$ such that $|d(x, v) - d(v, y)| \leq 1$ and v is connected by at most one edge to the shortest path from x to y and $c(v) = d(v, x)$.*

Proof. Let $v \in V$ with $|I(v)| \geq 2$ and $x \in I(v)$ with $d(v, x) = c(v)$. Assume that all $x' \in I(v) \setminus \{x\}$ are at distance $d(x', v) \leq c(v) - 2$ from v. Consider the shortest path $v \to v_1 \to v_2 \to \ldots \to x$ to x. In this case v can reduce its private cost by $v : [v_1 \to v_2]$ since this swap improves v's distance to x by 1 but increases the distances to every node in $I(v) \setminus \{x\}$ by at most 1. But this contradicts G being in a MAX-equilibrium.

We now consider all pairs $(x_i, y_i) \in I(v) \times I(v)$ for that hold $d(v, x_i) = c(v)$ and $d(v, y_i) \geq c(v) - 1$. Let us assume that v is connected to each shortest path from x_i to y_i by at least two edges that do not lie on that path. (See Figure 2 for a visualization.) Thus, v is not located on the shortest path from x_i to y_i for all i. This implies that in the graph $G \setminus \{v\}$ for each pair (x_i, y_i) there exists a connected component containing both nodes x_i, y_i. Since each two nodes at distance exactly $c(v)$ form such a pair, all nodes of $I(v)$ at distance exactly $c(v)$ must be located in the same connected component, which gives for every pair (x_i, y_i) that both nodes are contained in the same component. Hence, all nodes $x' \in I(v)$ at distance $d(x', v) \geq c(v) - 1$ from v are in the same connected component and by the two edges distance constraint, there must be a path $v \to v_1 \to v_2$ that is a subpath of every path from v to every node x_i and y_i. Hence, v can perform the improving swap $v : [v_1 \to v_2]$ (cf. Figure 2). This swap decreases the distance to all nodes x_i, y_i by one and increases each distance to other nodes (i.e., nodes $w \in I(v)$ with $d(w, v) \leq c(v) - 2$) by at most one and hence contradicts G being in a MAX-equilibrium. □

Definition 1 (MAX-arrangement). *Let $v_0 \in V$ and $v_1 \in I(v_0)$ such that $d(v_0, v_1) = c(v_0)$. Consider a sequence of nodes $v_0, \ldots, v_m$ with $v_i \in I(v_{i-1})$, $i = 1, \ldots, m$, with private costs $c(v_i) > 3$ for $i = 0, \ldots, m-1$ and $c(v_m) = 3$. We call this sequence a* MAX-arrangement *if for all $i = 2, \ldots, m$ it holds (see Figure 4 for a visualization):*

$$v_i = \underset{v_i \in I(v_{i-1})}{\operatorname{argmax}} \left\{ d(v_{i-2}, v_i) \;\middle|\; \begin{array}{l} v_{i-1} \textit{ is connected by} \leq 1 \textit{ edge to the} \\ \textit{shortest path from } v_{i-2} \textit{ to } v_i \end{array} \right\}$$

The key property of a MAX-arrangement is stated by the following two lemmas: consider a node v_i in a MAX-arrangement, then (1) its MAX-arrangement successor node

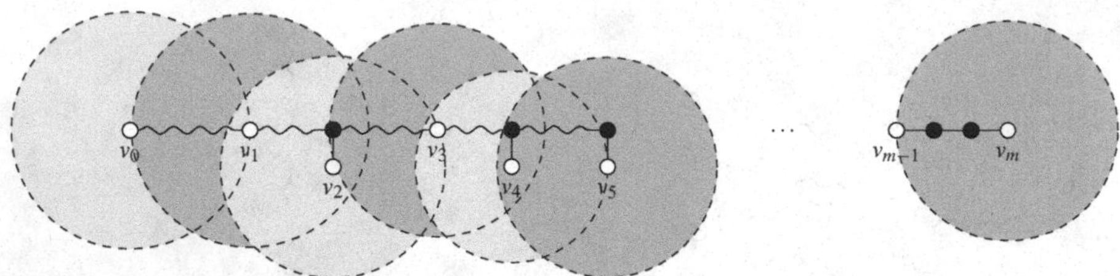

Fig. 4. Visualization of a MAX-arrangement. The radius of a circle around a node corresponds to the node's private cost. Curled lines denote shortest paths.

v_{i+1} cannot have a "much lower" private cost than v_i and (2) in the connection graph the shortest path from v_i to v_{i+1} can overlap by at most one edge with the shortest path to v_i's MAX-arrangement predecessor node.

Lemma 2. *For each two successive nodes v_i, v_{i+1} ($0 \leq i < m$) in a MAX-arrangement $v_0, \ldots, v_m$ it holds $d(v_i, v_{i+1}) \geq c(v_i) - 1$ and hence $c(v_{i+1}) \geq c(v_i) - 1$.*

Proof. Consider a node v_i, $0 \leq i < m$, in the MAX-arrangement. Then by Lemma 1, there exist $x, y \in I(v_i)$ with $d(v_i, x) = c(v_i)$ and $c(v_i) \geq d(v_i, y) \geq c(v_i) - 1$ such that v_i is connected by at most one edge to the shortest path from x to y. At least one of these nodes is a valid candidate for the next MAX-arrangement node v_{i+1} (even if neither x nor y is v_{i+1}, we get a distance lower bound) and we get $d(v_i, v_{i+1}) \geq \min\{d(v_i, x), d(v_i, y)\} \geq c(v_i) - 1$. This gives, $c(v_{i+1}) \geq c(v_i) - 1$. □

Lemma 3 (Increasing Distance). *Let I be a set of interests and $G = (V, E)$ a corresponding tree in a MAX-equilibrium with $v_0, \ldots, v_k$ a MAX-arrangement. Then the distances to v_0 are monotonically increasing, i.e., $d(v_0, v_i) \leq d(v_0, v_{i+1})$ for $i = 1, \ldots, k-1$.*

Proof. By $c(v_1) \geq 3$ we get with Remark 1 that $|I(v_1)| \geq 2$. Hence by Lemma 1, there exists a node v_2, such that the paths v_1 to v_0 and v_1 to v_2 overlap by at most one edge. By construction of the MAX-arrangement the distance $d(v_0, v_2)$ is maximal among all distances from v_0 to nodes $v \in I(v_1)$ and hence we get $d(v_0, v_1) \leq d(v_0, v_2)$.

Assume that there is a node v_i with smallest index $i \geq 2$ in the MAX-arrangement for which the claim does not hold. That is $d(v_0, v_{i-1}) \leq d(v_0, v_i) > d(v_0, v_{i+1})$. Denote by x the most distant node from v_0 that is on all shortest paths from v_0 to v_{i-1}, v_0 to v_i, and v_0 to v_{i+1}. (Such a node x exists since especially v_0 fulfills the restrictions.) By the choice of i and since all these paths cross node x, we get:

$$d(x, v_{i-1}) \leq d(x, v_i) > d(x, v_{i+1}) \tag{1}$$

By definition of the MAX-arrangement, v_i is connected by at most one edge to the shortest path from v_{i-1} to v_{i+1}. Hence, x must be a node on the path from v_{i-1} to v_{i+1}. First note that x cannot be v_i or a neighbor of v_i, since for those cases with (1) we get $d(x, v_{i+1}) < d(x, v_i) \leq 1$. Further, x must lie on the shortest path from v_{i-1} to v_i, since otherwise x would lie on the shortest path from v_i to v_{i+1} which implies by $d(v_{i-1}, v_i) \geq 3$ that $d(x, v_i) < d(x, v_{i-1})$. This gives $d(x, v_i) \leq d(x, v_{i+1})$ and is a contradiction. □

Lemma 4. *Let I be a set of interests and $G = (V,E)$ a corresponding tree in a MAX-equilibrium. Consider a MAX-arrangement $v_0,\dots,v_m$. Then, no edge in E is used more than two times by the shortest path visiting the nodes $v_0,\dots,v_m$ in the given order.*

Proof. We label the nodes of G by their distances to v_0. This is, for every $v \in V$ we define a *level* by $\text{level}(v) := d(v_0, v)$. We consider an arbitrary node v_k with $k \in \{1,\dots,m-1\}$ and the corresponding shortest path $v_k =: w_0 \to w_1 \to \dots \to w_t := v_{k+1}$ to node v_{k+1}. By definition, v_k is connected by at most one edge to the shortest path from v_{k-1} to v_{k+1} (see Figure 3). By Lemma 3 we have $\text{level}(v_{k-1}) \le \text{level}(v_k) \le \text{level}(v_{k+1})$. Hence, for $i = 2,\dots,t-1$ we get $\text{level}(w_i) < \text{level}(w_{i+1})$. This is, at most one edge (explicitly edge $\{w_0, w_1\}$) of the shortest path from v_0 to v_k is used a second time by the shortest path traversal from v_k to v_{k+1}. By Lemma 2 we have $t \ge c(v_k) - 1 \ge 3$ and get $\text{level}(v_k) < \text{level}(v_{k+1})$. □

Now we prove that given a node v_0 of a MAX-equilibrium tree with $c(v_0) > 3$, there exists a MAX-arrangement starting with v_0 and closing with a node with private cost 3. With the previous results about MAX-arrangements, this leads to the upper bound.

Lemma 5. *Let I be a set of interests and $G = (V,E)$ a corresponding tree in a MAX-equilibrium. Then for $v_0, v_1 \in V$ with $d(v_0, v_1) > 3$ and $v_1 \in I(v_0)$ there exists a MAX-arrangement starting with v_0. And for every such MAX-arrangement it holds that the shortest path that visits all nodes of the MAX-arrangement in the given order uses at least $(c(v_0))^2 + c(v_0) - 6)/4$ different edges of G.*

Proof. Existence: v_0, v_1 obviously fulfill the conditions of a MAX-arrangement. Thus, it suffices to show that, given the beginning of a MAX-arrangement $v_0,\dots,v_i$ with $c(v_j) > 3, j = 0,\dots,i-1$, we can either find a next node v_{i+1} that suffices the conditions or otherwise $c(v_i) = 3$. Assume $c(v_i) > 3$. Then, by Lemma 1 there exist $x, y \in I(v_i)$ with $d(v_i, x) = c(v_i)$ and $c(v_i) \ge d(v_i, y) \ge c(v_i) - 1$ such that v_i is connected by at most one edge to the shortest path from x to y. Since $c(v_i) > 3$, also $c(x) \ge 3$ and $c(y) \ge 3$ hold. Now, for at least one node (x or y) we have that this node is most distant to v_{i-1}, it is not v_{i-2}, and thus it fulfills the conditions for a MAX-arrangement.

Traversal: We now can apply the previous lemmas for providing the minimal length of such a MAX-arrangement: Lemma 2 states that by construction we always have $c(v_{i+1}) \ge c(v_i) - 1$. Lemma 3 implies that no node can be contained more than once in a MAX-arrangement. By the arguments above, we can always find a new node for the MAX-arrangement until we reach a node w with $c(w) = 3$. Hence, the MAX-arrangement contains at least $c(v_0) - 2$ nodes. Since the distance between two succeeding nodes in the MAX-arrangement decreases by at most one per node, a traversal of this MAX-arrangement consists of at least $\sum_{i=3}^{c(v_0)} i = (c(v_0)^2 + c(v_0) - 6)/2$ edges. From these, by Lemma 4, at least $(c(v_0))^2 + c(v_0) - 6)/4$ edges are different. □

Theorem 1 (Restated) *Let I be a set of interests and $G = (V,E)$ a corresponding tree in a MAX-equilibrium, $n := |V|$. Then, for all $v \in V$ we have $c(v) \in \mathcal{O}(\sqrt{n})$.*

Proof. W.l.o.g. we may assume that there is at least one $\{v, v'\} \in I$ with $d(v, v') \ge 3$. Let nodes $v_0, v_1 \in V$, $v_1 \in I(v_0)$ have maximal distance among all nodes,

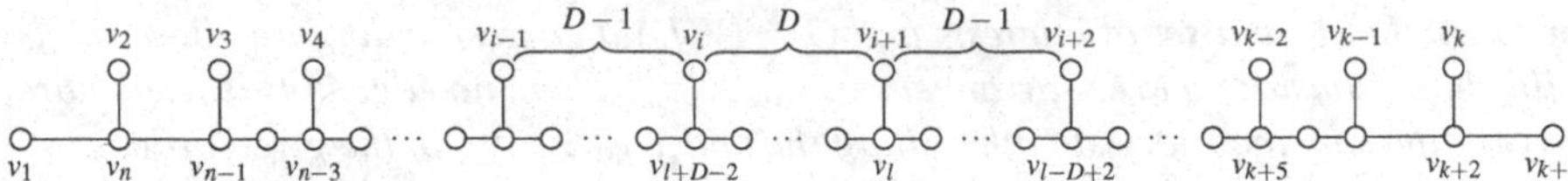

Fig. 5. Tree $G=(V,E)$ in a MAX-equilibrium with private cost $\Omega(D)$ for node v_i, with $D := \sqrt{|V|-2}+1$, $k := 2D-3$, and $l = n - \sum_{i=1}^{D} i$.

$D := d(v_0, v_1) = c(v_0)$. Then, by Lemma 5 we can find a MAX-arrangement $v_0, \ldots, v_m$ whose traversal uses at least $(D^2+D-6)/4$ different edges. Since our tree has exactly $n-1$ edges, we get $(D^2+D-6)/4 \leq n-1$ as an upper bound for the size of every MAX-arrangement and hence the private cost upper bound is $D \in \mathcal{O}(\sqrt{n})$. □

2.2 The Private Cost Upper Bound Is Tight

Next, we show that the upper bound of $\mathcal{O}(\sqrt{|V|})$ for the private costs is tight by constructing a MAX-equilibrium instance with one player having private cost $\Omega(\sqrt{|V|})$.

Remark 2. For a connection graph with nodes $V := \{v_1, \ldots, v_n\}$, let $I := \{\{v_i, v_{i+1}\} | i = 1, \ldots, n-1\} \cup \{\{v_n, v_1\}\}$ be interests such that (V, I) is a circle. Then, a node v_i with degree one in G cannot perform any swap if and only if it holds $|d(v_{i-1}, v_i) - d(v_i, v_{i+1})| \leq 1$ and v_i is connected by one edge to the shortest path from v_{i-1} to v_{i+1} (cf. Lemma 1).

Theorem 2. *There exists a set of interests and a corresponding tree $G=(V,E)$ in a MAX-equilibrium with a node $v_i \in V$ that has private cost $c(v_i) \in \Omega\left(\sqrt{|V|}\right)$.*

Sketch. We consider interests $I := \{\{v_i, v_{i+1}\} | i = 1, \ldots, n-1\} \cup \{\{v_n, v_1\}\}$ and the connection graph as stated in Figure 5. (For the proof see the full version [5].) □

2.3 Existence of MAX-equilibria and the Price of Anarchy

In this section we compute the *price of stability* (PoS) and the *price of anarchy* (PoA). Let the *social optimum* represent an instance with the smallest social cost of any tree over all nodes (which is not necessarily in a MAX-equilibrium). Then, the PoS denotes the ratio between the minimum social cost of a MAX-equilibrium and the cost of a social optimum. Whereas the PoA denotes the ratio between the worst social cost of a MAX-equilibrium and the cost of a social optimum.

In the full version [5], we provide a simple approximation algorithm that generates for any interest graph a MAX-equilibrium tree whose social cost is at most twice as high as an optimal solution, yielding the following lemma.

Lemma 6. *For every set of interests I there exists a corresponding tree $G=(V,E)$ in a MAX-equilibrium that causes social cost $c(G) \leq 2n$, $n := |V|$.*

Theorem 3. *The price of stability for I-BNCG is at most* 2.

Proof. Let I be a set of interests over nodes V, $n := |V|$. Then each connection graph that is a tree induces social cost of at least n. By Lemma 6 there exists a connection graph in a MAX-equilibrium with social cost of at most $2n$. Thus, the price of stability is at most $2n/n = 2$. □

Lemma 7. *There exist interest graphs over n nodes with a corresponding MAX-equilibrium tree that causes social cost of* $\Omega(n \cdot \sqrt{n})$.

Sketch. Consider $I := \{\{v_i, v_{i+1}\} | i = 1, \ldots, \lfloor n/2 \rfloor - 1\} \cup \{\{v_i, v_1\} | i = \lfloor n/2 \rfloor, \ldots, n\} \cup \{\{v_{n/2-1}, v_i\} | i = \lfloor n/2 \rfloor, \ldots, n\}$ and construct a similar graph as in Figure 5 but with $\lfloor n/2 \rfloor$ nodes at position of v_i, each with private cost $\Omega(\sqrt{n})$. (See full version [5].) □

Theorem 4. *The price of anarchy for I-BNCG is* $\Theta(\sqrt{n})$.

Proof. Theorem 1 provides an upper bound of $\mathcal{O}(\sqrt{n})$ for the private cost of every node in a tree in a MAX-equilibrium with n nodes. By this, $\mathcal{O}(n \cdot \sqrt{n})$ is an upper bound for the social cost of every MAX-equilibrium. Further, by Lemma 7 we get $\Omega(n \cdot \sqrt{n})$ as a worst case lower bound for the social cost of a graph in a MAX-equilibrium. For the cost of a social optimum, we get $\Theta(n)$. (Each social optimum incurs cost of at least n and at most $2n$.) Hence, we get $\Theta(n \cdot \sqrt{n}/n) = \Theta(\sqrt{n})$ for the price of anarchy. □

2.4 The Price of Anarchy for I-BNCG on General Graphs

Theorem 5. *The price of anarchy for I-BNCG with general connection graphs is* $\Theta(n)$.

Proof. First, note that the social cost of every instance are upper bounded by n^2 and lower bounded by n. Second, we provide an interest graph over n nodes ($n \equiv 0 \mod 6$) and a corresponding MAX-equilibrium graph $G = (V, E)$ with social cost $\Omega(n^2)$ (see Figure 6). To this, we connect $n/2$ nodes to a ring (ring nodes) and connect one additional (satellite) node to each of them. Each of the ring nodes is interested in its three adjacent nodes in G, whereas each satellite node is interested in its neighbor at the ring and in both satellite nodes at distance exactly $n/6 + 2$. This is an equilibrium and all $n/2$ satellite nodes have a private cost of $n/6 + 2$, i.e., the price of anarchy is $\Omega(n)$. □

3 Further Structural Properties of Equilibria

By Lemma 5 we achieved an upper bound for any MAX-arrangement (see Definition 1) contributed only by the property that the network is connected. Here, we introduce a second upper bound for a MAX-arrangement that is given by the size of a *maximum independent set* (MIS) in the interest graph. Having such an MIS of size M, we can bound the maximum private costs by $\mathcal{O}(M)$, which yields improved bounds for specific families of interest graphs. Particularly, this gives asymptotically same upper bounds for complete interest graphs on trees as those explicitly constructed by Alon et al. [2].

Theorem 6. *Let I be a set of interests and $G = (V, E)$ a corresponding tree in a MAX-equilibrium. Let M be the size of a maximum independent set in (V, I). Then for every MAX-arrangement $v_0, \ldots, v_m$: The length of this MAX-arrangement is at most $2 \cdot M$.*

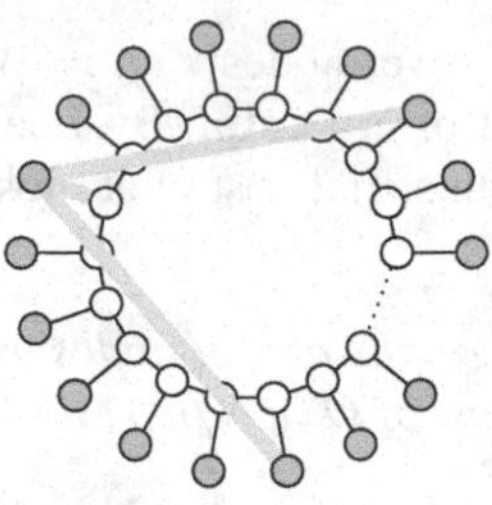

Fig. 6. MAX-equilibrium graph with social cost of $\Omega(n^2)$. Each white node is interested in its three neighbors. Each gray node is interested in its white neighbor and the two gray nodes at distance $n/6+2$.

Proof. We prove that the nodes of $v_0, \ldots, v_{m-1}$ with even index form an independent set in the interest graph (V,I). Consider an even index i and assume for contradiction that there is an even index $k < i$ such that $v_k \in I(v_i)$. By Lemma 3 we get $d(v_k, v_{k+1}) \leq d(v_k, v_{k+2})$. If $v_{k+2} \neq v_i$ with Lemma 2 and $c(v_j) > 3$ for all v_j in the MAX-arrangement we get $d(v_k, v_i) > d(v_k, v_{k+2}) + 1 \geq c(v_k)$. But this is a contradiction.

Thus, consider the case $v_{k+2} = v_i$. Since v_{k+1} is connected by at most one edge to the shortest path from v_k to v_{k+2} and $d(v_{k+1}, v_{k+2}) \geq 3$ we get $v_{k+2} \notin I(v_k)$. Otherwise we either get the same contradiction as before, or v_{k+1} would contradict to be the most distant node in $I(v_k)$ that fulfills the MAX-arrangement conditions. Hence, the nodes with even index form an independent set in (V,I), which gives $m \leq 2 \cdot M$ □

Corollary 1. *Let I be a set of interests and $G = (V,E)$ a corresponding tree in a MAX-equilibrium, $n := |V|$. Let the size M of any MIS in (V,I) be limited by $\sqrt{n}$. Then, for $v \in V$ we have $c(v) \in \mathcal{O}(M)$.*

Proof. W.l.o.g. we assume that there is a node with private cost greater than 3. Hence, there is a MAX-arrangement $v_0, \ldots, v_m, v_{m+1}$ with $c(v_i) > 3, i = 1, \ldots, m$ and $c(v_{m+1}) = 3$. By Theorem 6 we get $m \leq 2M$. Analog to Theorem 1, we get the upper bound. □

Corollary 2. *Let I be a set of interests and $G = (V,E)$ a corresponding tree in a MAX-equilibrium. If (V,I) is a complete graph, then $c(v) \in \mathcal{O}(1)$ for all $v \in V$.*

In the full version [5], we provide a scenario and a corresponding cyclic invocation sequence over all nodes, with each node performing a best-response improving swap (if possible), such that the nodes never reach a MAX-equilibrium. This gives:

Remark 3. I-BNCG is no potential game as defined by Monderer and Shapley [12].

4 Outlook and Future Work

In this paper, we presented tight worst case bounds for the private costs as well as for the social cost of any MAX-equilibrium on tree networks. Furthermore, we drew an interesting connection between the size of an MIS in the interest graph and upper bounds on the private/social costs. In comparison with MAX-equilibria on general graphs, we could show that the price of anarchy can perform much worse if the connection graph is not acyclic. However, it remains an open question whether the price of anarchy on general connection graphs with complete interests could perform better than $\mathcal{O}(n)$. For this, so far there is only a worst case lower bound of $\Omega(\sqrt{n})$ (by Alon et al. [2]) for the graph diameter in a MAX-equilibrium, yielding a lower bound for the price of anarchy. Techniques similar to our MAX-arrangement-technique may allow deeper insights into the nature of MAX-equilibria in that scenario. Apart from this, finding good upper bounds on the social cost of an AVG-equilibrium remains a challenging problem (in the full version [5], we give a lower bound of $\Omega(n)$ for the private costs).

Even if the existence of a MAX-equilibrium is always ensured (which we proved for trees), it remains an open question whether the dynamics ever reaches an equilibrium. We could state examples, where the network *never* converges to a MAX-equilibrium. It seems an interesting question whether we can guarantee the convergence by additional policies, e.g., by restricting the order in which nodes perform their swaps. And in case of a guaranteed convergence, how many swaps would it take to reach an equilibrium?

Currently, we only considered static interest graphs. But in practice, interests of network participants might change over time. Introducing a time model and considering certain (possibly restricted) changes of the interest graph seems a natural way to generalize our model, yielding an interesting online problem.

References

1. Albers, S., Eilts, S., Even-Dar, E., Mansour, Y., Roditty, L.: On nash equilibria for a network creation game. In: 17th SODA, pp. 89–98. ACM (2006)
2. Alon, N., Demaine, E.D., Hajiaghayi, M.T., Leighton, T.: Basic network creation games. In: 22nd SPAA, pp. 106–113. ACM (2010)
3. Anshelevich, E., Dasgupta, A., Kleinberg, J., Tardos, É., Wexler, T., Roughgarden, T.: The price of stability for network design with fair cost allocation. In: 45th FOCS, pp. 295–304. IEEE (2004)
4. Anshelevich, E., Dasgupta, A., Tardos, É., Wexler, T.: Near-optimal network design with selfish agents. In: 35th STOC, pp. 511–520. ACM (2003)
5. Cord-Landwehr, A., Hüllmann, M., Kling, P., Setzer, A.: Basic network creation games with communication interests. CoRR, arxiv.org/abs/1207.5419 (2012)
6. Demaine, E.D., Hajiaghayi, M.T., Mahini, H., Zadimoghaddam, M.: The price of anarchy in network creation games. In: 26th PODC, pp. 292–298. ACM (2007)
7. Fabrikant, A., Luthra, A., Maneva, E., Papadimitriou, C.H., Shenker, S.: On a network creation game. In: 22nd PODC, pp. 347–351. ACM (2003)
8. Halevi, Y., Mansour, Y.: A Network Creation Game with Nonuniform Interests. In: Deng, X., Graham, F.C. (eds.) WINE 2007. LNCS, vol. 4858, pp. 287–292. Springer, Heidelberg (2007)
9. Koutsoupias, E., Papadimitriou, C.: Worst-Case Equilibria. In: Meinel, C., Tison, S. (eds.) STACS 1999. LNCS, vol. 1563, pp. 404–413. Springer, Heidelberg (1999)
10. Lenzner, P.: On Dynamics in Basic Network Creation Games. In: Persiano, G. (ed.) SAGT 2011. LNCS, vol. 6982, pp. 254–265. Springer, Heidelberg (2011)
11. Mihalák, M., Schlegel, J.C.: The Price of Anarchy in Network Creation Games Is (Mostly) Constant. In: Kontogiannis, S., Koutsoupias, E., Spirakis, P.G. (eds.) SAGT 2010. LNCS, vol. 6386, pp. 276–287. Springer, Heidelberg (2010)
12. Monderer, D., Shapley, L.S.: Potential Games. Games and Economic Behavior 14(1), 124–143 (1996)

Common Knowledge and State-Dependent Equilibria

Nuh Aygun Dalkiran[1], Moshe Hoffman[2], Ramamohan Paturi[3], Daniel Ricketts[3], and Andrea Vattani[3]

[1] Bilkent University, Department of Economics
dalkiran@bilkent.edu.tr

[2] The Rady School of Management & Department of Computer Science and Engineering, UC San Diego
mohoffman@ucsd.edu

[3] Department of Computer Science and Engineering, UC San Diego
{paturi,daricket,avattani}@ucsd.edu

Abstract. Many puzzling social behaviors, such as avoiding eye contact, using innuendos, and insignificant events that trigger revolutions, seem to relate to common knowledge and coordination, but the exact relationship has yet to be formalized. Herein, we present such a formalization. We state necessary and sufficient conditions for what we call state-dependent equilibria – equilibria where players play different strategies in different states of the world. In particular, if everybody behaves a certain way (e.g. does not revolt) in the usual state of the world, then in order for players to be able to behave a different way (e.g. revolt) in another state of the world, it is both necessary and sufficient for it to be common p-believed that it is not the usual state of the world, where common p-belief is a relaxation of common knowledge introduced by Monderer and Samet [16]. Our framework applies to many player r-coordination games – a generalization of coordination games that we introduce – and common (r,p)-beliefs – a generalization of common p-beliefs that we introduce. We then apply these theorems to two particular signaling structures to obtain novel results.

1 Introduction

In the popular parable "The Emperor's New Clothes" [2], a gathering of adults pretends to be impressed by the Emperor's dazzling new suit despite the fact that he is actually naked. It is not until an innocent child cries out "But he has nothing on at all!" that the Emperor's position of authority and respect is questioned. This is a metaphor for a number of common political situations in which the populace knows the current regime is inept but takes no action against it until some seemingly insignificant event occurs, such as the child's cry. In fact, in Tunisia, despite years of political repression and poverty, it was not until the previously unknown street vendor Mohamed Bouazizi set himself on fire that citizens rose up in protest. Common knowledge – everyone knows

M. Serna (Ed.): SAGT 2012, LNCS 7615, pp. 84–95, 2012.

that everyone knows that... – might offer such an explanation for this strange phenomenon: while the boy's cry and the self immolation of Mohamed Bouazizi do not teach anyone that the government is inept, they make it commonly known that the government is inept. Likewise, common knowledge has been proffered as an explanation for many other puzzling social behaviors: it is common to avoid eye-contact when caught in an inappropriate act, despite the fact that looking away, if anything, increases the conspicuousness of a shameful deed. Nevertheless, even Capuchin monkeys look away when they ignore a request to help an ally in a tussle [21]. And few adults after a nice date are fooled by the inquiry "Would you like to come upstairs for a drink?" yet all but the most audacious avoid the explicit request [22].

Many authors have aptly noted that common knowledge plays an important role in these puzzling social behaviors [10, 4, 6, 22]. Avoiding eye contact prevents common knowledge that you were noticed, using innuendos enables a speaker to request something inappropriate without making the request commonly known, and prohibiting public displays of criticism of the government while not preventing people from realizing the flaws of their government, prevent the flaws from being commonly known. Authors have argued that common knowledge is important in these situations because common knowledge is needed for coordination. But without formal arguments, many important questions still remain, such as: what exactly needs to be "commonly known" in order to "coordinate"? What exactly will happen in the absence of common knowledge? Miscoordination? When common knowledge is lacking, but almost present, e.g. if everyone is pretty sure that everyone is pretty sure... will this have the same effect as common knowledge? Such details, which may seem pedantic, are crucial for answering practical questions such as: if I cannot think up an innuendo, will an appropriately placed cough midsentence do the trick? Why is it that sometimes we use innuendos and sometimes we go out of our way to state the obvious?

We will formalize the role of common knowledge in coordination, which will enable us to address each of these questions. The crucial step in our formalism is based on the insight of Rubinstein [23]. Rubinstein considers coordination games – games in which players make choices such that they would like to mimic the choice that others make. Rubinstein supposes that players coordinate on a particular action A in a given situation. He then supposes that the situation changes and asks whether the players can coordinate on a different action instead. He shows that unless it is commonly known that the situation has changed, players still must coordinate on A. The intuition is clear: even if one player knows that circumstances have changed, if he thinks the other player does not know this, then he expects the other player to play as if circumstances have not changed. Since it is a coordination game, he best responds by playing as if circumstances have not changed. Likewise, even if both players know that circumstances have changed, and both players know that both players know this, but one player does not realize the second player has this second degree of knowledge, then this player will expect the other player to play as if circumstances have not changed.

By the above argument, he best responds by playing as if circumstances have not changed. The same logic continues indefinitely.

Rubinstein presents a particular instance in which the above logic holds. The contribution of our paper is to show that this logic holds quite generally, for any two player coordination game, and in fact, for a generalization to many players. And moreover, we show that common knowledge is not just necessary for changing behaviors when circumstances change, but common knowledge is also sufficient. We hope that this will lead to a deeper understanding of these puzzling social behaviors, as well as some novel predictions.

OUR RESULTS. In this paper, we introduce *state-dependent equilibria*, which we define as equilibrium strategies in which players take different actions when the circumstances change. This notion allows us to address the questions that were left unanswered by the informal discussions of common knowledge and coordination. In particular, we characterize the conditions under which rational players are able to play state-dependent equilibria.

We begin by considering two-player coordination games. We show that it is not quite common knowledge that determines the existence of state-dependent equilibria but rather a relaxation of common knowledge. This notion corresponds with common p-beliefs, as developed by Monderer and Samet [16]: each believes with probability at least p that each believe with probability at least p.... In our framework, we show that p depends on the precise payoffs of the game and corresponds to the risk dominance of Harsanyi and Selten [14].

We then introduce a natural n-player generalization of coordination games that we call *r-coordination games* in which coordination on an action is successful if at least some fraction r of the players take that action. Accordingly, we also develop a generalization of common p-beliefs for this setting.

In order to derive our results, we provide a unifying theoretical framework for analyzing our games. Our framework gives tight necessary and sufficient conditions on the players' beliefs under which a state-dependent equilibrium exists. These conditions depend on the payoffs of the game (in particular on the risk dominance) and, in the case of r-coordination games, on the threshold fraction r required for successful coordination.

Our final contribution is to apply this framework both to simple but puzzling social behavior and to more complex distributed phenomena that arise in biology, economics, and sociology. The first application is eye-contact. We offer a post hoc explanation for why we avoid eye-contact when caught in an inappropriate act. For the second and third applications, we show how our results can be applied to situations in which the true state of the world is observed by all players with arbitrarily small noise, as in the global games literature [18, 20, 19]. This yields some novel predictions about social behaviors, such as which cues can be used to instigate a revolution, and when a researcher's reputation can be resilient to substandard work.

Due to space constraints, the proofs of our claims will appear in the full version of the paper.

1.1 Related Work

The concept of common knowledge was first formalized in multi-modal logic in 1969 by Lewis [15]. Aumann later put common knowledge in a set-theoretic framework [3].

In 1989, Rubinstein used common knowledge to analyze a problem related to the coordinated attack problem in computer science [23]. This problem, called the Electronic Mail Game, was the first example that common knowledge is very different than any finite order of knowledge. Rubinstein showed that the lack of common knowledge prevents players from switching strategies (i.e. prevents the existence of state-dependent equilibria) in the Electronic Mail Game. See [17] for a retrospective on the Electronic Mail Game. Our results show that common knowledge is not just necessary but also sufficient and holds for any coordination game and even r-coordination games.

Carlsson and Van Damme showed that when players have noisy signals about the payoffs in a coordination game, as the noise vanishes, the unique equilibrium in the game becomes the risk dominant equilibrium [5]. Morris and Shin applied this result to bank runs and currency crisis, showing that there is a unique underlying value at which currencies collapse and bank runs occur, in contrast to previous models, which permitted multiple equilibria and prevented comparative static analysis [18, 20, 19]. In some of our applications, we use similar signaling structures, but the uncertainty does not affect the payoffs. We find circumstances under which no state-dependent equilibria exist.

Monderer and Samet developed an approximate notion of common knowledge called common p-beliefs, which is relevant in our framework. We will draw heavily on their definitions and results [16]. Others have discussed the role of common knowledge in social puzzles, albeit less formally than in the aforementioned literature. Chwe discusses the role of common knowledge in public rituals [6]. Pinker *et al* discusses the role of common knowledge in innuendos [22]. Binmore and Friedell discuss the role of common knowledge in eye contact [9, 4]. In our paper we formalize the role of common knowledge in many of these social puzzles.

The role of common knowledge has been studied in the fields of distributed computing and artificial intelligence [11, 7, 12]. This line of work suggests that knowledge is an important abstraction for distributed systems and for the design and analysis of distributed protocols, in particular for achieving consistent simultaneous actions. Fagin and Halpern [13, 8] present an abstract model for knowledge and probability in which they assign to each agent-state pair a probability space to be used when computing the probability that a formula is true. A complexity-theoretic version of Aumann's celebrated Agreement Theorem is provided in [1].

2 Preliminaries

We will adopt the set-theoretic formulation of common knowledge introduced by Aumann [3]. In this model, there is a set Ω of "states of the world". Each player i has some information regarding the true state of the world. This information is given by a partition Π_i of Ω. In particular, for $\omega \in \Omega$, $\Pi_i(\omega)$ is the set of states indistinguishable from ω to player i – that is, when ω occurs, player i knows that one of the states in $\Pi_i(\omega)$ occurred but not which one. Finally, there is a probability distribution μ over Ω, representing the (common) prior belief of the players over the states of the world. These parameters all together constitute the information structure.

Definition 1 (Information structure). *An information structure is a tuple* $\mathcal{I} = (N, \Omega, \mu, \{\Pi_i\}_{i \in N})$ *where* N *is the set of players (with* $n := |N|$*),* Ω *is the set of possible states of the world,* μ *is a strictly positive common prior probability distribution over* Ω*, and* Π_i *is the information partition of player* i*.* $\Pi_i(\omega)$ *gives the set of states indistinguishable from* ω *to player* i*.*

A (Bayesian) game is now defined by an information structure, a set of possible actions for each player and a state-dependent utility for each player.

Definition 2 (Bayesian game). *A Bayesian game* Γ *is a tuple* $(\mathcal{I}, \{A_i\}_{i \in N}, \{u_i\}_{i \in N})$ *where* $\mathcal{I} = (N, \Omega, \mu, \{\Pi_i\}_{i \in N})$ *is an information structure,* A_i *is the (finite) set of possible actions that player* i *can take,* $u_i : A_1 \times \ldots \times A_n \times \Omega \to \mathcal{R}$ *is the utility for player* i *given the state of the world and the actions of all players.*

A strategy profile prescribes the action (possibly randomized) that each player takes at each state of the world.

Definition 3 (Strategy profile). *A strategy profile is a function* $\sigma = (\sigma_1, \ldots, \sigma_n) : \Omega \to A_1 \times \ldots \times A_n$ *that specifies what action each player takes in each state of the world.*

Since a player cannot distinguish between states belonging to the same partition, it is enforced that if a player i plays some strategy $\sigma = \sigma_i(\omega)$ at some state $\omega \in \Omega$, it must be the case that i plays σ at all states $\omega' \in \Pi_i(\omega)$. We can now recall the definition of Bayesian Nash equilibrium.

Definition 4 (Bayesian Nash equilibrium). *A strategy profile* $\sigma = (\sigma_1, \ldots, \sigma_n) : \Omega \to A_1 \times \ldots \times A_n$ *is a Bayesian Nash equilibrium (BNE) of* Γ *if for all* $i \in N$*,*

1. $\sigma_i(\omega) = \sigma_i(\omega')$ *whenever* $\omega \in \Pi_i(\omega')$*.*
2. $\int_{\omega \in \Omega} u_i(\sigma_i(\omega), \sigma_{-i}(\omega)) \mathrm{d}\mu(\omega) \geq \int_{\omega \in \Omega} u_i(\sigma'_i(\omega), \sigma_{-i}(\omega)) \mathrm{d}\mu(\omega)$ *for all* σ' *satisfying property 1.*

We now introduce our key definition of *state-dependent equilibria*, which we define as equilibrium strategies in which players take different actions when the circumstances change. This notion allows us to address the questions that were left unanswered by the informal discussions of common knowledge and coordination.

Definition 5 (State-dependent BNE). *We say that a Bayesian Nash equilibrium* σ^* *is state-dependent if for some* $\omega, \omega' \in \Omega$*,* $i \in N$*, we have that* $\sigma^*_i(\omega) = A$ *and* $\sigma^*_i(\omega') = B$*.*

We now define the notion of p-belief, introduced by Monderer and Samet [16], which extends the notion of common knowledge by Aumann [3]. Let p be a number between 0 and 1. We say that a player i p-believes the event E at state of the world ω if the subjective probability that i assigns to E at ω is at least p. That is, whenever ω is the true state of the world, i believes that an event in E occurred with probability at least p. Henceforth, we will use short expressions such as "i p-believes E at ω" to refer to this concept.

We denote by $\mathcal{B}_i^p(E)$ the set of all states of the world at which player i p-believes E.

Definition 6 (p-belief [16]). *For any $0 \leq p \leq 1$, we say that player i p-believes E at ω if $\mu(E \mid \Pi_i(\omega)) \geq p$. We will denote by $\mathcal{B}_i^p(E)$ the event that i p-believes E, i.e. $\mathcal{B}_i^p(E) = \{\omega \mid \mu(E \mid \Pi_i(\omega)) \geq p\}$.*

Observe that by definition of $\mathcal{B}_i^p(E)$, the notation $\omega \in \mathcal{B}_i^p(E)$ indicates that whenever ω occurs, player i believes with probability at least p that the event E occurred. An event E is then defined p-evident if whenever it occurs, each player i believes with probability at least p that it indeed occurred.

Definition 7 (evident p-belief [16]). *An event E is evident p-belief if for all $i \in N$ we have $E \subseteq \mathcal{B}_i^p(E)$.*

The following concept extends the notion of common knowledge.

Definition 8 (common p-belief [16]). *An event C is common p-belief at state ω if there exists an evident p-belief event E such that $\omega \in E$, and for all $i \in N$, $E \subseteq \mathcal{B}_i^p(C)$.*

Monderer and Samet provide a nice example that illustrates this concept: suppose the true state is either E or F with equal probability. The true state is announced and each of two players independently hears the announcement with probability $1 - \epsilon$, $0 < \epsilon < 1/2$. Then if E is the true state and both hear the announcement then E is common p-belief for all $p < 1 - \epsilon$ even though it is not common knowledge.

3 Two Player Framework

In this section we consider the classic 2-player, 2-strategy symmetric coordination game. The payoffs are as follows:

	A	B
A	a, a	b, c
B	c, b	d, d

Assumption 1 (Coordination game). *We make the following standard assumption on the parameters of a symmetric coordination game: $a > c$ and $d > b$.*

Throughout this paper, we will use $p^* = \frac{d-b}{d-b+a-c}$. This value is called risk-dominance [14]. Note that if player i believes with probability exactly p^* that the other player will play A at ω, then player i will be indifferent between playing A and B at ω.

For convenience, we will use the following definitions throughout this section.

Definition 9. *Given any strategy profile* σ*, we let* $A_i(\sigma) = \{\omega | \sigma_i(\omega) = A\}$ *and* $B_i(\sigma) = \{\omega | \sigma_i(\omega) = B\}$*, i.e. the set of states where player* i *plays* A *and* B *respectively.*

We now state our main result for the 2-player case. The main question we ask is when is it possible for the two players to coordinate on different actions in different states of the world. We answer this question in terms of the existence of evident p-belief events (where p depends on the payoff matrix) showing that such events are necessary and sufficient.

Theorem 1. *There exists a state-dependent Bayesian Nash equilibrium* σ^* *if and only if there exists a non-empty evident* p^**-belief event* E *and a non-empty evident* $(1-p^*)$*-belief event* F *such that* $E \cap F = \emptyset$*.*

While evident knowledge is both necessary and sufficient for state-dependent equilibria, our theorem further allows us to specify how the strategies must depend on these evident events, which we express in the following corollary:

Corollary 1. *A strategy profile* σ^* *is a state-dependent Bayesian Nash equilibrium if and only if there exists a non-empty evident* p^**-belief event* E *and a non-empty evident* $(1-p^*)$*-belief event* F *such that* $\mathcal{B}_i^{p^*}(E) \cap \mathcal{B}_i^{1-p^*}(F) = \emptyset$ *and* $\mathcal{B}_i^{p^*}(E) \cup \mathcal{B}_i^{1-p^*}(F) = \Omega$ *for all* i*, in which case* $A_i(\sigma^*) = \mathcal{B}_i^{p^*}(E)$ *and* $B_i(\sigma^*) = \mathcal{B}_i^{1-p^*}(F)$ *for all* i*.*

Our next corollary states the relationship between state-dependent equilibria and common knowledge.

Corollary 2. *If* σ^* *is a Bayesian Nash equilibrium such that* $\sigma_i^*(\omega) = A$ *and* $\sigma_i^*(\omega') = B$*, then* $\neg\omega'$ *is common* p^**-belief at* ω *and* $\neg\omega$ *is common* $(1-p^*)$*-belief at* ω'*.*

4 Application: A Rationale for Avoiding Eye-Contact

Two Charedi men, Michael and Dave, go to a bar, and each spots the other, purposely looking away before meeting eyes. Why?

Suppose that the next day they have to decide whether to tell the Rabbi. If one expects the other to tell, he is better off also admitting to his actions. On the other hand, if one does not expect the other to tell, then he is better off also not admitting to his transgression. The payoffs can be interpreted as the coordination game from the two-player framework by interpreting A as the act of not telling the Rabbi, B as the act of telling the Rabbi.

We make the reasonable assumption that if at least one of the men stays home, neither tells the Rabbi that he saw the other player at the bar (since he in fact did not). We will use our framework from section 3 to show that (a) there is always an equilibrium in which they both tell the Rabbi if they make eye-contact at the bar, and (b) under mild assumptions, if they do not make eye contact, neither will tell the Rabbi.

THE MODEL. We now specify the information structure: we suppose that in one state of the world, at least one of them stays home ($\mathcal{H}$) while in another state of the world, Dave enters the bar, and Michael is already sitting at the bar. When Dave walks in, Michael is either staring at the bartender, in which case he would not see Dave, or looking at the door, in which case he would. As soon as Dave enters, he sees Michael, so he quickly turns around and walks out. Dave turns around before or after noticing if Michael saw him.

The set of possible states of the world is given by $\Omega = \{\mathcal{H}, (\mathcal{M}, \mathcal{D}), (\mathcal{M}', \mathcal{D}), (\mathcal{M}, \mathcal{D}'), (\mathcal{M}', \mathcal{D}')\}$. We interpret the states of the world as follows: $\mathcal{H}$ is the state where Dave *does not* go to the bar and stays at ($\mathcal{H}$)ome. $\mathcal{M}$ is the event that Michael goes to the bar and sees Dave, and $\mathcal{D}$ is the event that Dave sees Michael. $(\mathcal{M}, \mathcal{D})$ is the state where Dave goes to the bar, Michael sees him, and Dave sees that Michael saw him (i.e. they make eye-contact). $(\mathcal{M}', \mathcal{D}')$ is the state where Dave goes to the bar, Michael is looking at the bartender, and Dave leaves the bar before checking if Michael saw him.

The information partitions are given as follows:

$$\Pi_{\mathcal{M}} = \{\{\mathcal{H}, (\mathcal{M}', \mathcal{D}'), (\mathcal{M}', \mathcal{D})\}, \{(\mathcal{M}, \mathcal{D}')\}, \{(\mathcal{M}, \mathcal{D})\}\}$$
$$\Pi_{\mathcal{D}} = \{\{\mathcal{H}\}, \{(\mathcal{M}', \mathcal{D}'), (\mathcal{M}, \mathcal{D}')\}, \{(\mathcal{M}', \mathcal{D})\}, \{(\mathcal{M}, \mathcal{D})\}\}$$

Observe that $(\mathcal{M}, \mathcal{D})$ is an evident p^*-belief event, that is, when eye contact happens, it becomes common knowledge between Michael and Dave as expected.

We use the following independent probabilities to deduce the priors over the state space: $p_{\mathcal{B}}$ is the probability that Dave goes to the bar i.e., he *does not* stay home; $p_{\mathcal{M}'}$ is the probability that Michael is looking at the bartender when Dave walks in; $p_{\mathcal{D}'}$ is the probability that, conditioned on Dave going to the bar, he leaves the bar without noticing Michael.

Our first claim is an almost trivial one which shows that there always exists an equilibrium in which they both tell the Rabbi if they make eye-contact.

Claim. There exists a Bayesian-Nash equilibrium of Γ such that $\sigma^*(\mathcal{H}) = (A, A)$ and $\sigma^*((\mathcal{M}, \mathcal{D})) = (B, B)$ for any $p_{\mathcal{B}}, p_{\mathcal{M}'}, p_{\mathcal{D}'}$.

Our next claim shows conditions under which if Michael and Dave do not make eye-contact, they must continue playing A if they play A on $\mathcal{H}$. That is, suppose Michael and Dave coordinate on (A, A) when Dave stays home; under what conditions is it the case that they can play (B, B) only at $(\mathcal{M}, \mathcal{D})$, i.e. only when they make eye-contact.

Claim. Suppose σ^* is a Bayesian-Nash equilibrium of Γ with $\sigma^*(\mathcal{H}) = (A, A)$. If $p_{\mathcal{M}'} > p^*$ and $\frac{p_{\mathcal{B}} p_{\mathcal{M}'}}{p_{\mathcal{B}} p_{\mathcal{M}'} + (1 - p_{\mathcal{B}})} < 1 - p^*$ then $\sigma^*(\omega) \neq (B, B)$ for all $\omega \neq (\mathcal{M}, \mathcal{D})$.

Now that we have formalized why someone might want to avoid eye contact, we can discuss when this is worthwhile. For instance, avoiding eye contact will not serve any purpose when it is very likely that they saw each other, e.g. if the bar had nobody else present and was very well lit (i.e. when $p_{\mathcal{M}'}$ and $p_{\mathcal{D}'}$ are small). Likewise, avoiding eye contact serves no purpose if, when it is commonly known that both parties see each other doing an act, neither is expected to play any differently than if neither transgressed (i.e. $\sigma^*(\mathcal{M}, \mathcal{D}) = (A, A)$). For example, the transgression is not perceived as related to the ensuing coordination game, e.g. if the two religious men have already discussed their secret abhorrence of the religion.

Moreover, avoiding eye contact only serves a purpose if there will be an ensuing coordination game (i.e. $a > c$). If in fact Michael would prefer to rat on Dave, regardless of whether Dave rats on Michael (e.g. because he knows the Rabbi will believe him, and he would like Dave to be excommunicated) then Dave does not help himself by avoiding Michael's eyes. In fact, to the extent that Dave thinks this might be the case, he might want to avoid eye contact, as it may make his presence more conspicuous to Michael.

Lastly, Michael may even purposely make eye contact, or yell out "hey Dave, is that you," if he in fact *wants* to switch from them both playing A to both playing B (which would be the case if $d > a$). For instance, this would be the case if Dave was looking for someone to leave the community with him and help him start a new life in the secular world.

5 *n*-Player Framework

We now introduce r-coordination games. Let Ω be all possible states of the world. There are n players, each of whom can take action A or B. A player's payoff for a particular action is a function of the fraction of players who play B. In particular, a player's payoffs are a function of whether the fraction of players who play B exceeds a threshold $\bar{r}$. Let r denote the fraction of players who play B. The payoffs are as follows.

$$u_i(A, r) = \begin{cases} a : r \leq \bar{r} \\ b : r > \bar{r} \end{cases} \qquad u_i(B, r) = \begin{cases} c : r \leq \bar{r} \\ d : r > \bar{r} \end{cases}$$

We again use assumption 1 on the values of the parameters, namely that $a > c$ and $d > b$. In this context, these assumptions on the payoff parameters generalize that of a 2 player coordination game in that a player best respond by playing A if and only if sufficiently many others play A.

We will also assume that n is sufficiently large such that a particular player's decision to play A or B does not affect whether r exceeds $\bar{r}$.

Furthermore, we will again use $p^* = \frac{d-b}{d-b+a-c}$. For n-players, p^* is a generalization of risk dominance. If player i believes with probability exactly p that at least $(1 - \bar{r})$ players will play A at ω, then player i will be indifferent between playing A and B at ω.

Note that this setup is a generalization of the two player setup. In particular, if there are two players, then we can let $\bar{r}$ be any value in $(1/2, 1)$ in order to obtain the two player model.

In Definitions 10, 11, and 12, we generalize p-beliefs, evident p-beliefs, and common p-beliefs to n players.

Definition 10 ((*r*, *p*)-belief). *For any* $0 \leq p \leq 1$ *and any* $0 \leq r \leq 1$, *we say that event* E *is* (r, p)*-belief at* ω *if* $|\{i \mid \omega \in \mathcal{B}_i^p(E)\}| \geq rn$. *We define* $\mathcal{B}^{r,p}(E) = \{\omega : |\{i \mid \omega \in \mathcal{B}_i^p(E)\}| \geq rn\}$ *as the event that at least a fraction of* r *players* p*-believes* E.

Definition 11 (evident (r, p)-belief). *An event E is evident (r, p)-belief if $E \subseteq \mathcal{B}^{r,p}(E)$.*

Definition 12 (common (r, p)-belief). *Given an event C, let $C^0 = \mathcal{B}^{r,p}(C)$ and inductively define $C^n = \mathcal{B}^{r,p}(\bigcap_{i<n} C^i)$ for all $n \geq 2$. Then C is common (r, p)-belief at ω if $\omega \in \bigcap_{n \geq 1} C^n$*

Note that common (r, p)-beliefs is identical to common p-beliefs when $n = 2$ and $r = 1$. The following theorem and corollaries are analogous to our two-player theorems and corollaries, despite the differing setup and proofs.

Theorem 2. *There exists a state-dependent Bayesian Nash equilibrium σ^* if and only if there exists a non-empty evident $(1 - \bar{r}, p^*)$-belief event E and a non-empty evident $(s, 1 - p^*)$-belief event F such that $E \cap F = \emptyset$ for some $s > \bar{r}$.*

Corollary 3. *A strategy profile σ^* is a state-dependent Bayesian Nash equilibrium if and only if there exists a non-empty evident $(1 - \bar{r}, p^*)$-belief event E and a non-empty evident $(s, 1 - p^*)$-belief event F for some $s > \bar{r}$ such that $\mathcal{B}_i^{p^*}(E) \cap \mathcal{B}_i^{1-p^*}(F) = \emptyset$ and $\mathcal{B}_i^{p^*}(E) \cup \mathcal{B}_i^{1-p^*}(F) = \Omega$ for all i, in which case $A_i(\sigma^*) = \mathcal{B}_i^{p^*}(E)$ and $B_i(\sigma^*) = \mathcal{B}_i^{1-p^*}(F)$ for all i.*

Corollary 4. *If σ^* is a Bayesian Nash equilibrium such that $|\{j \mid \sigma_j^*(\omega) = A\}| \geq 1 - \bar{r}$ and $|\{j \mid \sigma_j^*(\omega') = B\}| > \bar{r}$, then $\neg\omega'$ is common $(1 - \bar{r}, p^*)$-belief at ω and $\neg\omega$ is common $(\bar{r}, 1 - p^*)$-belief at ω'.*

6 n-Player Application: The Emperor's Clothes

Suppose that John Doe is on his way to being the next game theorist superstar. He finally comes out with his first paper, and superficially it is a spectacular paper. However, the paper offers no real insight, a fact that John attempts to hide with mathematical complexity. And this is fairly clear to nearly everyone in the field. Nevertheless, editors start requesting the paper, departments start offering him positions, conferences start asking him to give the keynote. Why?

Presumably, no one wants to be the lone person in the field who disrespects the superstar. For example, nobody wants to be the only person *not* to invite John to a conference or a special journal issue; he might end up with a powerful enemy, even if John's research is not good. However, if everyone in the field disrespects John Doe, then everyone benefits from doing likewise, since no one wants his keynote speaker to be unpopular or his new recruit never to be invited to conferences. Thus, we can model this as a r-coordination game where A is the act of showing John Doe respect (e.g. inviting him to a conference), and B is an act of disrespect.[1]

We make the assumption that if in fact John Doe's research were as great as people expected, then everyone would treat him with respect. Furthermore, we assume

[1] Note that "The Emperor's New Clothes" can be seen as a metaphor for this story. John Doe is analogous to the Emperor and his colleagues are analogous to the citizens who do not, initially, publicly disrespect the obviously flawed superstar.

that if a person can detect that John's research is bad, he can only approximately estimate how many others can detect this as well. We will show (Theorem 3) that, under mild conditions, if John's research is bad, no matter what fraction of people in the field can detect that his research has no insight, he will still be treated with respect. However, if people know exactly what fraction of the field know that John's research is bad, and that fraction is sufficiently high, then it is possible to treat John with disrespect (subsequent Claim). This is in stark contrast with the case where the error in a person's estimate is arbitrarily small.

THE MODEL. We model the information structure as follows: we assume that if John's research is in fact bad, then $1-\epsilon$ of the population can detect that it is bad. Everyone who can detect that it is bad has some impression of how easy it is for others to detect how bad it is; namely, they each get a signal θ_i which is independently drawn from $\mathcal{U}[\epsilon-\delta, \epsilon+\delta]$. After observing his private signal, but not ϵ, player i can choose to play A or B. As in the general setup, the payoff from each action is a function of the fraction of players who play B. Let r denote the fraction of players who play B. The payoffs are as in Section 5.

We can interpret Theorem 3 as follows. Suppose players disrespect John if their private signal θ_i of the true state ϵ is smaller than some (arbitrarily small) threshold $\bar{\epsilon}$. Then, if the fraction $\bar{r}$ of players needed to coordinate on B is larger than the risk-dominance p^*, this set of strategies is not an equilibrium. Note that the condition on $\bar{r}$ does *not* depend on $\bar{\epsilon}$. Another way of interpreting our results is the following. Even if many believe that many believe that many believe...that John's research is bad (for finitely many iterations), John will still be respected. Whereas, if it is common knowledge (subsequent Claim), e.g. if it is publicly announced how bad John's research is, he will no longer be respected.

Theorem 3. *Let* $\epsilon \sim \mathcal{U}[0,1]$ *and* $\theta_i \sim^{iid} \mathcal{U}[\epsilon-\delta, \epsilon+\delta]$ *for all* i *and for some* $\delta > 0$. *Let* σ^* *be a strategy profile such that* $\sigma_i^*(\theta_i) = B$ *when* $\theta_i \leq \bar{\epsilon}$ *and* $\sigma_i^*(\theta_i) = A$ *when* $\theta_i > \bar{\epsilon}$ *for some* $\bar{\epsilon} \in [\delta, 1-\delta]$. *Then for* $\delta \to 0$, σ^* *is not a Bayesian Nash equilibrium if* $\bar{r} > p^*$.

We contrast this result with the scenario in which the exact value of ϵ is observed by those who can detect that John's research is bad (i.e. $\theta_i = \epsilon$). The following claim can be easily established.

Claim. The strategy profile σ^* is a Bayesian Nash equilibrium if $\sigma_i^*(\theta_i) = A$ if $\epsilon \leq 1-\bar{r}$ and $\sigma_i^*(\theta_i) = B$ otherwise.

Acknowledgments. We would like to thank Ehud Kalai, Aviad Heifetz, Mehmet Ekmekci, Emir Kamenica, Balasz Szentes and Asher Wolinsky. This research was partially supported by grant RFP-12–11 from the Foundational Questions in Evolutionary Biology Fund, by the National Science Foundation under Grant No. 0905645 and by the Army Research Office grant number W911NF-11-1-0363. Any opinions, findings, and conclusions or recommendations expressed in this material are those of the author(s) and do not necessarily reflect the views of the National Science Foundation.

References

[1] Aaronson, S.: The complexity of agreement. In: Proceedings of the Thirty-Seventh Annual ACM Symposium on Theory of Computing, STOC 2005, pp. 634–643. ACM, New York (2005)

[2] Andersen, H.C.: The Emperor's New Clothes. Fairy Tales Told for Children - First Collection (1837)

[3] Aumann, R.J.: Agreeing to Disagree. The Annals of Statistics 4(6), 1236–1239 (1976)

[4] Binmore, K.: Game theory - a very short introduction. Clarendon Press (2007)

[5] Carlsson, H., van Damme, E.: Global Games and Equilibrium Selection. Econometrica 61(5), 989–1018 (1993)

[6] Chwe, M.S.-Y.: Rational Ritual: Culture, Coordination, and Common Knowledge. Princeton University Press (2003)

[7] Dwork, C., Moses, Y.: Knowledge and common knowledge in a byzantine environment: Crash failures. Information and Computation 88(2), 156–186 (1990)

[8] Fagin, R., Halpern, J.Y.: Reasoning about knowledge and probability. J. ACM 41(2), 340–367 (1994)

[9] Friedell, M.F.: On the structure of shared awareness. Behavioral Science 14(1), 28–39 (1969)

[10] Geanakoplos, J.: Common knowledge. Journal of Economic Perspectives 6(4), 53–82 (1992)

[11] Halpern, J.Y., Moses, Y.: Knowledge and common knowledge in a distributed environment. J. ACM 37(3), 549–587 (1990)

[12] Halpern, J.Y., Moses, Y.: A guide to completeness and complexity for modal logics of knowledge and belief. Artif. Intell. 54(3), 319–379 (1992)

[13] Halpern, J.Y., Tuttle, M.R.: Knowledge, probability, and adversaries. J. ACM 40(4), 917–960 (1993)

[14] Harsanyi, J.C., Selten, R.: A General Theory of Equilibrium Selection in Games. MIT Press Books, vol. 1. The MIT Press (1988)

[15] Lewis, D.: Convention: A Philosophical Study. Harvard University Press, Cambridge (1969)

[16] Monderer, D., Samet, D.: Approximating common knowledge with common beliefs. Games and Economic Behavior 1(2), 170–190 (1989)

[17] Morris, S.: Coordination, communication and common knowledge: A retrospective on electronic mail game. Oxford Review of Economic Policy 18(4) (February 2002)

[18] Morris, S., Shin, H.S., Yale University. Cowles Foundation for Research in Economics: Global games: theory and applications. Cowles Foundation discussion paper. Cowles Foundation for Research in Economics (2000)

[19] Morris, S., Shin, H.S.: Unique equilibrium in a model of self-fulfilling currency attacks. American Economic Review 88(3), 587–597 (1998)

[20] Morris, S., Shin, H.S.: Rethinking multiple equilibria in macroeconomic modeling. In: NBER Macroeconomics Annual 2000, vol. 15, NBER Chapters, pp. 139–182. National Bureau of Economic Research, Inc. (June 2001)

[21] Perry, S., Manson, J.: Manipulative Monkeys: The Capuchins of Lomas Barbudal. Harvard University Press (2008)

[22] Pinker, S., Nowak, M.A., Lee, J.J.: The logic of indirect speech. PNAS 105(3), 833–838 (2008)

[23] Rubinstein, A.: The electronic mail game: Strategic behavior under "almost common knowledge". American Economic Review 79(3), 385–391 (1989)

Approximating the Minmax Value of Three-Player Games within a Constant is as Hard as Detecting Planted Cliques

Kord Eickmeyer[1], Kristoffer Arnstfelt Hansen[2,⋆], and Elad Verbin[2]

[1] National Institute of Informatics, Tokyo
eickmeye@nii.ac.jp
[2] Aarhus University
arnsfelt@cs.au.dk, elad.verbin@gmail.com

Abstract. We consider the problem of approximating the minmax value of a multi-player game in strategic form. We argue that in three-player games with 0-1 payoffs, approximating the minmax value within an additive constant smaller than $\xi/2$, where $\xi = \frac{3-\sqrt{5}}{2} \approx 0.382$, is not possible by a polynomial time algorithm. This is based on assuming hardness of a version of the so-called planted clique problem in Erdős-Rényi random graphs, namely that of *detecting* a planted clique. Our results are stated as reductions from a promise graph problem to the problem of approximating the minmax value, and we use the detection problem for planted cliques to argue for its hardness. We present two reductions: a randomised many-one reduction and a deterministic Turing reduction. The latter, which may be seen as a derandomisation of the former, may be used to argue for hardness of approximating the minmax value based on a hardness assumption about *deterministic* algorithms. Our technique for derandomisation is general enough to also apply to related work about ϵ-Nash equilibria.

1 Introduction

We consider games in strategic form between 3 players. These are given by a finite strategy space for each player, S_1,S_2, and S_3 (also called the *pure strategies*), together with utility functions $u_1, u_2, u_3 : S_1 \times S_2 \times S_3 \to \mathbb{R}$. We can identify the strategy spaces with the sets $[n_1]$,$[n_2]$, and $[n_3]$, where $n_i = |S_i|$. We shall refer to this as a $n_1 \times n_2 \times n_3$ game. In this paper only the utilities for Player 1 are relevant.

Let Δ_1,Δ_2, and Δ_3 be the sets of probability distributions over S_1,S_2, and S_3 respectively; these are also called *mixed strategies*. The minmax value (also known as the threat value) for Player 1 is given by:

⋆ Hansen and Verbin acknowledge support from the Danish National Research Foundation and The National Science Foundation of China (under the grant 61061130540) for the Sino-Danish Center for the Theory of Interactive Computation, within which this work was performed. They also acknowledge support from the Center for Research in Foundations of Electronic Markets (CFEM), supported by the Danish Strategic Research Council.

M. Serna (Ed.): SAGT 2012, LNCS 7615, pp. 96–107, 2012.

$$\min_{(\sigma_2,\sigma_3)\in\Delta_2\times\Delta_3}\max_{\sigma_1\in\Delta_1}\mathop{\mathrm{E}}_{a_i\sim\sigma_i}[u_1(a_1,a_2,a_3)]$$

A strategy profile (σ_2, σ_3) for Player 2 and Player 3 for which this values is obtained is called an optimal minmax profile. It is not hard to see that Player 1 may always obtain the maximum by a pure strategy, i.e., the minmax value is equal to:

$$\min_{(\sigma_2,\sigma_3)\in\Delta_2\times\Delta_3}\max_{a_1\in S_1}\mathop{\mathrm{E}}_{\substack{a_2\sim\sigma_2\\a_3\sim\sigma_3}}[u_1(a_1,a_2,a_3)] \tag{1}$$

The corresponding notion of minmax value in finite two-player games is a fundamental notion of game theory. Minmax values have been studied much less in multi-player player games, but are arguably also here of fundamental interest. In particular the minmax value of such games is crucial for the statements as well as proofs of the so-called *folk theorems* that characterise the Nash equilibria of *repeated games.* The problem of *computing* the minmax value of a multi-player game was first considered only recently by Borgs et al. [1], exactly in the context of studying computational aspects of the folk theorem. In particular they show that approximating the minmax value of a three-player game within a specific inverse polynomial additive error is NP hard.

Here, to be able to talk meaningfully about approximation within an additive error, we assume that all payoffs have been *normalised* to be in the interval between 0 and 1. The question of approximating the minmax value was considered further by Hansen et al. [2]. Using a "padding" construction it was observed that the NP hardness result of Borgs et al. extends to any inverse polynomial additive error. This was complemented by a quasipolynomial approximation algorithm obtaining an approximation to within an arbitrary $\epsilon > 0$, which was obtained using a result of Lipton and Young [3], stating that in an $n \times n$ matrix game with payoffs between 0 and 1, each player can guarantee a payoff within any $\epsilon > 0$ of the value of the game using strategies that simply consist of a uniform choice from a multiset of $\lceil \ln n/(2\epsilon^2)\rceil$ pure strategies. We summarise these results by the following theorem.

Theorem 1 ([1,2]). *For any constant $\epsilon > 0$ it is NP hard to approximate the minmax value of an $n \times n \times n$ game with 0-1 payoffs within additive error $1/n^\epsilon$. On the other hand, there is an algorithm that, given $\epsilon > 0$ and a $n \times n \times n$ game with payoffs between 0 and 1, approximates the minmax value from above with additive error at most ϵ in time $n^{O(\log(n)/\epsilon^2)}$*

This naturally raises the question of whether it is possible to approximate the minmax value within any constant $\epsilon > 0$ in polynomial time, or even whether it is possible to approximate the minmax value within *some* nontrivial additive constant $0 < \epsilon < 1/2$ in polynomial time. Due to the quasipolynomial time algorithm above, it is unlikely that the theory of NP completeness can shed light on this question.

A similar situation is present for the problem of computing a Nash equilibrium in two-player bimatrix games. Celebrated recent results [4,5] show that this problem is complete for the complexity class PPAD. On the other hand several works provide algorithms for computing an ϵ-Nash equilibrium. An ϵ-Nash equilibrium in a $n \times n$ bimatrix game with payoffs between 0 and 1 can be computed in time $n^{O(\log(n)/\epsilon^2)}$ [6], by an algorithm similar to the one described above for the minmax value. As for polynomial time algorithms, several algorithms have been devised for decreasing additive error ϵ (see e.g. [7] for references). The current best such algorithm achieves $\epsilon = 0.3393$ [7]. How well a Nash equilibrium can be approximated in the sense of ϵ-Nash equilibria is a major open question. Having a polynomial time algorithm, polynomial also in $1/\epsilon$, or in other words having a fully polynomial time approximation scheme (FPTAS), would imply that every problem in the class PPAD would be solvable in polynomial time [5]. Currently there is no evidence for or against the existence of a polynomial time algorithm for any fixed $\epsilon > 0$, or in other words a polynomial time approximation scheme (PTAS) for computing ϵ-Nash equilibria.

The Planted Clique Problem. Our result depends on assuming hardness of the so-called planted clique problem (more precisely, the *detection* variant). Let $G_{n,p}$ denote the distribution of Erdős-Rényi random graphs on n vertices where each potential edge is included in the graph independently at random with probability p. Most frequently the case of $p = 1/2$ is considered, but we will be interested in having $p > 0$ be a small constant. This choice is made in order to get a conclusion as strong as possible from our proof.

It is well known that in almost every graph from $G_{n,p}$ the largest clique is of size $2\log_{1/p} n - O(\log\log n)$ [8]. The hidden clique problem is defined using the distribution $G_{n,p,k}$ [9,10] of graphs on n vertices defined as follows: A graph G is picked according to $G_{n,p}$, then a set of k vertices is chosen uniformly at random, independent of G, and connected to form a clique. Thus apart from the planted k-clique the graph is completely random. The (*search* variant of the) planted clique problem is then defined as follows: Given a graph G chosen at random from $G_{n,p,k}$, find a k-clique in the graph G. Note that when the parameter k is significantly larger than $2\log_{1/p} n$, the planted clique is with high probability the unique maximum clique in the graph, and thus it also makes sense to talk about finding *the planted* clique, with high probability. Furthermore, by guessing $2\log_{1/p} n$ of the vertices of the planted clique and determinig their common neighbours, such a clique can be found in quasipolynomial time.

The planted clique problem is known as a difficult combinatorial problem. Indeed the current best polynomial time algorithms for solving the planted clique problem [11,12] are only known to work when $k = \Omega(\sqrt{n})$. We may compare this with the observation due to Kučera [10] that for $k \geq C\sqrt{n\log n}$ when C is a suitably large constant, the vertices of the clique would almost surely be the vertices of largest degree, and hence easy to find. The planted clique problem has also been proposed as a basis for a cryptographic one-way function [13]. For this application, however, the size of the planted clique is $k = (1+\epsilon)\log_{1/p} n$, which is smaller than the expected size of the largest clique. Recently, Feldman et al. gave further evidence towards the computational hardness of planted clique by proving lower bounds for a broad class of algorithms they call *statistical algorithms* [14].

The planted clique *detection* problem is defined as follows: Given a graph G chosen at random from either (i) $G_{n,p}$, or (ii) $G_{n,p,k}$, decide which is the case.

It is an interesting open question whether the above game-theoretic problems and the planted clique problem actually have the same complexity (which is likely to be between P and NP). As a first step, in Section 3 we give a reduction from optimal Nash equilibria to a variant of the minmax value in three-player games.

1.1 Our Results

We show a relationship between the task of approximating the minmax value in a three-player game and the planted clique detection problem. Our result builds heavily on the ideas of the work of Hazan and Krauthgamer in [15] (see also [16]). These are described in the next section.

In our results we prove hardness of approximating the minmax value, and aim to obtain a conclusion as strong as possible, while maintaining a reasonable hardness assumption.

We will actually state our results using the following promise[1] graph problem Gap-DBS, parametrized by numbers $0 < c_1 < c_2$ and $\eta > 0$. Let $G = (V_1, V_2, E)$ be a bipartite graph. For $S \subseteq V_1$, $T \subseteq V_2$ the *density* of the subgraph induced by S and T is given by $\mathrm{d}(S,T) = \frac{|E(S,T)|}{|S||T|}$. Note that if we let A denote the adjacency matrix of G and let u_S and u_T be the probability vectors that are uniform on the sets S and T, then we have $\mathrm{d}(S,T) = u_S^{\mathsf{T}} A u_T$.

> GAP DENSE BIPARTITE SUBGRAPH (GAP-DBS)
> *Input:* Bipartite graph $G = (V_1, V_2, E)$, $|V_1| = |V_2| = n$
> *Promise:* Either
> (i) There exist $S \subseteq V_1$, $T \subseteq V_2$, $|S| = |T| = c_2 \ln n$, such that $\mathrm{d}(S,T) \geq 1 - \eta$, or
> (ii) For all $S \subseteq V_1$, $T \subseteq V_2$, $|S| = |T| = c_1 \ln n$, it holds that $\mathrm{d}(S,T) \leq \eta$.
>
> *Problem:* Decide which of these is the case

We also introduce the following gap problem for the minmax value of three-player games with 0-1 payoffs, parametrised by numbers $0 \leq \alpha < \beta \leq 1$

> GAP THREE-PLAYER MINMAX (GAP-MINMAX)
> *Input:* $n \times n \times n$ game G with 0-1 payoffs
> *Promise:* The minmax value for Player 1 in G is either at most α, or at least β.
> *Problem:* Decide which of these is the case

[1] Clearly if there exist sets S and T with $|S| = |T| = c_2 \ln n$ and $\mathrm{d}(S,T) \geq 1 - \eta$, there exist subsets $S' \subseteq S$ and $T' \subseteq T$ with $|S'| = |T'| = c_1 \ln n$ and $\mathrm{d}(S',T') \geq 1 - \eta$ as well.

We are now ready to state our results. Throughout the paper $\xi = \frac{3-\sqrt{5}}{2} \approx 0.382$ is the smaller of the two roots of $x^2 - 3x + 1 = 0$, which is also known as $1 - \varphi$, where φ is the conjugate golden ratio.

Theorem 2. *There exist reductions from the Gap-DBS problem to the Gap-minmax problem as follows.*

1. *For every $0 < \eta < 0.1$ and $0 < c_1 < c_2$ satisfying $\frac{c_2}{c_1} > \frac{2\ln(1/\eta)}{(1-\eta)\eta^2}$ there is a randomised many-one reduction from the Gap-DBS problem to the Gap-minmax problem with parameters $(\eta, \xi - \eta/5)$.*
2. *For every $0 < \eta < 0.1$ and $0 < c_1 < c_2$ satisfying $\frac{c_2}{c_1} > 1/\eta$ there is a deterministic Turing reduction from the Gap-DBS problem to to the Gap-minmax problem with parameters $(\eta, \xi - \eta/5)$.*

We prove the two parts of this theorem as two separate theorems, stated as Theorem 8 and Theorem 12. We note that, interestingly, the constant ξ has previously turned up as the additive error $\xi + \delta$, for arbitrary $\delta > 0$, obtained by an approximation algorithm for computing ϵ-Nash equilibria [17].

One can view the second reduction in Theorem 2 as a derandomisation of the first one, at the cost of turning the many-one reduction into a Turing reduction. On the other hand the required ratio between c_1 and c_2 is actually much smaller.

We will use the planted clique problem to argue that the Gap-DBS is hard for certain settings of parameters (c_1, c_2, η). For this we use similar arguments as in [15,16]. Given a graph H that is an input to the planted clique detection problem, we let A be the adjacency matrix of H and let G be the bipartite graph that also has A as adjacency matrix. We wish to have the following property: If H was chosen from $G_{n,p,k}$, then with high probability G belongs to case (i) of the Gap-DBS problem, and if H was instead chosen from $G_{n,p}$ then with high probability G belongs to case (ii) of the Gap-DBS problem. This can indeed be obtained with an appropriate assumption about the clique detection problem. We have the following statement, whose proof we omit due to space limitations.

Proposition 3. *For any $\eta > 0$ there exist $p > 0$ and $c_1 > 0$ such that for $k = c_2 \ln n$, with $c_2 > c_1$, Gap-DBS with parameters (c_1, c_2, η) is as hard as the hidden clique detection problem for $G_{n,p,k}$.*

The choice of $c_2 > c_1$ of interest for us will be dictated by the choice of reduction we wish to use from Theorem 2, and in turn dictates the precise hardness assumption for the planted clique detection problem needed. However we find it natural to assume that the planted clique detection problem is hard for $G_{n,p,k}$ for any $p > 0$ and any $k = c_2 \ln n$, $c_2 > 2/\ln\frac{1}{p}$, i.e., with k strictly greater than the largest clique in $G_{n,p}$. Thus our results can be stated as follows.

Theorem 4. *For every $\epsilon > 0$, there is no randomised polynomial time algorithm that with high probability approximates the minmax value of an $n \times n \times n$ game with payoffs between 0 and 1 within an additive error $\xi/2 - \epsilon$, unless there exist $p > 0$ and $c_2 > 2/\ln\frac{1}{p}$ and a randomised polynomial time algorithm that solves the planted clique detection problem for $G_{n,p,k}$ with high probability, for $k = c_2 \ln n$.*

Theorem 5. *For every $\epsilon > 0$, there is no polynomial time algorithm approximating the minmax value of an $n \times n \times n$ game with payoffs between 0 and 1 within an additive error $\xi/2 - \epsilon$, unless there exist $0 < c_1 < c_2$ satisfying $c_2 > c_1/\eta$ and a (deterministic) polynomial time algorithm that solves the Gap-DBS problem with parameters (η, c_1, c_2), for $\eta = 5/3\epsilon$.*

1.2 Related Work

The problem of computing a Nash equilibrium in a bimatrix is PPAD complete. However, there are many different properties such that asking for a Nash equilibrium that satisfies the property is an NP hard problem [18,19]. In particular it is NP hard to compute a Nash equilibrium maximizing the social welfare, i.e., maximizing the sum of the two players' payoffs.

Hazan and Krauthgamer [15], motivated by the question of whether there is a PTAS for computing ϵ-Nash equilibria, considered an "ϵ-Nash" variant of the problem of maximizing social welfare, namely that of computing an ϵ-Nash equilibrium whose social welfare is no less than the maximal social welfare achievable by a Nash equilibrium, minus ϵ. In order to describe all the results in the following, say that an ϵ-Nash equilibrium is δ*-good* if its social welfare is no less than the maximal social welfare achievable by a Nash equilibrium, minus δ.

Remark 6. For the notion introduced by Hazan and Krauthgamer, Minder and Vilkenchik [16] use the terminology "ϵ-best ϵ-Nash equilibrium". However we feel this is somewhat of a misnomer, since the social welfare is compared to the largest achievable by a Nash equilibrium rather than an ϵ-Nash equilibrium. Indeed, a simple example[2] shows that for any $\epsilon > 0$ one may have a game where the (unique) Nash equilibrium has social welfare ϵ, but there exist an ϵ-Nash equilibrium of social welfare 1. For this reason we will instead call it "ϵ-good". In fact, let us generalise the notion and say that an ϵ-Nash equilibrium is δ*-good* if its social welfare is no less than the maximal social welfare achievable by a Nash equilibrium, minus δ.

Hazan and Krauthgamer gave a randomised polynomial time reduction from the planted clique problem to the problem of computing an ϵ-good ϵ-Nash equilibrium. More precisely, they show there are constants $\epsilon, c > 0$ such that if there is a polynomial time algorithm that computes in a two-player bimatrix game an ϵ-good ϵ-Nash equilibrium, then there is a randomised polynomial time algorithm that solves the planted clique problem in $G_{n,1/2}$ for $k = c \log_2 n$ with high probability.

This result was sharpened by Minder and Vilenchik [16], making the constant c smaller. In particular they obtain $c = 3 + \delta$, for arbitrary $\delta > 0$ (here $\delta > 0$ dictates an upper bound on ϵ), and for the similar problem of detecting a planted clique they obtain $c = 2 + \delta$. Essentially the goal of Minder and Vilenchik was the opposite of ours. Namely, viewing their result as arguing for hardness, their goal

[2] Consider just the bimatrix game given by the two 1×2 matrices $[1\ 0]$ for the row player and $[0\ \epsilon]$ for the column player.

was to obtain an assumption as weak as possible, while maintaining a nontrivial conclusion.

Austrin et al. [20] considered the other goal of obtaining strong hardness conclusions for computing δ-good ϵ-Nash equilibria (as well as ϵ-Nash versions of computing second equilibria and small support equilibria, and approximating pure Bayes Nash equilibria), assuming hardness for the planted clique problem. For this reason their work is the most relevant to use for comparing with our results. With the goal of obtaining strong hardness conclusions for computing δ-good ϵ-Nash equilibria in mind, one now needs to consider both of the parameters, ϵ and δ, and their relationship. Austrin et al. consider the extreme cases for both of these parameters individually and obtain the following results.

Theorem 7 (Austrin et al.).

1. *For any $\eta > 0$ there exists $\delta = \Omega(\eta^2)$ such that computing a δ-good ϵ-Nash equilibrium is as hard as the planted clique problem, for $\epsilon = 1/2 - \eta$.*
2. *For any $\eta > 0$ there exists $\epsilon = \Omega(\eta^2)$ such that computing a δ-good ϵ-Nash equilibrium is as hard as the planted clique problem, for $\delta = 2 - \eta$.*

Furthermore Austrin et al. give a simple polynomial time algorithm that computes a $\frac{1}{2}$-Nash equilibrium with social welfare at least as large as any Nash equilibrium, showing that the first part of Theorem 7 is tight. Clearly the second part is tight as well. On the other hand it appears that the tightness of these results were possible due to the focus on a single parameter at a time, and the exact trade-off possible between these two parameters still seems unclear.[3]

The reductions in [15,16] are randomised reductions, and we remark that our derandomisation technique can be used for these reductions as well.

2 The Reductions

We collect the utilities for Player 1 in matrices, one for each pure strategy. Thus we define $n_2 \times n_3$ matrices $A^{(1)}, \ldots, A^{(n_1)}$ by $a^{(i)}_{j,k} = u_1(i, j, k)$. In this notation, if Player 1 plays the pure strategy i and Player 2 and Player 3 play by mixed strategies x and y, the expected payoff to Player 1 is given by $x^\mathsf{T} A^{(i)} y$.

2.1 The Randomised Reduction

In this section we present a randomised reduction from approximate planted clique to minmax value in three-player games. To be precise, we prove the following result:

[3] While the statements of Theorem 7 are given using asymptotic notation, the proofs provide concrete (albeit not particularly optimised) constants. For instance the proof of the first part gives $\delta = 1/288$ for $\epsilon = 1/4$, and the proof of the second part gives $\epsilon = 1/288$ for $\delta = 3/2$.

Theorem 8. *Let $0 < \eta < 0.1$ and $0 < c_1 < c_2$ and such that $\frac{c_2}{c_1} > \frac{2\ln(1/\eta)}{(1-\eta)\eta^2}$. Then there is a randomised polynomial time many-one reduction which, given as input the adjacency matrix $A \in \{0,1\}^{n\times n}$ of a bipartite graph G, outputs a three-player game G_A such that with high probability*

- *if there are subsets $S, T \subseteq [n]$ of size at least $c_2 \ln n$ such that $\mathrm{d}(S,T) \geq 1-\eta$, then $\mathrm{minmax}_1 G_A \leq \eta$.*
- *if $\mathrm{d}(S,T) < \eta$ for every $S, T \subseteq [n]$ of size at least $c_1 \ln n$, then $\mathrm{minmax}_1 G_A > \xi - \frac{\eta}{5}$.*

We will need the following lemma, whose proof is an easy application of the Chernoff bound.

Lemma 9. *Let $0 < \delta < 1$, and $k_1 = c_1 \ln n$, $k_2 = c_2 \ln n$, where $0 < c_1 < c_2$ satisfy $c_2 > \frac{2\ln(1/\delta)}{(1-\delta)\delta^2} \cdot c_1$. Let $D \subseteq [n]$ be a fixed subset of size $|D| = k_2$. Then there is a constant c such that if we we choose at random $m = n^c$ subsets $S_1, \ldots, S_m \subseteq [n]$, by letting $j \in S_i$ with probability $1-\delta$, independently for every i and j, with probability at least $1 - n^{-\Omega(1)}$ the sets satisfy the following properties.*

(a) For all i, $|S_i \cap D| \geq (1-\delta)^2 k_2$.
(b) For every set $S \subseteq [n]$ of size $|S| = k_1$, there exists i such that $S_i \cap S = \emptyset$.

Proof (of Thm. 8). We use Lemma 9 with c_1 and c_2 as in the problem description and $\delta = 1 - \sqrt{1-\eta} = \eta/2 + O(\eta^2)$. Let m be as in the lemma. The reduction first guesses $2m$ subsets $S_1^{(\mathrm{r})}, \ldots, S_m^{(\mathrm{r})}, S_1^{(\mathrm{c})}, \ldots, S_m^{(\mathrm{c})}$ at random as in the lemma. It then outputs a three-player game G_A as follows:

- Players 2 and 3 have n strategies each.
- Player 1 has $2m+1$ strategies given by matrices B, $R^{(1)}, \ldots, R^{(m)}$, and $S^{(1)}, \ldots, S^{(m)}$. The matrix B is defined as $B = 1 - A$, and $R^{(k)}$ and $C^{(k)}$ for $k = 1, \ldots, m$, are given by

$$(R^{(k)})_{ij} = \begin{cases} 1 & \text{if } i \notin S_k^{\mathrm{r}} \\ 0 & \text{if } i \in S_k^{\mathrm{r}} \end{cases} \quad \text{and} \quad (C^{(k)})_{ij} = \begin{cases} 1 & \text{if } j \notin S_k^{\mathrm{c}} \\ 0 & \text{if } j \in S_k^{\mathrm{c}} \end{cases}$$

We claim that this game satisfies our assumptions.

For the first part, let $S, T \subseteq [n]$ be sets of size at least $c_2 \log n$ such that $\mathrm{d}(S,T) \geq 1-\eta$. By choosing appropriate subsets, we may assume that, in fact, $|S| = |T| = c_2 \log n$. Furthermore, by Lemma 9, with high probability $|S_i^{\mathrm{c}} \cap T| \geq (1-\delta)^2 c_2 \ln n$. Thus if players 2 and 3 play strategies u_S and u_T, respectively, Player 1 will receive payoff at most $1 - (1-\delta)^2 = \delta(2-\delta) = (1-\sqrt{1-\eta})(1+\sqrt{1-\eta}) = \eta$ by playing any of the strategies corresponding to $R^{(k)}$ and $C^{(k)}$, while playing the strategy corresponding to B will give Player 1 payoff $1 - \mathrm{d}(S,T) < \eta$.

For the second part, we assume to the contrary that G has density $\mathrm{d}(S,T) < \eta$ for all sets S, T of size at least $c_1 \ln n$, but $\min\max G_A \leq a$. Let (σ_2, σ_3) be an optimal strategy profile, i.e., $\max\{\sigma_2^\mathsf{T} B\sigma_3, \sigma_2^\mathsf{T} R^{(k)}\sigma_3, \sigma_2^\mathsf{T} C^{(k)}\sigma_3\} \leq a$. We first show that on any support of size at most k_1 each of σ_2 and σ_3 places probability at most a: Suppose $S \subseteq [n]$ and $|S| \leq k_1$ with $\Pr_{\sigma_2}[S] = p$. Then by switching to an appropriate set action corresponding to $R^{(k)}$, Player 1 might increase his payoff to at least p. Thus $p \leq a$. The proof for σ_3 is the same, replacing $R^{(k)}$ with $C^{(k)}$. We set, with foresight, $a = \xi - \frac{\eta}{5}$, $b = 1 - \xi - \frac{\eta}{2}$, and $c = 1 - \eta$. Direct calculations show that for $0 < \eta < 0.1$, these values satisfy

$$a < b < c < 1 \qquad (1-a)b > a \qquad \text{and} \qquad (1-a)c > b\ . \tag{2}$$

We show that there exist sets S and T of size at least $c_1 \ln n$ such that $u_S^\mathsf{T} A u_T \geq 1 - c$: Define $T = \{i \mid \sigma_2^\mathsf{T} B e_i \leq b\}$, and let $p = \Pr_{\sigma_3}[T]$. Then $a \geq \sigma_2^\mathsf{T} B\sigma_3 \geq (1-p)b$, and therefore $(1-p)b < a$, which means $1 - p < a/b$. But we have $1 - a > a/b$, which then implies $p > a$, and therefore $|T| \geq c_1 \ln n$ as argued above. Furthermore, by definition of T we have $\sigma_2^\mathsf{T} B u_T \leq b$. Next, define $S = \{i \mid {e_i}^\mathsf{T} B u_T \leq c\}$, and let $p = \Pr_{\sigma_2}[S]$. Similarly to before we then have $b \geq \sigma_2^\mathsf{T} B u_T \geq (1-p)c$ which means $(1-p)c < b$, and thus $1 - p < b/c$. But we have $1 - a > b/c$, which then implies $p > a$, and again we obtain that $|S| \geq c_1 \ln n$. Furthermore, by definition of S and $B = 1 - A$ we have $u_S^\mathsf{T} A u_T \geq 1 - c = \eta$.

Remark 10. We remark that the above analysis is tight, namely that in the case when $\mathrm{d}(S,T) < \eta$ for every $S, T \subseteq [n]$ of size at least $c_1 \ln n$, it is not possible to prove a lower bound on the minmax value better than ξ in the game constructed.

2.2 Derandomisation

In this section we derandomise our result in Theorem 8, at the price of turning our many-one reduction into a Turing reduction.

Recall that randomness was needed by our reduction for the construction of the sets $S_i^{(r)}$ and $S_i^{(c)}$. We now show how these sets can be constructed explicitly, giving a derandomised analogue of Lemma 9:

Lemma 11. *Let $0 < k_1 < k_2 < n \in \mathbb{N}$. Then there are families $A^{(1)}, \ldots, A^{(r)}$ of subsets of $[n]$ such that*

- *there are $r = 2^{O(k_2)} \log n$ families, and each family is of size $s = \binom{k_2}{k_1}$,*
- *for every set $M \subseteq [n]$ of size k_2, there is an index $j \in [r]$ such that $\left|A_i^{(j)} \cap M\right| = k_2 - k_1$, for all $i \in [s]$ and*
- *for every set $M \subseteq [n]$ of size k_1 and every $j \in [r]$, there is an index $i \in [s]$ such that $A_i^{(j)} \cap M = \emptyset$.*

These sets can be constructed in time polynomial in n and r. In particular, if $k_2 = O(\log n)$ then both r and s are polynomial in n, and the families of subsets can be constructed in time polynomial in n.

Proof. In [21], Alon et al. gave a construction of a family $H = \{f_1, \ldots, f_r\}$ of *perfect hash functions* from $[n]$ to $[k_2]$. This means

- each f_j is a function from $[n]$ to $[k_2]$ and
- for each $M \subseteq [n]$ of size k_2, at least one of the f_j is injective on M.

Moreover, $r = 2^{O(k_2)} \log n$ and the functions can be constructed in time polynomial in n and r.

Let $s = \binom{k_2}{k_1} \leq 2^{k_2}$ and let $M_1, \ldots, M_s$ be an enumeration of the subsets of $[k_2]$ of size k_1. Define $A_i^{(j)} := \{x \in [n] \mid f_j(x) \notin M_i\}$. These subsets meet the size restrictions claimed in the lemma and are readily seen to be constructable in time poly(n, r).

Now, let $M \subseteq [n]$ be of size k_2, and suppose f_j is injective on M. Then $A_i^{(j)} \cap M = \{x \in M \mid f_j(x) \notin M_i\}$, and because f_j is a bijection between M and $[k_2]$, this set has size $k_2 - k_1$ for all $i \in [s]$.

Furthermore, if $M \subseteq [n]$ is of size k_1, then $|f_j(M)| \leq k_1$ for all $j \in [r]$. Thus for each j there is an i such that $f_j(M) \subseteq M_i$, which implies $A_i^{(j)} \cap M = \emptyset$.

Our derandomised reduction now looks as follows:

Theorem 12. *For $0 < \eta < 0.1$ and $0 < c_1 < c_2$ and such that $\frac{c_2}{c_1} > \frac{1}{\eta}$, there is a polynomial-time Turing reduction from Gap-DBS to Gap-Minmax with a gap $(\eta, \xi - \eta/5)$.*

Proof. The reduction works as in the randomised case, the main difference being that instead of guessing sets $S_i^{(r)}$ and $S_i^{(c)}$ at random, we construct (polynomially many) set families $A^{(1)}, \ldots, A^{(r)}$ using the construction in Lemma 11 with $k_{1/2} = c_{1/2} \ln n$. We then use each pair of such families to construct a game $G_A^{(j_1,j_2)}$ as in the proof of Theorem 8; using the family $A^{(j_1)}$ for the row strategies and $A^{(j_2)}$ for the column strategies. We show that

- if $\mathrm{d}(S,T) \geq 1 - \eta$ for some sets S, T of size at least $c_2 \ln n$, then $\mathrm{minmax}_1 G_A^{(j_1,j_2)} \leq \eta$, for *some* j_1 and j_2, and
- if $\mathrm{d}(S,T) \leq \eta$ for all sets S, T of size at least $c_1 \ln n$, then $\mathrm{minmax}_1 G_A^{(j_1,j_2)} \geq \xi - \eta/5$, for *all* j_1, j_2.

The proof works as in the randomised case: For the first part, we note that by Lemma 11, for some j_1, j_2 and all i we have $\left|A_i^{(j_1)} \cap S\right| = k_2 - k_1 \geq (1-\eta)k_2$ and $\left|A_i^{(j_2)} \cap T\right| = k_2 - k_1 \geq (1-\eta)k_2$, and therefore $\mathrm{minmax}_1 G_A^{(j_1,j_2)} \leq \eta$ in this case. The second part is unchanged from the randomised case.

3 A Reduction from Optimal NE to Minmax

The following reduction gives evidence to the fact that computing the minmax-value in three-player games is at least as hard as finding ϵ-Nash equilibria with high average payoff.

Theorem 13. *There is a polynomial time reduction which, given payoff matrices $R, C \in [0,1]^{m \times n}$ specifying a game $\mathcal{G}$ in which the players have m and n strategies respectively, and $\alpha \in [0,1]$, $\epsilon > 0$, outputs payoff matrices for Player 1 in a three-player game $\mathcal{H}$ such that:*

– *If $\mathcal{G}$ has an ϵ-Nash equilibrium with average payoff* $> 1-\alpha$, *then* $\text{minmax}_1\mathcal{H} \leq \alpha$.
– *If $\mathcal{G}$ has no 2ϵ-Nash equilibrium with average payoff* $> 1 - \alpha - \epsilon$, *then* $\text{minmax}_1\mathcal{H} > \alpha + \epsilon$.

Proof. Player 1 has $m + n + 1$ strategies, Player 2 has m strategies and Player 3 has n strategies. We group Player 1's strategies into three categories:

1. one strategy called v which has payoff matrix $1 - (R + C)/2$,
2. for each $\tilde{\imath} \in [m]$ a strategy $a_{\tilde{\imath}}$ with payoff matrix $\alpha - \epsilon + (R_{\tilde{\imath}j} - R_{ij})_{i,j}$,
3. for each $\tilde{\jmath} \in [n]$ a strategy $b_{\tilde{\jmath}}$ with payoff matrix $\alpha - \epsilon + (C_{i\tilde{\jmath}} - C_{ij})_{i,j}$.

Let $\sigma_2 \in \Delta_m$ and $\sigma_3 \in \Delta_n$ be mixed strategies for players 2 and 3. Then

1. the expected payoff for Player 1 when playing strategy v is one minus the social welfare of the game specified by R and C if players 2 and 3 play the strategy profile (σ_2, σ_3),
2. the expected payoff when playing $a_{\tilde{\imath}}$ is $\alpha - \epsilon$ plus Player 2's gain when defecting to strategy $\tilde{\imath}$,
3. the expected payoff when playing $b_{\tilde{\jmath}}$ is $\alpha - \epsilon$ plus Player 3's gain when defecting to strategy $\tilde{\jmath}$.

In particular, if $\sigma_2 \in \Delta_m$ and $\sigma_3 \in \Delta_n$ are an ϵ-Nash equilibrium with average payoff $> 1 - \alpha$, then no strategy for Player 1 in $\mathcal{H}$ will have expected payoff $> \alpha$, if players 2 and 3 play according to σ_2 and σ_3, so $\text{minmax}_1\mathcal{H} \leq \alpha$.

On the other hand, suppose that $\mathcal{G}$ has no 2ϵ-Nash equilibrium with average payoff $> 1 - \alpha - \epsilon$. Let σ_2 and σ_3 be strategies for players 2 and 3 in $\mathcal{H}$. If Player 1 receives payoff $< \alpha + \epsilon$ when responding to σ_2 and σ_3 with strategy v, then the average payoff of (σ_2, σ_3), as a pair of strategies in $\mathcal{G}$, will be at least $1 - \alpha - \epsilon$. By our assumption on $\mathcal{G}$, (σ_2, σ_3) can not be an 2ϵ-Nash equilibrium, i.e., one of the players can gain more than 2ϵ by deviating. But then one of the strategies $a_{\tilde{\imath}}, b_{\tilde{\jmath}}$ will give Player 1 an expected payoff of at least $\alpha + \epsilon$ in $\mathcal{H}$. Therefore $\text{minmax}_1\mathcal{H} > \alpha + \epsilon$ in this case.

4 Conclusion

We have considered a promise graph problem, which is hard assuming standard hardness assumptions on detecting planted cliques in random graphs. We have shown that the problem of approximating the minmax value in three-player games with 0-1 payoffs is at least as hard as this promise graph problem, by giving both a randomised many-one reduction and a deterministic Turing reduction. We believe this gives a satisfactory answer (in the negative) to the question of whether the minmax value in three-player games can be approximated in polynomial time within any additive error $\epsilon > 0$. We leave open the problem of whether the minmax value of three-player games can be approximated within *some* nontrivial additive error $0 < \epsilon < 1/2$ in polynomial time.

References

1. Borgs, C., Chayes, J., Immorlica, N., Kalai, A.T., Mirrokni, V., Papadimitriou, C.: The myth of the folk theorem. Games and Economic Behavior 70(1), 34–43 (2010); Special Issue In Honor of Ehud Kalai
2. Hansen, K.A., Hansen, T.D., Miltersen, P.B., Sørensen, T.B.: Approximability and Parameterized Complexity of Minmax Values. In: Papadimitriou, C., Zhang, S. (eds.) WINE 2008. LNCS, vol. 5385, pp. 684–695. Springer, Heidelberg (2008)
3. Lipton, R.J., Young, N.E.: Simple strategies for large zero-sum games with applications to complexity theory. In: STOC 1994, pp. 734–740. ACM Press (1994)
4. Daskalakis, C., Goldberg, P.W., Papadimitriou, C.H.: The complexity of computing a Nash equilibrium. SIAM Journal on Computing 39(1), 195–259 (2009)
5. Chen, X., Deng, X., Teng, S.H.: Settling the complexity of computing two-player Nash equilibria. J. ACM 56, 14:1–14:57 (2009)
6. Lipton, R.J., Markakis, E., Mehta, A.: Playing large games using simple strategies. In: EC 2003, pp. 36–41. ACM (2003)
7. Tsaknakis, H., Spirakis, P.G.: An optimization approach for approximate nash equilibria. Internet Mathematics 5(4), 365–382 (2008)
8. Bollobás, B.: Random Graphs. Cambridge University Press (2001)
9. Jerrum, M.: Large cliques elude the metropolis process. Random Structures & Algorithms 3(4), 347–359 (1992)
10. Kučera, L.: Expected complexity of graph partitioning problems. Discrete Appl. Math. 57, 193–212 (1995)
11. Alon, N., Krivelevich, M., Sudakov, B.: Finding a large hidden clique in a random graph. Random Structures & Algorithms 13(3-4), 457–466 (1998)
12. Feige, U., Krauthgamer, R.: Finding and certifying a large hidden clique in a semirandom graph. Random Structures & Algorithms 16, 195–208 (2000)
13. Juels, A., Peinado, M.: Hiding cliques for cryptographic security. Designs, Codes and Cryptography 20, 269–280 (2000)
14. Feldman, V., Grigorescu, E., Reyzin, L., Vempala, S.S., Xiao, Y.: Statistical algorithms and a lower bound for planted clique. Technical Report 064, ECCC (2012)
15. Hazan, E., Krauthgamer, R.: How hard is it to approximate the best nash equilibrium? SIAM Journal on Computing 40(1), 79–91 (2011)
16. Minder, L., Vilenchik, D.: Small Clique Detection and Approximate Nash Equilibria. In: Dinur, I., Jansen, K., Naor, J., Rolim, J. (eds.) APPROX and RANDOM 2009. LNCS, vol. 5687, pp. 673–685. Springer, Heidelberg (2009)
17. Daskalakis, C., Mehta, A., Papadimitriou, C.H.: Progress in approximate nash equilibria. In: EC 2007, pp. 355–358. ACM (2007)
18. Conitzer, V., Sandholm, T.: Complexity results about nash equilibria. In: IJCAI 2003, pp. 765–771. Morgan Kaufmann (2003)
19. Gilboa, I., Zemel, E.: Nash and correlated equilibria: Some complexity considerations. Games and Economic Behavior 1(1), 80–93 (1989)
20. Austrin, P., Braverman, M., Chlamtáč, E.: Inapproximability of NP-Complete Variants of Nash Equilibrium. In: Goldberg, L.A., Jansen, K., Ravi, R., Rolim, J.D.P. (eds.) APPROX/RANDOM 2011. LNCS, vol. 6845, pp. 13–25. Springer, Heidelberg (2011)
21. Alon, N., Yuster, R., Zwick, U.: Color-coding. J. ACM 42, 844–856 (1995)

Approximate Well-Supported Nash Equilibria Below Two-Thirds*

John Fearnley[1], Paul W. Goldberg[1],
Rahul Savani[1], and Troels Bjerre Sørensen[2]

[1] Department of Computer Science, University of Liverpool, UK
[2] Department of Computer Science, University of Warwick, UK

Abstract. In an ϵ-Nash equilibrium, a player can gain at most ϵ by changing his behaviour. Recent work has addressed the question of how best to compute ϵ-Nash equilibria, and for what values of ϵ a polynomial-time algorithm exists. An *ϵ-well-supported* Nash equilibrium (ϵ-WSNE) has the additional requirement that any strategy that is used with non-zero probability by a player must have payoff at most ϵ less than a best response. A recent algorithm of Kontogiannis and Spirakis shows how to compute a 2/3-WSNE in polynomial time, for bimatrix games. Here we introduce a new technique that leads to an improvement to the worst-case approximation guarantee.

1 Introduction

In a bimatrix game, a Nash equilibrium is a pair of strategies in which both players only assign probability to best responses. The apparent hardness of computing an exact Nash equilibrium [5,4] has led to work on computing approximate Nash equilibria, and two notions of approximate Nash equilibria have been developed. The first, and more widely studied, notion is of an *ϵ-approximate Nash equilibrium* (ϵ-Nash), where each player is required to achieve an expected payoff that is within ϵ of a best response. A line of work [7,6,2] has investigated the best ϵ that can be guaranteed in polynomial time. The current best result in this setting is a polynomial time algorithm that finds a 0.3393-Nash equilibrium [12].

However, ϵ-Nash equilibria have a drawback: since they only require that the expected payoff is within ϵ of a pure best response, it is possible that a player could be required to place probability on a strategy that is arbitrarily far from being a best response. This issue is addressed by the second notion of an approximate Nash equilibrium. An *ϵ-well supported approximate Nash equilibrium* (ϵ-WSNE), requires that both players only place probability on strategies that have payoff within ϵ of a pure best response. This is a stronger notion of equilibrium, because every ϵ-WSNE is an ϵ-Nash, but the converse is not true.

* This work is supported by by EPSRC grant EP/H046623/1 "Synthesis and Verification in Markov Game Structures", and EPSRC grants EP/G069239/1 and EP/G069034/1 "Efficient Decentralised Approaches in Algorithmic Game Theory." A full version of this paper is available at `http://arxiv.org/abs/1204.0707`

M. Serna (Ed.): SAGT 2012, LNCS 7615, pp. 108–119, 2012.

In contrast to ϵ-Nash, there has been relatively little work ϵ-WSNE. The first result on the subject gave a $\frac{5}{6}$ additive approximation [7], but this only holds if a certain a graph-theoretic conjecture is true. The best-known polynomial-time additive approximation algorithm was given by Kontogiannis and Spirakis, and achieves a $\frac{2}{3}$-approximation [10]. We will call this algorithm the KS algorithm. In [9], which is an earlier conference version of [10], the authors presented an algorithm that they claimed was polynomial-time and achieves a ϕ-WSNE, where $\phi = \frac{\sqrt{11}}{2} - 1 \approx 0.6583$, but this was later withdrawn, and instead the polynomial-time $\frac{2}{3}$-approximation algorithm was presented in [10]. It has also been shown that there is a PTAS for ϵ-WSNE if and only if there is a PTAS for ϵ-Nash [4].

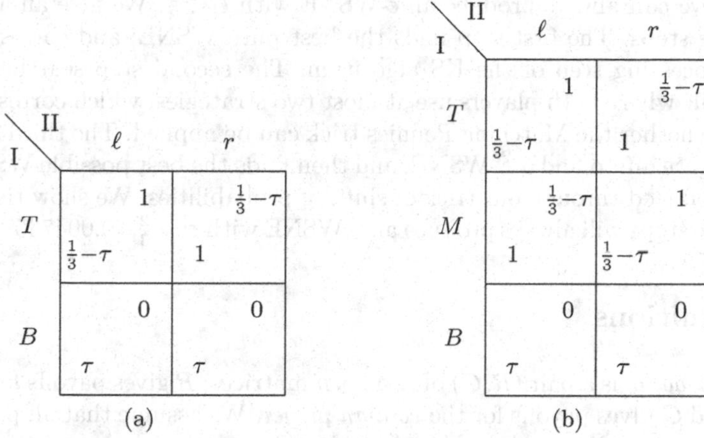

Fig. 1. Two examples that approach the worst case for the KS algorithm

Our approach. We build on the KS algorithm for finding a $\frac{2}{3}$-WSNE. Figure 1a gives a game where the KS algorithm produces a $\frac{2}{3}$-WSNE. The KS algorithm begins by checking there is a *pure* $\frac{2}{3}$-WSNE. In Figure 1a, there is a pure $\frac{2}{3}$-WSNE when $\tau = 0$, but not when $\tau > 0$, because any pure profile where both payoffs are at least $\frac{1}{3}$ is a $\frac{2}{3}$-WSNE. If no pure $\frac{2}{3}$-WSNE exists, the algorithm solves the zero-sum game $(D, -D)$, where $D = \frac{1}{2}(R - C)$, and gives the solution as a WSNE in the original game. In Figure 1a, if τ is small, then the solution to the zero-sum game has the row player playing B, and the column player mixing equally between ℓ and r. The *regret* for the row player is the difference between the payoff of a best response, and the lowest payoff of a row used by the row player. In our example, the row player's regret is the difference between the payoff of B and the payoff of T, and we can see that as $\tau \to 0$, the row player's regret approaches $\frac{2}{3}$. Since we have a ϵ-WSNE only if both players have regret smaller than ϵ, the quality of the WSNE approaches the worst-case bound of $\frac{2}{3}$.

Notice that in Figure 1a we can improve things for the row player by transferring some of the *column* player's probability from r to ℓ. The row player's regret is reduced, and the column player's regret is the same. However, consider Figure 1b. Once again this is approximately worst-case for the KS algorithm; the column player again mixes ℓ and r, while the row player uses row B, again getting regret of about $\frac{2}{3}$. This game is designed to prevent the trick of shifting some of the column player's probability so as to reduce the row player's regret.

In this case however, there is a new trick, which is to focus on rows T and M, and columns ℓ and r, where the payoffs are similar to the Matching Pennies game. By mixing uniformly on these strategies, the players both obtain average payoffs more than $\frac{1}{3}$, so that their regret in the entire game must be less than $\frac{2}{3}$.

Our main result is to show that one of these tricks can always be applied, and that we can always produce an ϵ-WSNE with $\epsilon < \frac{2}{3}$. We give an algorithm with three steps. The first step finds the best pure WSNE, and corresponds to the preprocessing step of the KS algorithm. The second step searches for the best WSNE where both players use at most two strategies, which corresponds to checking whether the Matching Pennies trick can be applied. The third step uses the KS algorithm to find a $\frac{2}{3}$-WSNE, and then finds the best possible WSNE that can be produced through our trick of shifting probabilities. We show that one of these three steps will always produce an ϵ-WSNE with $\epsilon = \frac{2}{3} - 0.004735 \approx 0.6619$.

2 Definitions

A *bimatrix game* is a pair (R, C) of two $n \times n$ matrices: R gives payoffs for the *row player*, and C gives payoffs for the *column player*. We assume that all payoffs are in the range $[0, 1]$. We use $[n] = \{1, 2, \dots n\}$ to denote the *pure strategies* for each player. To play the game, both players simultaneously select a pure strategy: the row player selects a row $i \in [n]$, and the column player selects a column $j \in [n]$. The row player then receives $R_{i,j}$, and the column player receives $C_{i,j}$.

A *mixed strategy* is a probability distribution over $[n]$. We denote a mixed strategy as a vector $\mathbf{x}$ of length n, such that $\mathbf{x}_i$ is the probability that the pure strategy i is played. The *support* of mixed strategy $\mathbf{x}$, denoted $\text{Supp}(\mathbf{x})$, is the set of pure strategies i with $x_i > 0$. If $\mathbf{x}$ and $\mathbf{y}$ are mixed strategies for the row and column player, respectively, then we call $(\mathbf{x}, \mathbf{y})$ a *mixed strategy profile*.

Let $\mathbf{y}$ be a mixed strategy for the column player. The *best responses* against $\mathbf{y}$ for the row player is the set of pure strategies that maximize the payoff against $\mathbf{y}$. More formally, a pure strategy $i \in [n]$ is a best response against $\mathbf{y}$ if, for all pure strategies $i' \in [n]$ we have: $\sum_{j \in [n]} \mathbf{y}_j \cdot R_{i,j} \geq \sum_{j \in [n]} \mathbf{y}_j \cdot R_{i',j}$. Column player best responses are defined analogously. A mixed strategy profile $(\mathbf{x}, \mathbf{y})$ is a *mixed Nash equilibrium* if every pure strategy in $\text{Supp}(\mathbf{x})$ is a best response against $\mathbf{y}$, and every pure strategy in $\text{Supp}(\mathbf{y})$ is a best response against $\mathbf{x}$. Nash [11] showed that all bimatrix games have a mixed Nash equilibrium.

An *approximate well-supported Nash equilibrium* weakens the requirements of a mixed Nash equilibrium. For a mixed strategy $\mathbf{y}$ of the column player, a pure strategy $i \in [n]$ is an ϵ*-best response* for the row player if, for all pure strategies $i' \in [n]$ we have: $\sum_{j\in[n]} \mathbf{y}_j \cdot R_{i,j} \geq \sum_{j\in[n]} \mathbf{y}_j \cdot R_{i',j} - \epsilon$. We define ϵ-best responses for the column player analogously. A mixed strategy profile $(\mathbf{x}, \mathbf{y})$ is an *ϵ-well-supported Nash equilibrium* (ϵ-WSNE) if every pure strategy in $\text{Supp}(\mathbf{x})$ is an ϵ-best response against $\mathbf{y}$, and every pure strategy in $\text{Supp}(\mathbf{y})$ is an ϵ-best response against $\mathbf{x}$.

3 Our Algorithm

We begin with an algorithm for finding the best WSNE on a given pair of supports. Let S_c and S_r be supports for the column and row player, respectively. We define an LP, which assumes that the row player uses a strategy with support S_r, and then finds a strategy on S_c that minimizes the row player's regret.

Definition 1. *Let $\mathbf{y}'$ be a mixed strategy for the column player. We define:*

Minimize: ϵ

Subject to:

$$R_{i'} \cdot \mathbf{y}' - R_i \cdot \mathbf{y}' \leq \epsilon \qquad i \in S_r,\ i' \in [n] \tag{1}$$

$$\mathbf{y}'_j = 0 \qquad j \notin S_c \tag{2}$$

A linear program for the row player can be defined symmetrically.

Let $(\mathbf{y}^*, \epsilon_\mathbf{y})$ be a solution of the LP given in Definition 1 (that is, $\mathbf{y}^*$ and $\epsilon_\mathbf{y}$ are the values of $\mathbf{y}'$ and ϵ that result) with parameters S_r and S_c, and let $(\mathbf{x}^*, \epsilon_\mathbf{x})$ be a solution of the corresponding LP for the row player. We define ϵ^* to be $\max(\epsilon_\mathbf{x}, \epsilon_\mathbf{y})$, and we have the following property.

Proposition 2. $(\mathbf{x}^*, \mathbf{y}^*)$ *is an ϵ^*-WSNE.*

More importantly, we can show that $(\mathbf{x}^*, \mathbf{y}^*)$ is at least as good, or better than, all well-supported Nash equilibria with support S_c and S_r.

Proposition 3. *For every ϵ-WSNE* $(\mathbf{x}, \mathbf{y})$ *with* $\text{Supp}(\mathbf{x}) = S_r$ *and* $\text{Supp}(\mathbf{y}) = S_c$*, we have $\epsilon^* \leq \epsilon$.*

Our algorithm for finding a WSNE consists of three distinct procedures.

(1) **Find the best pure WSNE.** The KS algorithm requires a preprocessing step that eliminates all pure $\frac{2}{3}$-WSNE, and this is a generalisation of that step. Suppose that the row player plays row i, and that the column player plays column j. Let: $\epsilon_r = \max_{i'}(R_{i',j}) - R_{i,j}$, and $\epsilon_c = \max_{j'}(C_{i,j'}) - C_{i,j}$. Thus i is an ϵ_r-best response against j, and that j is an ϵ_c-best response against i. Therefore, (i, j) is a $\max(\epsilon_r, \epsilon_c)$-WSNE. We can find the best pure WSNE by checking all $O(n^2)$ possible pairs of pure strategies. Let ϵ_p be the best approximation guarantee that is found by this procedure.

(2) **Find the best WSNE with 2×2 support.** We can use the linear program from Definition 1 to implement this procedure. For each of the $O(n^4)$ possible 2×2 supports, we solve the LPs to find a WSNE. Proposition 3 implies that this WSNE is at least as good as the best WSNE on those supports. Let ϵ_m be the best approximation guarantee that is found by this procedure.

(3) **Find an improvement over the KS algorithm.** The KS algorithm constructs a zero-sum game $(D, -D)$, where $D = \frac{1}{2}(R - C)$, and solves it. Kontogiannis and Spirakis showed that, if there is no pure $\frac{2}{3}$-WSNE, the min-max strategies for the zero-sum game are always a $\frac{2}{3}$-WSNE in the original game [10]. To find an improvement over the KS algorithm, we take the mixed strategy pair $(\mathbf{x}, \mathbf{y})$ that is produced by the KS algorithm, and we use the linear program from Definition 1 with parameters $S_r = \text{Supp}(\mathbf{x})$ and $S_c = \text{Supp}(\mathbf{y})$. Let $(\mathbf{x}^*, \mathbf{y}^*)$ be the mixed strategy profile returned by the LPs, and let ϵ_i be the smallest value such that $(\mathbf{x}^*, \mathbf{y}^*)$ is a ϵ_i-WSNE.

We take the smallest of ϵ_p, ϵ_m, and ϵ_i, and return the corresponding WSNE.

4 Outline

We want to show that our algorithm finds a $(\frac{2}{3} - z)$-WSNE, for some $z > 0$. The precise value of z will be determined during the proof, so for now we treat z as a parameter. At a high level, we will show that if $\epsilon_p > \frac{2}{3} - z$, and if $\epsilon_m > \frac{2}{3} - z$, then we must have $\epsilon_i \leq \frac{2}{3} - z$. Recall that Procedure (3) takes the mixed strategy profile $(\mathbf{x}, \mathbf{y})$, and finds the best WSNE on the supports of $\mathbf{x}$ and $\mathbf{y}$. Our approach is to use the assumptions that $\epsilon_p > \frac{2}{3} - z$ and $\epsilon_m > \frac{2}{3} - z$ to construct $(\mathbf{x}', \mathbf{y}')$, which is a specific $(\frac{2}{3} - z)$-WSNE on the supports of $\mathbf{x}$ and $\mathbf{y}$. The existence of $(\mathbf{x}', \mathbf{y}')$ then implies that Procedure (3) must produce at least a $(\frac{2}{3} - z)$-WSNE.

In our proof, we focus on how the mixed strategy $\mathbf{y}'$ can be constructed from $\mathbf{y}$. However, all of our arguments can be applied symmetrically in order to construct $\mathbf{x}'$ from $\mathbf{x}$. Our approach is to take the strategy $\mathbf{y}$ and to improve it. If $\mathbf{x}$ is not a $(\frac{2}{3} - z)$-best response against $\mathbf{y}$, then there must be at least one row i such that $R_i \cdot \mathbf{y} > \frac{2}{3} - z$. We call these *bad rows*, and the goal of our construction is to improve all bad rows, so that we can find a $(\frac{2}{3} - z)$-WSNE. We will first define a strategy $\mathbf{y}^{imp}$, which improves a specific bad row. Then, we define $\mathbf{y}'$ to be a convex combination of $\mathbf{y}$ and $\mathbf{y}^{imp}$. Formally, we will define $\mathbf{y}' = \mathbf{y}(t)$, where $t \in [0, 1]$, and $\mathbf{y}(t) := (1 - t) \cdot \mathbf{y} + t \cdot \mathbf{y}^{imp}$.

For the remainder of the proof, we will be concerned with finding a value of z for which the following property holds.

Definition 4. *$P(z)$ is the property of (non-negative real value) z that there exists $t \in [0, 1]$ such that, for all row player strategies $\mathbf{x}'$ with* $\text{Supp}(\mathbf{x}') = \text{Supp}(\mathbf{x})$, *$\mathbf{x}'$ is a $(\frac{2}{3} - z)$-best response against $\mathbf{y}(t)$.*

Since all of our arguments can also be applied to the row player, if $P(z)$ holds then there must exist a t such that $(\mathbf{x}(t), \mathbf{y}(t))$ is a $(\frac{2}{3} - z)$-WSNE. Our goal is to find the largest value of z for which $P(z)$ holds in all bimatrix games. Once we have determined the appropriate z, we will have then shown that our algorithm will always find a $(\frac{2}{3} - z)$-WSNE for all possible input games.

In the final part of our proof, we will develop a test that represents a sufficient condition for $P(z)$ to hold in all bimatrix games. If the test is passed then $P(z)$ holds in all bimatrix games, but we do not prove that $P(z)$ does not hold when the test is failed. Our test is monotone in z, and so to complete our proof, we use binary search to find the largest z for which the test tells us that $P(z)$ holds. We find that the test is passed when $z = 0.004735$, but failed when $z = 0.004736$. Thus, we arrive at our main result.

Theorem 5. *The algorithm given in Section 3 finds a* $(\frac{2}{3} - 0.004735)$*-WSNE.*

5 The Proof

5.1 Re-analysing the KS Algorithm

The original KS algorithm uses a preprocessing step that checks for a *pure* $\frac{2}{3}$-WSNE, and stops if one is found. In our version we initially check for a pure $\frac{2}{3} - z$-WSNE, a stronger requirement that leaves more input games that have to be handled by the rest of the algorithm. The results we establish for the rest of the algorithm are given in terms of the column player's strategy; corresponding results hold when the row player is considered.

Proposition 6. *Assume that* $\epsilon_p > \frac{2}{3} - z$*, and let* $(\mathbf{x}, \mathbf{y})$ *be the WSNE returned by the KS algorithm. If the row player has regret larger than* $\frac{2}{3} - z$ *in* $(\mathbf{x}, \mathbf{y})$*, then for all rows* i' *we have both of the following:*

$$R_{i'} \cdot \mathbf{y} \leq \frac{2}{3} + 2z, \qquad\qquad R_{i'} \cdot \mathbf{y} - C_{i'} \cdot \mathbf{y} \leq 3z.$$

This proposition shows that, under our new assumptions the KS algorithm will now produce a mixed strategy pair $(\mathbf{x}, \mathbf{y})$ that is a $(\frac{2}{3} + 2z)$-WSNE. The main goal of our proof is to show that the probabilities in $\mathbf{x}$ and $\mathbf{y}$ can be rearranged to construct a $(\frac{2}{3} - z)$-WSNE. From this point onwards, we only focus on improving the strategy $\mathbf{y}$, with the understanding that all of our techniques can be applied in the same way to improve the strategy $\mathbf{x}$.

Our improvement procedure must consider the rows i whose payoff lies in the range $\frac{2}{3} - z < R_i \cdot \mathbf{y} \leq \frac{2}{3} + 2z$. We call these rows *bad* rows, because they are the rows that must be improved to produce a $(\frac{2}{3} - z)$-WSNE. We classify the bad rows according to how bad they are.

Definition 7. *A row* i *is* q-bad *if* $R_i \cdot \mathbf{y} = \frac{2}{3} + 2z - qz$.

It can be seen from Proposition 6 that every row is q bad for some $q \geq 0$, and we are particularly interested in the q-bad rows with $0 \leq q < 3$.

5.2 The Structure of a q-Bad Row

To define our improvement procedure, we must understand the structure of a q-bad row. If i is a q-bad row, then we can apply the second inequality of Proposition 6 to obtain:

$$C_i \cdot \mathbf{y} \geq \frac{2}{3} - z - qz. \tag{3}$$

Now consider a q-bad row i with $q < 3$. We can deduce the following three properties about row i.

- Definition 7 tells us that $R_i \cdot \mathbf{y}$ is close to $\frac{2}{3}$.
- Equation (3) tells us that $C_i \cdot \mathbf{y}$ is close to $\frac{2}{3}$.
- The fact that $\epsilon_p > \frac{2}{3} - z$ implies that, for each column j, we must either have $R_{i,j} < \frac{1}{3} + z$ or $C_{i,j} < \frac{1}{3} + z$, because otherwise (i, j) would be a pure $(\frac{2}{3} - z)$-WSNE.

In order to satisfy all three of these conditions simultaneously, the row i must have a very particular form, which the rows T and M in Figure 1b show: approximately half of the probability assigned by $\mathbf{y}$ must be given to columns j where $R_{i,j}$ is close to 1 and $C_{i,j}$ is close to $\frac{1}{3}$, and the other (approximately) half of the probability assigned by $\mathbf{y}$ must be given to columns j where $R_{i,j}$ is close to $\frac{1}{3}$ and $C_{i,j}$ is close to 1.

Building on this observation, we split the columns of each row i into three sets. We define the set B_i of *big* columns to be $B_i = \{j : R_{i,j} \geq \frac{2}{3} + 2z\}$, and the set S_i of *small* columns to be $S_i = \{j : C_{i,j} \geq \frac{2}{3} + 2z\}$. Finally, we have the set of *other* columns $O_i = \{1, 2, \ldots, n\} \setminus (B_i \cup S_i)$, which contains all columns that are neither big nor small. We can then formalise our observations by giving inequalities about the amount of probability that $\mathbf{y}$ can assign to these sets.

Proposition 8. *If i is a q-bad row then:*

$$\sum_{j \in O_i} \mathbf{y}_j \leq \frac{2qz}{\frac{1}{3} - 2z},$$

$$\sum_{j \in B_i} \mathbf{y}_j \geq \frac{\frac{1}{3} + z - qz - (\frac{1}{3} + z) \sum_{j \in O_i} \mathbf{y}_j}{\frac{2}{3} - z},$$

$$\sum_{j \in S_i} \mathbf{y}_j \geq \frac{\frac{1}{3} - 2z - qz - (\frac{1}{3} + z) \sum_{j \in O_i} \mathbf{y}_j}{\frac{2}{3} - z}.$$

The first inequality is obtained by an application of Markov's inequality. The second two can be proved by substituting bounds for B_i, S_i, and O_i into Definition 7 and Equation 3. The inequalities show that, if $q = 0$, then $\mathbf{y}$ must give a roughly equal split between the big and small columns. As q increases, our inequalities become weaker, and the split may become more lopsided.

5.3 The Improved Strategies $\mathbf{y}^{imp}$ and $\mathbf{y}(t)$

We now define an improved version of $\mathbf{y}$. We start by constructing $\mathbf{y}^{imp}$, which will improve the *worst* bad row. That is, we choose $\bar{\imath}$ to be the index of a row in $\arg\max_i(R_i \cdot \mathbf{y})$, and therefore $\bar{\imath}$ is a $\bar{q}$-bad row such that there is no q-bad row with $q < \bar{q}$. We fix $\bar{\imath}$ and $\bar{q}$ to be these choices for the rest of this paper. If $\bar{q} \geq 3$, then $\mathbf{y}$ does not need to be improved. Therefore, we can assume that $\bar{q} < 3$.

We aim to improve row $\bar{\imath}$ by moving the probability assigned to $B_{\bar{\imath}}$ to $S_{\bar{\imath}}$. This is a generalisation of shifting probability from the first column to the second column in Figure 1a. Formally, we define the strategy $\mathbf{y}^{imp}$, for each j with $1 \leq j \leq n$, as:

$$\mathbf{y}_j^{imp} = \begin{cases} 0 & \text{if } j \in B_{\bar{\imath}}, \\ \mathbf{y}_j + \frac{\mathbf{y}_j \cdot \sum_{k \in B_{\bar{\imath}}} \mathbf{y}_k}{\sum_{k \in S_{\bar{\imath}}} \mathbf{y}_k} & \text{if } j \in S_{\bar{\imath}}, \\ \mathbf{y}_j & \text{otherwise.} \end{cases}$$

The strategy $\mathbf{y}^{imp}$ improves the specific bad row $\bar{\imath}$, but other rows may not improve, or even get worse in $\mathbf{y}^{imp}$. Therefore, we propose that $\mathbf{y}$ should be gradually improved towards $\mathbf{y}^{imp}$. More formally, for the parameter $t \in [0, 1]$, we define the strategy $\mathbf{y}(t)$ to be $(1-t) \cdot \mathbf{y} + t \cdot \mathbf{y}^{imp}$.

5.4 An Upper Bound on $R_i \cdot \mathbf{y}^{imp}$

Recall that $P(z)$ checks whether there exists a t such that all row player strategies with support Supp($\mathbf{x}$) are $(\frac{2}{3}-z)$-best responses against $\mathbf{y}(t)$. In order to perform this test, we check whether there exists a t such that $R_i \cdot \mathbf{y}(t) \leq \frac{2}{3} - z$, for all rows i. Thus, eventually, we will need an upper bound on $R_i \cdot \mathbf{y}(t)$ for each row i. Since $\mathbf{y}(t)$ is a convex combination of $\mathbf{y}$ and $\mathbf{y}^{imp}$, we begin the construction of our test by finding an upper bound on $R_i \cdot \mathbf{y}^{imp}$.

The strategy $\mathbf{y}^{imp}$ is defined by moving all probability from $B_{\bar{\imath}}$ to $S_{\bar{\imath}}$. We are interested in the effect that this can have on a q-bad row $i \neq \bar{\imath}$. If we consider the partition of the columns in $\bar{\imath}$ into $(B_{\bar{\imath}}, S_{\bar{\imath}}, O_{\bar{\imath}})$, and the partition of the columns in i into (B_i, S_i, O_i), then we have a decomposition into nine possible intersections:

Row $\bar{\imath}$	$B_{\bar{\imath}}$			$S_{\bar{\imath}}$			$O_{\bar{\imath}}$		
Row i	B_i	S_i	O_i	B_i	S_i	O_i	B_i	S_i	O_i

We cannot know the precise amount of probability that $\mathbf{y}$ assigns to each of the sets in the decomposition. However, Proposition 8 gives useful constraints on the probabilities allocated to the sets used in the decomposition. We will use these inequalities to write down a linear program that characterises $R_i \cdot \mathbf{y}^{imp}$.

The LP will have one variable for each of the sets in the decomposition. The idea is that each variable should represent the amount of probability that $\mathbf{y}$ assigns to that set. Thus, we have nine variables: d_{bb}, d_{bs}, d_{bo}, and so on, where the variable d_{bb} represents $\sum_{j \in B_{\bar{\imath}} \cap B_i} \mathbf{y}_j$, the variable d_{bs} represents $\sum_{j \in B_{\bar{\imath}} \cap S_i} \mathbf{y}_j$, and so on. For convenience, we use $\sum d_{b*}$ as a shorthand for $d_{bb} + d_{bs} + d_{bo}$, and $\sum d_{*b}$ as a shorthand $d_{bb} + d_{sb} + d_{ob}$. We also use $\sum d_{s*}$, $\sum d_{*s}$, $\sum d_{o*}$, and $\sum d_{*o}$, which have analogous definitions. Finally, we use $\sum d_{**}$ as a shorthand for $\sum d_{b*} + \sum d_{s*} + \sum d_{o*}$.

The LP is shown in Figure 2; the constraints that variables d_{ij} are non-negative, and should sum to 1 are not shown. The LP takes three parameters: z,

$\bar{q}$, and q. The inequalities of this LP are taken directly from Proposition 8, and each inequality appears twice: once for row $\bar{\imath}$, and once for row i. The objective function is intended to capture $R_i \cdot \mathbf{y}^{imp}$, and it the auxiliary function:

$$\phi(z,q) = \left(1 + \frac{\frac{1}{3} + z + qz + \frac{2qz}{\frac{1}{3}-2z}}{\frac{1}{3} - 2z - qz - (\frac{1}{3}+z)\frac{2qz}{\frac{1}{3}-2z}}\right).$$

If $\mathrm{s}(z, \bar{q}, q)$ is the solution of this LP, then we have the following proposition.

Proposition 9. *For every q-bad row i we have $R_i \cdot \mathbf{y}^{imp} \leq \mathrm{s}(z, \bar{q}, q)$.*

Maximize:

$$\phi(z,\bar{q})\left(d_{sb} + (\frac{1}{3}+z)\cdot d_{ss} + (\frac{2}{3}+2z)\cdot d_{so}\right) + d_{ob} + (\frac{1}{3}+z)\cdot d_{os} + (\frac{2}{3}+2z)\cdot d_{oo}$$

Subject to:

$$\sum d_{b*} \geq \frac{\frac{1}{3} + z - \bar{q}z - (\frac{1}{3}+z)(\sum d_{o*})}{\frac{2}{3} - z} \quad (4)$$

$$\sum d_{*b} \geq \frac{\frac{1}{3} + z - qz - (\frac{1}{3}+z)(\sum d_{*o})}{\frac{2}{3} - z} \quad (5)$$

$$\sum d_{s*} \geq \frac{\frac{1}{3} - 2z - \bar{q}z - (\frac{1}{3}+z)(\sum d_{o*})}{\frac{2}{3} - z} \quad (6)$$

$$\sum d_{*s} \geq \frac{\frac{1}{3} - 2z - qz - (\frac{1}{3}+z)(\sum d_{*o})}{\frac{2}{3} - z} \quad (7)$$

$$\sum d_{o*} \leq \frac{2\bar{q}z}{\frac{1}{3} - 2z} \quad (8)$$

$$\sum d_{*o} \leq \frac{2qz}{\frac{1}{3} - 2z} \quad (9)$$

Fig. 2. A linear program that gives an upper bound on $R_i \cdot \mathbf{y}^{imp}$

5.5 Applying the Matching Pennies Argument

Recall that ϵ_m is computed in stage 2 of our algorithm, and is the quality of the best WSNE with 2×2 support. So far, we have not used the assumption that $\epsilon_m > \frac{2}{3} - z$. In this section we will see how this assumption can be used to strengthen our LP. We define a matching pennies sub-game as follows.

Definition 10 (Matching Pennies). *Let i and i' be two rows, and let j and j' be two columns. If $j \in B_i \cap S_{i'}$ and $j' \in B_{i'} \cap S_i$, then we say that i, i', j, and j' form a matching pennies sub-game.*

An example of a matching pennies sub-game is given by l, r, T, and M in Figure 1b, because we have $l \in B_M \cap S_T$, and we have $r \in B_T \cap S_M$. In this example, we can obtain an exact Nash equilibrium by making the row player mix uniformly between T and M, and making the column player mix uniformly between l and r. However, in general we can only expect to obtain an $(\frac{2}{3} - z)$-WSNE using this technique.

Proposition 11. *If there is a matching pennies sub-game, then we can construct a $(\frac{2}{3} - z)$-WSNE with a 2×2 support.*

Thus, we can assume that our game does not contain a matching pennies sub-game, because otherwise Procedure (2) would have found a $(\frac{2}{3} - z)$-WSNE. Note that, by definition, if the game does not contain a matching pennies sub-game, then for all rows i we must have either $B_{\bar{i}} \cap S_i = \emptyset$, or $B_i \cap S_{\bar{i}} = \emptyset$.

We can use this observation to strengthen our LP. We define two LPs, each of which is constructed by adding an extra constraint to our existing LP. In the first LP we add the constraint $d_{bs} = 0$, and in the second LP we add the constraint $d_{sb} = 0$. We refer to the solutions of these two LPs as $\mathrm{s}_1(z, \bar{q}, q)$ and $\mathrm{s}_2(z, \bar{q}, q)$ respectively. We then obtain the following strengthening of Proposition 9.

Proposition 12. *For each q-bad row i we either have $R_i \cdot \mathbf{y}^{imp} \le \mathrm{s}_1(z, \bar{q}, q)$, or we have $R_i \cdot \mathbf{y}^{imp} \le \mathrm{s}_2(z, \bar{q}, q)$.*

5.6 A Linear Upper Bound for Our LPs

Now we can finally obtain our bound for $R_i \cdot \mathbf{y}^{imp}$, by proving an upper bound for $\mathrm{s}_k(z, \bar{q}, q)$. It is not difficult to show that s_k is monotonically increasing in $\bar{q}$. Since $\bar{q} < 3$, we can therefore argue that $\mathrm{s}_k(z, \bar{q}, q) \le \mathrm{s}_k(z, 3, q)$. Then, using standard techniques from sensitivity analysis in linear programming, it is possible to bound $\mathrm{s}_k(z, 3, q)$ by a linear function.

Proposition 13. *We can compute $c_{z,k}$ and $d_{z,k}$ so that $\mathrm{s}_k(z, 3, q) \le c_{z,k} + d_{z,k} \cdot q$.*

To obtain our final upper bound on $R_i \cdot \mathbf{y}^{imp}$, we simply take the maximum over the two LPs. That is, we set $c_z = \max(c_{z,1}, c_{z,2})$ and $d_z = \max(d_{z,1}, d_{z,2})$. This then leads to our final upper bound for $R_i \cdot \mathbf{y}^{imp}$.

Proposition 14. *We have $R_i \cdot \mathbf{y}^{imp} \le c_z + d_z \cdot q$, for every q-bad row i.*

5.7 The Test for $P(z)$

Finally, we can describe the test that determines whether $P(z)$ holds in all bimatrix games. The test constructs a point t_z^*, and then checks whether $R_i \cdot \mathbf{y}(t_z^*) \le \frac{2}{3} - z$ holds for all rows i.

We begin by defining t_z^*, which is the smallest value of t for which, if i is a 0-bad row, then $R_i \cdot \mathbf{y}(t) \le \frac{2}{3} - z$. By definition we have that $R_i \cdot \mathbf{y} = \frac{2}{3} + 2z$, and we also know that $R_i \cdot \mathbf{y}^{imp} \le c_z + d_z \cdot 0$. Therefore t_z^* is the solution of:

$$(\frac{2}{3} + 2z) \cdot (1 - t_z^*) + c_z \cdot t_z^* = \frac{2}{3} - z.$$

This can be seen graphically in Figure 3a. The line in the figure starts at $\frac{2}{3} + z$ when $t = 0$, and ends at c_z when $t = 1$. The point t_z^* is the value of t at which this line crosses $\frac{2}{3} - z$. We can solve the equation to obtain the following formula:

$$t_z^* = \frac{3z}{\frac{2}{3} + 2z - c_z}. \tag{10}$$

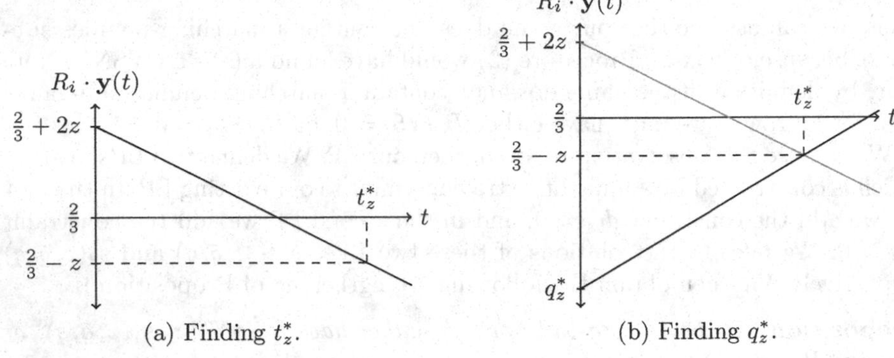

(a) Finding t_z^*. (b) Finding q_z^*.

Fig. 3. Diagrams that show how t_z^* and q_z^* are found

Next, we define a constant q_z^*. For each row i, there is a trivial bound of:

$$R_i \cdot \mathbf{y}^{imp} \leq 1. \tag{11}$$

Note that if q is large, then this bound will be better than our bound of $c_z + d_z \cdot q$. The next step of our procedure is to find q_z^*, which is the smallest value of q such that, using this trivial bound (11), we can conclude that $R_i \cdot \mathbf{y}(t_z^*) \leq \frac{2}{3} - z$. Formally, we define q_z^* to be the solution of:

$$(\frac{2}{3} + 2z - q_z^* z) \cdot (1 - t_z^*) + t_z^* = \frac{2}{3} - z.$$

This can be seen diagrammatically in Figure 3b: we fix a line that passes through 1 when $t = 1$, and $\frac{2}{3} - z$ when $t = t_z^*$. Then, q_z^* is defined to be the point at which this line meets the y-axis of the graph, where $t = 0$. Solving the equation gives the following formula for q_z^*.

$$q_z^* = \frac{(2z - \frac{1}{3}) \cdot t_z^* - 3z}{zt_z^* - z} \tag{12}$$

For rows i that are q-bad with $q \geq q_z^*$, we can apply the trivial bound (11) to argue that $R_i \cdot \mathbf{y}(t_z^*) \leq \frac{2}{3} - z$. Therefore, we need only be concerned with rows i that are q-bad with $0 \leq q < q_z^*$. The next proposition gives a simple test that can be used to check whether all such rows will have the property $R_i \cdot \mathbf{y}(t_z^*) \leq \frac{2}{3} - z$.

Proposition 15. *If $c_z + d_z \cdot q_z^* \leq 1$, then $R_i \cdot \mathbf{y}(t_z^*) \leq \frac{2}{3} - z$ for all rows i.*

Thus, our test for checking whether $P(z)$ holds in all bimatrix games can be summarised as follows. First we compute the constants c_z and d_z. Then we use these to compute t_z^* and q_z^*. Finally, we check whether $c_z + d_z \cdot q_z^* \leq 1$. If the inequality holds, then Proposition 15 implies that $P(z)$ is true. To complete the proof of Theorem 5, it suffices to note that our test proves that $P(z)$ holds in all bimatrix games for $z = 0.004735$.

6 Conclusions

In Section 3, we presented a polynomial-time algorithm for computing a $(\frac{2}{3} - z)$-WSNE, where $z = 0.004735$. We do not believe that our analysis is tight, as it uses several restrictions that our algorithm does not face. For example, $\mathbf{y}(t)$ uses the same support as the strategy returned by the KS algorithm, whereas the LP given in Definition 1 can return a subset of this support. Another example is that in the analysis we only consider 2×2 subgames in which players mix uniformly, whereas Procedure 2 considers all mixtures.

An interesting open question is the following. Does every bimatrix game possess a $\frac{1}{2}$-WSNE, where both players use at most two strategies? This is known to be true with high probability in random games [1], but not known in general.

References

1. Bárány, I., Vempala, S., Vetta, A.: Nash equilibria in random games. Random Struct. Algorithms 31(4), 391–405 (2007)
2. Bosse, H., Byrka, J., Markakis, E.: New algorithms for approximate Nash equilibria in bimatrix games. Theoretical Computer Science 411(1), 164–173 (2010)
3. Bradley, S.P., Hax, A.C., Magnanti, T.L.: Applied Mathematical Programming. Addison-Wesley (1977), `http://web.mit.edu/15.053/www/`
4. Chen, X., Deng, X., Teng, S.-H.: Settling the complexity of computing two-player Nash equilibria. Journal of the ACM 56(3),14:1–14:57 (2009)
5. Daskalakis, C., Goldberg, P.W., Papadimitriou, C.H.: The complexity of computing a Nash equilibrium. SIAM Journal on Computing 39(1), 195–259 (2009)
6. Daskalakis, C., Mehta, A., Papadimitriou, C.H.: Progress in approximate Nash equilibria. In: Proceedings of ACM-EC, pp. 355–358 (2007)
7. Daskalakis, C., Mehta, A., Papadimitriou, C.H.: A note on approximate Nash equilibria. Theoretical Computer Science 410(17), 1581–1588 (2009)
8. Jansen, B., de Jong, J.J., Roos, C., Terlaky, T.: Sensitivity analysis in linear programming: just be careful! European Journal of Operational Research 101(1), 15–28 (1997)
9. Kontogiannis, S.C., Spirakis, P.G.: Efficient Algorithms for Constant Well Supported Approximate Equilibria in Bimatrix Games. In: Arge, L., Cachin, C., Jurdziński, T., Tarlecki, A. (eds.) ICALP 2007. LNCS, vol. 4596, pp. 595–606. Springer, Heidelberg (2007)
10. Kontogiannis, S.C., Spirakis, P.G.: Well supported approximate equilibria in bimatrix games. Algorithmica 57(4), 653–667 (2010)
11. Nash, J.: Non-cooperative games. The Annals of Mathematics 54(2), 286–295 (1951)
12. Tsaknakis, H., Spirakis, P.G.: An optimization approach for approximate Nash equilibria. Internet Mathematics 5(4), 365–382 (2008)

Mechanisms and Impossibilities for Truthful, Envy-Free Allocations*

Michal Feldman[1,2] and John Lai[1]

[1] Harvard School of Engineering and Applied Sciences, Cambridge MA 02138, USA
jklai@post.harvard.edu

[2] Hebrew University of Jerusalem, Jerusalem, Israel
michal.feldman@huji.ac.il

Abstract. We study mechanisms for combinatorial auctions that are simultaneously incentive compatible (IC), envy free (EF) and efficient in settings with *capacitated* valuations — a subclass of subadditive valuations introduced by Cohen et al. [4]. Capacitated agents have valuations which are additive up to a publicly known capacity. The main result of Cohen et al. [4] is the assertion that the Vickrey-Clarke-Groves mechanism with Clarke pivot payments is EF (and clearly IC and efficient) in the case of homogeneous capacities. The main open problem raised by Cohen et al. [4] is whether the existence result extends beyond homogeneous capacities. We resolve the open problem, establishing that no mechanism exists that is simultaneously IC, EF and efficient for capacitated agents with heterogeneous capacities. In addition, we establish the existence of IC, EF, and efficient mechanisms in the special cases of capacitated agents with heterogeneous capacities, where (i) there are only two items; or (ii) the individual item values are binary. Finally, we show that the last existence result does not extend to the stronger notion of *Walrasian* mechanisms, i.e. mechanisms whose allocation and payments correspond to a Walrasian equilibrium.

1 Introduction

A combinatorial auction mechanism takes as input agents' valuations for bundles of items and computes an allocation and payment for each agent. Incentive compatibility (IC) and envy freeness (EF) are two desirable properties of combinatorial auction mechanisms. IC ensures that agents cannot gain by misreporting their private information [11], while EF imposes a notion of fairness on the outcome of the auction. Specifically, EF requires that no agent prefers the allocation and payment of another agent to her own [5, 6, 13, 14, 18].

IC is desirable for various reasons. IC mechanisms create incentives for the agents to report their true values, and as a result, the computed allocation may better optimize the objective of the auctioneer. In addition, IC mechanisms are considered fair in the sense that they do not advantage more sophisticated agents. This is, however, a very weak notion of fairness, and it is well known

* A full version of this paper including all proofs is available on the authors' websites.

M. Serna (Ed.): SAGT 2012, LNCS 7615, pp. 120–131, 2012.

that IC mechanisms may not adhere to very basic fairness requirements [1]. In particular, IC mechanisms may produce outcomes which are not EF. This may be problematic in certain settings, such as government run spectrum auctions, since the participants, after observing the outcome, may question the fairness of the auction, and perceive others as being favored by the mechanism. Recent experiments show that people place extremely high value on fairness. For example, Rafaeli et al. [17] show that people care about fairness in queues even more than the actual delay they experience. If outcomes are EF, in contrast, then no agent views other agents' outcomes as preferable.

EF outcomes can be thought of as a relaxation of the outcomes of a *Walrasian equilibrium*. In a Walrasian equilibrium, we have item prices such that every agent receives a bundle that maximizes her utility (i.e., valuation for the bundle minus the sum of the prices of the bundle's items), and the market clears (i.e., every unsold item has a price of zero).[1] If a Walrasian equilibrium exists, then the corresponding outcome is efficient [2] and valuations that are *gross-substitutes* (which is a subclass of subadditive valuations) always admit a Walrasian equilribrium [9].

While every Walrasian equilibrium outcome is clearly EF, the other direction does not hold. In contrast to a Walrasian equilibrium outcome, envy free outcomes assign (arbitrary) bundle prices, which may not correspond to item prices. If the allocation and payments of a mechanism correspond to a Walrasian equilibrium outcome, we say that the mechanism is *Walrasian*.

In this paper, we focus on combinatorial auction mechanisms that are simultaneously IC, EF and efficient; i.e, maximize social welfare. We also consider how our results are affected by replacing the EF requirement with the stronger Walrasian requirement. Because we focus on efficient allocations, the problem of finding IC+(EF or Walrasian) mechanisms reduces to finding payment rules which are IC+(EF or Walrasian), except, possibly, for cases where there may be multiple efficient allocations as in Section 4.

Notably, without the additional EF (or Walrasian) requirement, the family of Vickrey-Clarke-Groves (VCG) mechanisms [3, 8] is known to be IC and efficient for arbitrary valuations. Moreover, the classic results of Green and Laffont [7] and Holmstrom [10] prove that for the efficient allocation and valuations that are connected domains (which include the valuations studied in this paper), any IC mechanism is a VCG mechanism. VCG mechanisms allocate according to an efficient allocation, and determine the payment for each agent in a way that reporting one's true valuations is a dominant strategy. VCG mechanisms are essentially a family of payment rules. The most common payment rule is known as the *Clarke pivot* rule, in which an agent's payment is the externality that the agent imposes on the other agents.

Similarly, without the additional IC requirement, Mu'alem [15] shows that the efficient allocation can always be supported by EF payments. In particular,

[1] We differentiate between a Walrasian equilibrium and a Walrasian equilibrium outcome since a Walrasian equilibrium requires specification of item prices while an outcome simply states the bundle and payment of each agent.

an allocation has supporting EF payments iff it is *locally efficient* — a weaker notion than global efficiency. Thus, an EF and efficient mechanism exists for arbitrary valuation functions.

Therefore, every efficient allocation can be supported by IC payments and can also be supported by EF payments. Unfortunately, it is not always the case that the set of IC payment rules shares a non-empty intersection with the set of EF payment rules, i.e., there may not be a payment rule that can simultaneously satisfy IC and EF. Most of the mechanism design literature focuses on mechanisms that are either IC or EF, but not much attention has been given to the combination of both properties.

One exception is the unit demand case, where each agent desires at most one item. Under these preferences, it is known that VCG with Clarke pivot payments is Walrasian [9, 12] (and is, therefore, clearly IC and EF). Another more recent systematic treatment of the problem is the work of Cohen et al. [4] which considers mechanisms that are IC, EF and efficient for various subadditive valuation classes. In particular, Cohen et al. [4] introduce the class of *capacitated* valuations, which is a natural generalization of unit-demand. Agents with capacitated valuations are associated with a publicly known capacity c and values for individual items. An agent's value for a bundle of items is the sum of the values for the c most valued items in the bundle. We refer to the case where all agents are capacitated and have the same capacity as *homogeneous capacities* and the general case where agents may have arbitrary capacities as the *heterogeneous capacities* case. Because the capacities are publicly known, these classes of valuations are connected and any IC and efficient mechanism must be a VCG mechanism. The results of Cohen et al. [4] are summarized in Figure 1. The main

	capacitated - heterogeneous	capacitated - homogeneous
IC + Walrasian	**NO [derived by right column]**	**NO** [Cohen et al. [4]] **NO** for binary valuations **[new]**
IC + EF	**NO [new: main result]** **YES** for $n = 2$ [Cohen et al. [4]] **YES** for $m = 2$ **[new]** **YES** for binary valuations **[new]**	**YES** [Cohen et al. [4]]

Fig. 1. This table specifies the existence of a particular type of mechanism (rows) for various families of valuation functions (columns). Efficiency is required in all entries. The results are divided between those that are established by Cohen et al. [4] and those that are established here, indicated as **[new]**.

result is that the VCG mechanism with Clarke pivot payments is EF for homogeneous capacities. For the broader class of heterogeneous capacities, Cohen et al. [4] show that the VCG mechanism with Clarke pivot payments is not EF, but it is left open whether there exists any mechanism that is simultaneously IC, EF, and efficient. This problem is the main open problem raised by Cohen et al. [4]. For the special case in which there are only two agents (with heterogeneous capacities), it is shown that a particular VCG mechanism (that does not use Clarke pivot payments) is always EF. They also show that under the additional requirement of *no positive transfers* (i.e., payments are weakly positive), no IC, EF, and efficient mechanism exists, even for two agents and two items.

In this paper, we resolve open problems raised in Cohen et al. [4], and establish several additional results for additional natural special cases. Our results are summarized in Figure 1, marked by [**new**]. Our main results are:

- We prove that for heterogeneous capacities, there is no mechanism that is IC, EF and efficient, even if no other requirement (such as no positive transfers) is imposed. To establish this impossibility, we take a computational approach which frames the problem of finding satisfactory VCG payments as a linear program. This result shows that homogeneous capacities is a maximal class that admits an IC, EF, and efficient mechanism. If the capacities are not homogeneous, then IC, EF, and efficient mechanisms no longer exist.
- We devise an IC, EF, and efficient mechanism for heterogeneous capacities in the special case of two items. This result complements the positive result of Cohen et al. [4] which establishes existence for the special case of two agents. Interestingly, the Clarke pivot payment is not EF in either of these cases. Moreover, the two cases rely on different payment rules.
- We then restrict attention to the interesting special case in which agents' valuations for individual items are binary; i.e., in $\{0, 1\}$. We refer to this class as the *binary valuations* class. This is a natural setting where each agent likes a subset of the items but still has a capacity. In this case, there exists a mechanism for heterogeneous capacities that is simultaneously IC, EF, and efficient. In particular, we show that that VCG with Clarke pivot payments is EF if ties in the efficient allocation are broken based on a lexicographic order that favors higher-capacity agents. The tie breaking method is shown to be critical; VCG with Clarke pivot payments is not EF if ties are broken arbitrarily (see Section 4). The proof involves viewing allocations as flows on a particular graph and using augmenting paths and flow decomposition. Similar techniques were used to prove the main result of Cohen et al. [4].
- Finally, we consider mechanisms that are IC, Walrasian, and efficient. We find that, while IC, EF and efficient mechanisms exist for binary valuations and heterogeneous capacities, this result does not extend to IC, Walrasian, and efficienct mechanisms. In particular, we show that there is no IC, Walrasian, and efficient mechanism even for binary valuations and homogeneous capacities.

2 Model and Preliminaries

Suppose we have a set $N = \{1, \ldots, n\}$ of agents and a set $G = \{1, \ldots, m\}$ of goods. We will index agents by i and j and goods by k. Each agent i is associated with a valuation function $v_i : 2^G \to \mathbb{R}_{\geq 0}$ that maps each bundle of goods to the agent's value for that bundle. A valuation profile $v = (v_1, \ldots, v_n)$ consists of a valuation function for each agent. We will often adopt the view of agent i and write a valuation profile as (v_i, v_{-i}), where v_{-i} denotes the valuations of all agents other than i. An allocation $a \in \mathcal{A}$ assigns a bundle of goods to each agent such that no good is given to more than one agent. Let a_i denote the bundle of items allocated to agent i under allocation a. We use the shorthand $v(a)$ to denote the *social welfare* of allocation a, i.e. $\sum_{i=1}^{n} v_i(a_i)$. An allocation is *efficient* if it maximizes social welfare amongst all allocations.

An allocation rule g maps a valuation profile to an allocation, and a payment rule p maps a valuation profile to a payment for each agent, with $g_i(v)$ and $p_i(v)$ denoting the bundle and payment of agent i, respectively. We assume quasi-linear utilities, i.e., the utility of agent i who receives bundle a_i and pays p_i is $v_i(a_i) - p_i$. A mechanism $M = (g, p)$ consists of an allocation rule and payment rule. The following properties of mechanisms are central to our study.

Definition 1. *A mechanism (g, p) is* efficient *if $g(v)$ is an efficient allocation for all v.*

Definition 2. *A mechanism (g, p) is* incentive-compatible *(IC) if there is no benefit to mis-reporting, i.e., for every agent i and every valuation profile (v_i, v_{-i}), $v_i(g_i(v_i, v_{-i})) - p_i(v_i, v_{-i}) \geq v_i(g_i(v_i', v_{-i})) - p_i(v_i', v_{-i})$.*

Definition 3. *A mechanism (g, p) is* envy-free *(EF) if no agent prefers the allocation and payment of another agent to her own, i.e., for every i, for every (v_i, v_{-i}), for every $j \neq i$, $v_i(g_i(v_i, v_{-i})) - p_i(v_i, v_{-i}) \geq v_i(g_j(v_i, v_{-i})) - p_j(v_i, v_{-i})$.*

Definition 4. *A mechanism (g, p) is* Walrasian *if the allocation and payments correspond to a Walrasian equilibrium outcome. In other words, there exists a price vector $(q_1, \ldots, q_m)$ such that:*

$$g_i(v) \in \arg\max_{S \subseteq G} \left(v_i(S) - \sum_{k \in S} q_k \right) \tag{1}$$

$$p_i(v) = \sum_{k \in g_i(v)} q_k \tag{2}$$

$$q_k = 0 \quad \text{if } k \text{ is unallocated in } g(v) \tag{3}$$

It is easy to verify that a Walrasian mechanism is also EF due to the first condition of Walrasian equilibrium, which stipulates that agents are allocated bundles which maximize their utility given the Walrasian item prices.

In this paper, we study mechanisms where g is an efficient allocation rule. Because we will be considering efficient allocations, it is convenient to introduce the following notation. Given a valuation profile v, Opt refers to an efficient allocation when all agents are considered. There may be multiple efficient allocations due to ties, but we point out where this distinction is important (e.g., in Section 4). Elsewhere, we assume that Opt is any efficient allocation. Opt^{-i} refers to an efficient allocation when agent i is excluded. Since Opt and Opt^{-i} are allocations, Opt_j and Opt_j^{-i} give the allocation of agent j in these allocations.

2.1 Characterization of IC and EF Mechanisms

When g is an efficient allocation, IC mechanisms are guaranteed to exist. In particular, Vickrey-Clarke-Groves mechanisms are IC.

Definition 5. *A* Vickrey-Clarke-Groves (VCG) mechanism *is a mechanism (g, p), where $g(v)$ is an efficient allocation and $p(v)$ takes on the following form,*

$$p_i(v) = h_i(v_{-i}) - \sum_{j \neq i} v_j(Opt_j),$$

where h_i can be any function of v_{-i}.

One of the most common choices of the h_i function is the *Clarke pivot* payment rule, given by

$$h_i(v_{-i}) = \sum_{j \neq i} v_j(\mathrm{Opt}_j^{-i}). \tag{4}$$

The obtained payment is then $p_i(v) = \sum_{j \neq i} v_j(\mathrm{Opt}_j^{-i}) - \sum_{j \neq i} v_j(\mathrm{Opt}_j)$, which can be interpreted as the externality that agent i imposes on the other agents.

It is well known that VCG mechanisms are IC from the classic results of Clarke [3] and Groves [8]. When the possible valuations of each agent form a connected domain (i.e., there is a path between any two possible valuations that stays within the set of possible valuations), VCG mechanisms are the only IC and efficient mechanisms [7, 10]. Therefore, when considering IC and efficient mechanisms for connected domains, the only flexibility one has is in the choice of the function $h_i(v_{-i})$.

If we consider VCG mechanisms, EF is equivalent to imposing a simple condition on the $h_i(v_{-i})$ functions. When clear in the context, we will often drop the input v_{-i} and simply refer to $h_i(v_{-i})$ using h_i.

Theorem 1. *[16] A VCG mechanism with efficient allocation Opt is EF iff for every valuation profile v and for every pair of agents i, j:*

$$h_i(v_{-i}) - h_j(v_{-j}) \leq v_j(Opt_j) - v_i(Opt_j). \tag{5}$$

Note that if there are multiple efficient allocations, then EF may depend on which efficient allocations are chosen by the mechanism. This turns out to be the case when we study binary valuations in Section 4. When the choice of efficient allocations is unimportant or when the efficient allocations are unique, the problem of finding IC, EF, and efficient mechanisms for connected domains reduces to finding h_i functions which satisfy (5).

2.2 Restricted Classes of Valuations

Following Cohen et al. [4], we consider the following classes of valuations. A valuation function is *superadditive* if for any sets $S, T \subseteq G$, $v_i(S) + v_i(T) \leq v_i(S \cup T)$. A valuation function v_i is *subadditive* if for any sets $S, T \subseteq G$, $v_i(S) +$

$v_i(T) \geq v_i(S \cup T)$. Pápai [16] proves that if valuations are superadditive, then VCG with Clarke pivot payments is EF (and trivially IC and efficient). In this paper, we focus on a subset of *subadditive* valuations. A valuation function is *capacitated* with capacity c if it is additive over items up to the capacity c. For sets of items with cardinality greater than c, the value is the sum of the c most valued items. In other words, if we let $top(v_i, S)$ denote the c most valued items in S with $top(v_i, S) = S$ if $|S| \leq c$, then

$$v_i(S) = \sum_{k \in top(v_i, S)} v_i(\{k\})$$

We refer to the case where all agents have the same capacity as the *homogeneous capacities* case, and the more general where capacities can differ as the *heterogeneous capacities* case. We assume that agent capacities are publicly known so that our valuations form a connected domain and VCG mechanisms are the only IC mechanisms.

3 General Capacitated Valuations

Cohen et al. [4] provide VCG payment rules which are EF for case of two capacitated agents and any number of items. We devise a mechanism for the complementary case, where there are two items and any number of capacitated agents. We also provide a negative result that shows that it is not possible to move beyond these special cases.

Theorem 2. *There exists an IC, EF, and efficient mechanism for two items and any number of capacitated agents.*

Theorem 3. *For capacitated valuations, where the number of items and the number of agents are both at least 3, there is no mechanism that is IC, EF, and efficient.*

The valuations in the proof of Theorem 3 involve agents with capacities 1 and 2, so it is not possible to further generalize the positive result for two items to any number of items but restricted capacities.

4 Binary Preferences

Up until now we assumed that agents' valuations for individual items are real numbers. In many real-life settings, however, bidders' preference structure is much simpler. In particular, consider a case where every agent has a set of *desired items*, which are items she is interested in getting. For example, a traveler who needs to express her preferred seats in an airplane would usually have in mind a set of desired seats (e.g., aisle seats). Such a preference structure can be represented by *binary valuations*, where an agent's valuation for every item is either 0 or 1. Moreover, in many situations agents simply do not know their valuations for items. In such cases, the binary valuation structure may serve as a good model, since agents, even if they cannot calculate their exact value for various items, can usually tell whether or not they want some item.

These examples motivate the study of IC, EF, and efficient mechanisms under this restricted preference structure. In particular, we ask whether the impossibility result from the previous section can be circumvented by considering the class of binary valuations (still under capacitated agents). This question is answered in the affirmative. Interestingly, in this case ties among efficient allocations cannot be broken arbitrarily. Only by breaking ties in a very certain way (which we will specify soon) can the desired result be achieved.

The last positive result, however, does not extend to IC, Walrasian, and efficient mechanisms, as even in the more restricted setting — that of agents with homogeneous capacities — there are simple examples that admit no IC, Walrasian, and efficient mechanism.

Theorem 4. *For capacitated agents, where $v_i(\{k\}) \in \{0,1\}$ for every i, k, there exists an IC, EF, and efficient mechanism.*

Before proceeding with the proof of Theorem 4, we establish some concepts and propositions that are needed in the proof. It will be useful to have in mind the following simple example.

Example 1. Suppose there are three agents, with agents 1 and 2 having capacity 1 and agent 3 having capacity 2. Agents 1 and 3 desire items b, c while agent 2 desires item a.

Because agent values are either 0 or 1, there may be many efficient allocations, and the particular efficient allocation chosen affects the envy-freeness of the resulting mechanism. We consider a *lexicographically-maximal* efficient allocation, where the sorting is done based on the agents' capacities. First, order the agents in a non-increasing order of capacities, arbitrarily breaking ties among agents with the same capacity. Next, compute an efficient allocation that is lexicographically-maximal (among all efficient allocations), according to the order above. i.e., find an efficient allocation such that there is no other efficient allocation that gives an agent with a lower index (i.e. higher capacity) greater value. We only consider allocations in which no agent receives more items than her capacity. This aids in obtaining EF yet is without loss with respect to efficiency because giving an agent more items than her capacity cannot increase welfare. In example 1, a lexicographically-maximal allocation gives agent 3 priority over agents 1 and 2 (since agent 3 has higher capacity). As a result, any lexicographically-maximal efficient allocation must give b, c to agent 3 and a to agent 2.

We show that a lexicographically-maximal efficient allocation, when combined with the Clarke-pivot rule, is IC and EF. Theorem 3.2 from Cohen et al. [4] shows that Clarke-pivot, when used with any efficient allocation, yields a payment rule where agents with higher capacity do not envy agents with lower capacity. As a result, to prove that our mechanism is EF, it remains to show that under

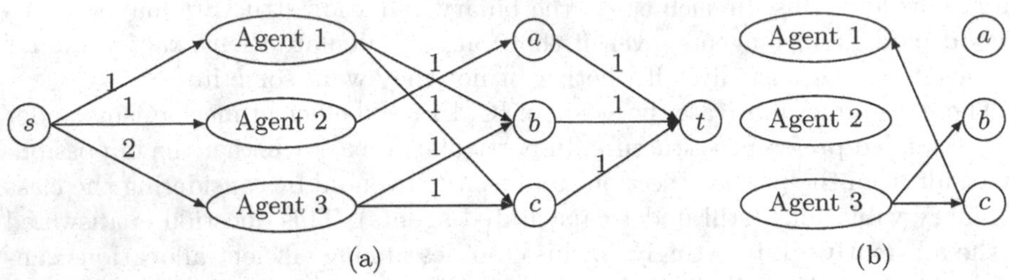

Fig. 2. (a) The graph $G(v)$ for the valuations in Example 1. (b) A graph representing the differences between Opt and D^{-3} for Example 1 (used in the proof of Theorem 4). Edges from agents to items indicate items an agent receives in Opt but not in D^{-3}. Edges from items to agents indicate items an agent receives in D^{-3} but not Opt. Here we assume that Opt allocates a to agent 2 and b, c to agent 3 while D^{-3} allocates a to agent 2 and b to agent 1.

a lexicographically-maximal efficient allocation and Clarke-pivot, agents with lower capacity do not envy agents with higher capacity.

For a given instance of valuations v, it will be useful to consider a directed graph $G(v)$ similar to Cohen et al. [4]. $G(v)$ contains a source, a node for each agent, a node for each item, and a sink. If an agent desires an item, $G(v)$ contains a directed edge from the agent to the item with capacity 1 (note not to confuse edge capacities in the graph representation with agents' capacities). The source is connected to each agent with a directed edge with capacity equal to the agent's capacity. Each item is connected to the sink with capacity 1. Figure 2(a) depicts this graph for Example 1. An allocation then corresponds to a feasible flow in $G(v)$ by connecting each agent to the items it is allocated and appropriately saturating the edges from the source to the agents and the items to the sink. Any integral flow also corresponds naturally to a feasible allocation.

Consider agents i and j, with agent i having strictly lower capacity than agent j. We wish to show that agent i will not envy agent j. A sufficient condition for this is $h_i - h_j \leq v_j(\text{Opt}_j) - v_i(\text{Opt}_j)$. In the remainder of this section, Opt refers to a lexicographically-maximal efficient allocation and Opt^{-i} refers to a lexicographically-maximal efficient allocation that excludes agent i. Consider the following procedure. Start with the lexicographically-maximal efficient allocation Opt. Remove agent i from this allocation by deallocating agent i (make all of the items allocated to agent i available). Call this allocation C^{-i}. C^{-i} necessarily has weakly less welfare than Opt^{-i} as it is a feasible allocation to the agents other than i. Consider $G(v_{-i})$, the directed graph that excludes agent i, and the flow on $G(v_{-i})$ corresponding to C^{-i}. We can find an allocation D^{-i} with $v(D^{-i}) = v(\text{Opt}^{-i})$ by adding augmenting paths to the flow on $G(v_{-i})$ corresponding to C^{-i}. Since all edge capacities are integer, it is without loss of generality to consider augmenting paths with net flow of 1. It is also without loss of generality to assume that each augmenting path only visits the sink once since any path that visits the sink multiple times contains a smaller augmenting path which visits the sink only once. The following propositions establish properties of these augmenting paths.

Proposition 1. *After each augmenting path, the total set of allocated items increases by exactly one item.*

Proposition 2. *The second to last node (i.e., the node prior to the sink) in each augmenting path is one of the items agent i was originally allocated in Opt.*

Proposition 3. *After adding an augmenting path, every agent other than i receives at least as many items as it did in Opt. Additionally, agent j will receive the same number of items as it did in Opt.*

We are now ready to prove Theorem 4.

Proof. Let Opt be a lexicographically-maximal efficient allocation, and let D^{-i} be the allocation formed by removing agent i and then adding augmenting paths to $G(v_{-i})$. Consider the following bipartite graph G_f and corresponding flow f that relates Opt and D^{-i}. The left hand side has nodes representing agents, and the right hand side has nodes representing items. There is an edge from an agent node to an item node if the agent receives the item in Opt but not in D^{-i}. There is an edge from an item node to an agent if the agent receives the item in D^{-i} but not in Opt. Let there be a flow of 1 on each edge in this graph. Figure 2(b) illustrates G_f and f for Example 1.

Proposition 3 establishes that the only source (node with greater outflow than inflow) is agent i, and that agent j has equal indegree and outdegree since it receives the same number of items in Opt and D^{-i}. Using flow decomposition, we can decompose f into paths and cycles. Each of the paths starts at agent i, with one path for each item agent i was allocated in Opt. By executing a path or cycle, we mean that for every agent to item edge we modify the current allocation by giving the item to the agent, and for every item to agent edge, we remove the item from the agent.

We now construct allocation E^{-j}, which will not allocate any items to agent j, starting from allocation D^{-i}. The items j receives in D^{-i} can be split into two sets. The first set consists of items it also received in Opt, and the second set consists of items it did not receive in Opt. Items in the second set will show up as an item to agent edge in G_f. The sum of the number of items in these two sets will be $v_j(\text{Opt}_j)$ (Proposition 3). For every item given to agent j in both Opt and D^{-i}, give the item to agent i. The remaining items that agent j receives in D^{-i} are part of either a cycle or a path in the flow decomposition of f. For every cycle that contains agent j, execute the cycle, and give the item agent j receives to agent i. This results in agent i receiving some item in Opt_j. For every path that contains agent j, execute the path, stopping at agent j. This results in agent i receiving an item that it desires.

After this process, every agent other than i, j receives the same exact number of items as in D^{-i}. Agent i receives $v_j(\text{Opt}_j)$ items, some of which are in Opt_j

and possibly undesired by agent i (the items j received in both Opt and D^{-i} and the items that were a part of cycles including agent j) and others which are desired by agent i (the items that were part of the paths starting with agent i and ending in agent j). Therefore, agent i receives a bundle that is Opt_j, with some items replaced by items the agent surely desires. As a result, $v_i(E^{-j}) \geq v_i(\text{Opt}_j)$. To complete the proof, we note that $v(E^{-j})$ is a lower bound on $v(\text{Opt}^{-j}) = h_j$ and verify the EF condition for agent i.

Example 1 demonstrates that the tie-breaking rule among efficient allocations is crucial, as some choices of efficient allocations do not yield EF Clarke pivot payments. The restriction to values in $\{0, 1\}$ is tight in sense that if agents have values in $\{r, s\}$ with $r, s > 0$, then VCG with Clarke pivot and lexicographically maximal allocations may no longer be EF. Our final result examines whether this positive result can be extended beyond EF to the stronger notion of Walrasian mechanisms. Notably, for the class of *unit-demand* valuations (homogeneously capacitated agents with capacity 1), VCG with Clarke pivot payments is Walrasian (even for real valuations) [9, 12]. We find that these results cannot be extended, even if we consider homogeneous capacities and binary valuations.

Theorem 5. *There exists no IC, Walrasian, and efficient mechanism for the class of homogeneously capacitated, binary valuations.*

5 Discussion and Open Problems

This work settles the main open question posed by Cohen et al. [4] regarding the existence of an IC, EF and efficient mechanism for valuation classes beyond homogeneous capacities. While there always exists an efficient IC mechanism, and similarly an efficient EF mechanism, there exists no mechanism that simultaneously satisfies both requirements when agents' capacities are heterogeneous. This result eliminates the hope for the existence of IC and EF mechanisms in the more general classes of submodular or subadditive valuations. The impossibility result is accompanied by two positive results, showing that existence of an IC and EF mechanism can be restored if either agents' valuations for individual items are binary or if there are only two items. The former result, however, does not extend to the stronger notion of a Walrasian mechanism, even if valuations are capacitated and binary. The natural future direction, given the impossibility result, is to resort to near-optimal outcomes. What is the best approximation to social welfare that can be achieved by a mechanism that is simultaneously EF and IC, for different valuation classes?

Acknowledgements. Michal Feldman is partially supported by the Israel Science Foundation (grant number 1219/09), by the Leon Recanati Fund of the Jerusalem School of Business Administration, the Google Inter-university center for Electronic Markets and Auctions, and the People Programme (Marie Curie Actions) of the European Union Seventh Framework Programme (FP7/2007-2013) under REA grant agreement number 274919. John Lai is supported by an NDSEG fellowship.

References

[1] Ausubel, L.M., Milgrom, P.: The lovely but lonely vickrey auction. In: Combinatorial Auctions, ch. 1. MIT Press (2006)

[2] Bikhchandani, S., Ostroy, J.: The package assignment model. Journal of Economic Theory 107(2), 377–406 (2002)

[3] Clarke, E.H.: Multipart pricing of public goods. Public Choice, 17–33 (1971)

[4] Cohen, E., Feldman, M., Fiat, A., Kaplan, H., Olonetsky, S.: Truth, Envy, and Truthful Market Clearing Bundle Pricing. In: Chen, N., Elkind, E., Koutsoupias, E. (eds.) WINE 2011. LNCS, vol. 7090, pp. 97–108. Springer, Heidelberg (2011)

[5] Dubins, L.E., Spanier, E.H.: How to cut a cake fairly. The American Mathetmatical Monthly 68(1), 1–17 (1961)

[6] Foley, D.K.: Resource allocation and the public sector. Yale Economic Studies (1967)

[7] Green, J., Laffont, J.: Characterization of satisfactory mechanisms for the revelation of preferences for public goods. Econometrica 45(2), 427–438 (1973)

[8] Groves, T.: Incentives in teams. Econometrica 41(4), 617–631 (1973)

[9] Gul, F., Stacchetti, E.: Walrasian equilibrium with gross substitutes. Journal of Economic Theory 87, 95–124 (1999)

[10] Holmstrom, B.: Groves schemes on restricted domains. Econometrica 47(5), 1137–1144 (1979)

[11] Hurwicz, L.: Optimality and informational efficiency in resource allocation processes. In: Arrow, K.J., Karlin, S., Suppes, P. (eds.) Mathematical Methods in the Social Sciences. Stanford University Press (1960)

[12] Leonard, H.B.: Elicitation of honest preferences for the assignment of individuals to positions. The Journal of Political Economy 91(3), 461–479 (1983)

[13] Maskin, E.S.: On the fair allocation of indivisible goods (1987)

[14] Moulin, H.: Fair Division and Collective Welfare. MIT Press (2004)

[15] Mu'alem, A.: On Multi-dimensional Envy-Free Mechanisms. In: Rossi, F., Tsoukias, A. (eds.) ADT 2009. LNCS, vol. 5783, pp. 120–131. Springer, Heidelberg (2009)

[16] Pápai, S.: Groves sealed bid auctions of heterogeneous objects with fair prices. Social Choice and Welfare 20(3), 371–385 (2003)

[17] Rafaeli, A., Kedmi, E., Vashdi, D., Barron, G.: Queues and fairness: A multiple study experimental investigation. Technical report, Technion-Israel Institute of Technology (2003)

[18] Svensson, L.: On the existence of fair allocations. Journal of Economics 43(3), 301–308 (1983)

Capacitated Network Design Games*

Michal Feldman[1] and Tom Ron[2]

[1] Hebrew University of Jerusalem and Harvard University
mfeldman@seas.harvard.edu
[2] Hebrew University of Jerusalem
rontom@gmail.com

Abstract. We study a *capacitated* symmetric network design game, where each of n agents wishes to construct a path from a network's source to its sink, and the cost of each edge is shared equally among its agents. The uncapacitated version of this problem has been introduced by Anshelevich *et al.* (2003) and has been extensively studied. We find that the consideration of edge capacities entails a significant effect on the quality of the obtained Nash equilibria (NE), under both the utilitarian and the egalitarian objective functions, as well as on the convergence rate to an equilibrium. The following results are established. First, we provide bounds for the price of anarchy (PoA) and the price of stability (PoS) measures with respect to the utilitarian (i.e., sum of costs) and egalitarian (i.e., maximum cost) objective functions. Our main result here is that, unlike the uncapacitated version, the network topology is a crucial factor in the quality of NE. Specifically, a network topology has a bounded PoA if and only if it is *series-parallel* (SP). Second, we show that the convergence rate of best-response dynamics (BRD) may be super linear (in the number of agents). This is in contrast to the uncapacitated version, where convergence is guaranteed within at most n iterations.

1 Introduction

The construction of large networks by strategic agents has been widely studied from a game-theoretic perspective in the last decade [3, 8, 9, 25]. For a motivating example, consider the construction and maintenance of large computer networks by independent economic agents with different, and often competing, self-interests. The game-theoretic perspective offers tools and insights that are fundamental to the understanding and analysis of these settings.

In a symmetric network design game, a network is given, where each edge is associated with some cost; and a set of n agents wish to buy some path from

* This work was partially supported by the Israel Science Foundation (grant number 1219/09), by the Leon Recanati Fund of the Jerusalem School of Business Administration, the Google Inter-university center for Electronic Markets and Auctions, and the People Programme (Marie Curie Actions) of the European Unions Seventh Framework Programme (FP7/2007-2013) under REA grant agreement number 274919. The authors wish to thank Eli Ben-Sasson and Irit Dinur for helpful discussion.

M. Serna (Ed.): SAGT 2012, LNCS 7615, pp. 132–143, 2012.

the network's source (s) to its sink (t). Every agent chooses an s-t path, and the cost of every edge is divided equally among the agents who use it. This is often called a *fair cost-sharing* method. The game theoretic twist is the assumption that each agent chooses its path strategically, so as to minimize its cost. It is well known that Nash equilibria of this game need not be efficient, where efficiency is usually defined with respect to either the sum of the agents' costs (referred to as the *utilitarian* or *sum-cost* objective) or to the maximum cost of any agent (referred to as the *egalitarian* or *max-cost* objective).

The efficiency loss is commonly quantified using the price of anarchy (PoA) [17, 23] and price of stability (PoS) [3] measures; the former refers to the ratio between the cost of the worst Nash equilibrium and the social optimum, whereas the latter refers to the ratio between the cost of the best Nash equilibrium and the social optimum. The network design game described above is fairly easy to analyze. The PoA is known to be tightly bounded by n with respect to the utilitarian objective function[1] [3]. It is not too difficult to see that the same bound holds with respect to the egalitarian objective. In addition, the PoA is independent of the network topology, as the worst case is obtained for two parallel links. The PoS, in contrast, is always equal to 1 (with respect to both objective functions), since in a symmetric network, the profile in which all agents share the shortest path from s to t is a Nash equilibrium. Finally, best-response dynamics (i.e., dynamics in which agents sequentially apply their best-response moves) exhibits a simple structure, where convergence to a NE is guaranteed within at most n steps.

Interestingly, as we shall soon see, a lot of the aforementioned results should be attributed to the assumption that the network edges are *uncapacitated*; i.e., it is assumed that edges may hold any number of agents. While this assumption has been employed by most of the studies on strategic network formation games, we claim that in real-life applications network links have a limit on the number of agents they can serve. To reflect this observation, we introduce *capacitated* network design games, in which every edge, in addition to its cost, is also associated with a *capacity* that specifies the number of agents it can hold. We study the quality of NE in these games (using both PoA and PoS measures) and the convergence rate of best-response dynamics. We are particularly interested in the effect of the *topology* of the underlying network on the obtained results.

In cases where edges are associated with capacities, a *feasibility* problem arises (i.e., whether there exists a solution that accommodates all the agents). However, as already hinted at by [3], if a feasible solution exists, the arguments used in the uncapacitated version can be applied to show that a pure NE exists and, moreover, every best-response dynamics converges to a pure NE. This observation motivates our study.

Our Contribution. For the PoA, the lower bound of n trivially carries over to the capacitated version; thus, one cannot expect for a bound better than n. The upper bound of n, however, does not carry over. In particular, we demonstrate

[1] While [3] consider an underlying directed graph, this bound carries over to the undirected case.

that the PoA can be arbitrarily high. As it turns out, however, the network topology plays a major role in the obtained PoA. A symmetric network topology G is said to be *PoA bounded* if for every symmetric network design game that is played on G, the PoA is bounded by n, independent of the edge costs and capacities. Our main result here is a full characterization of PoA-bounded network topologies. Specifically, we show that a symmetric network topology is PoA bounded if and only if it is a *series-parallel* (SP) network; i.e., a network that is built inductively by series and parallel compositions of SP networks. This result holds with respect to both the sum-cost and max-cost objectives. Moreover, for parallel-link networks, we show that the PoA (with respect to both the sum-cost and max-cost objectives) is essentially bounded by the maximum edge capacity in the network, and this is tight.

This separation between the graph topology and the assignment of edge costs and capacities reflects a separation between the underlying infrastructure and the edge characteristics. While the infrastructure is often stable over time, the edge characteristics may be modified over short time periods. A PoA bounded topology ensures that, no matter how edge characteristics evolve, the cost of a NE will never exceed n. Such topologies should be desired by network designers, who wish to guarantee the efficiency in their network despite the fact they do not control the actions of the individual users. Notably, within the class of SP networks, the worst case is obtained already for parallel links.

In contrast to the PoA, the PoS with respect to the sum-cost objective is not affected by the network topology. In particular, we provide a lower bound of $H(n)$ (i.e., the harmonic nth number) for the PoS on parallel-link graphs, and show that for every symmetric network the PoS is upper bounded by $H(n)$.

As for the max-cost objective function, for SP graphs the upper bound of n that is established for the PoA trivially carries over to the PoS, and a matching lower bound is established. For general graphs, we establish an upper bound of $n \log n$. Closing the gap between n and $\log n$ for the PoS in general graphs remains an open problem.

Most of our results for the PoA and PoS bounds are summarized in Table 1, where they are also contrasted with the corresponding results in the uncapacitated version (specified in brackets). These results suggest that the departure from the classic assumption of uncapacitated edges brings in significant differences in the quality of equilibria.

Additionally, we study the convergence rate of best-response dynamics (BRD) to a NE. Here too, the consideration of capacities introduces additional complexity that reveals itself through a slower conversion rate. While BRD in the uncapacitated version is guaranteed to converge within at most n iterations, we establish a lower bound of $\Omega(n^{3/2})$ for convergence in capacitated games. Moreover, this lower bound is obtained already in the simplest graphs; i.e., graphs that are composed of parallel links.

Finally, we note that while the feasibility problem in capacitated games is equivalent to a maximum flow computation, and thus can be solved in polynomial

Table 1. Summary of our results. The values in brackets correspond to the bounds for uncapacitated games. All the results, except for the PoS w.r.t max-cost for general networks are tight.

		Parallel links	SP	General
sum-cost (sc)	**PoA**	n (n)	n (n)	unbounded (n)
	PoS	$\log n(1)$	$\log n(1)$	$\log n(1)$
max-cost (mc)	**PoA**	n (n)	n (n)	unbounded (n)
	PoS	n (1)	n (1)	$n \log n(1)$

time, the optimization version of the problem is NP-complete (this can be easily verified through a reduction from 0-1 knapsack [14]).

Related Work. Various models of network design and formation games have been extensively studied in the last decade from a game-theoretic perspective [3–7, 18], with a great emphasis on the PoA and PoS measures. The PoA in network design games has been also studied with respect to the *strong equilibrium* solution concept by Epstein *et al.* [8], Andelman *et al.* [2] and Albers [1].

The role that network topology plays in game-theoretic settings has been studied in various models. In the model of network routing, it has been shown by Roughgarden and Tardos [25] that the PoA is independent of the network topology. In contrast, the network topology seems to matter a lot in other settings. Some prominent examples include the following. Milchtaich [21] showed that the *Pareto efficiency* of equilibria in network routing games (with a continuum of agents) strongly depends on the network topology. In addition, topological characterizations for symmetric network games have been also provided for other equilibrium properties, including (Nash and strong) equilibrium existence (see Milchtaich [20], Epstein *et al.* [8, 9], and Holzman and Law-Yone [15, 16]), and equilibrium uniqueness (see Milchtaich [19]).

Best-response dynamics (BRD) and its convergence rate has been the subject of intensive research recently. Since every congestion game is a potential game [22, 24], BRD always converge to a pure NE. However, they may in general take exponential number of steps depending on the number of agents, as established by Fabrikant *et al.* [11]. Anshelevich *et al.* [3] established that BRD may take exponential number of steps to converge in network design games, but is polynomial for the special case of two agents. Notably, as shall be discussed in Sect. 5, the exponential convergence rate does not apply in our setting. BRD convergence has been also studied in scheduling and routing games (see Even-Dar *et al.* [10], Fotakis [13], and Feldman and Tamir [12]).

2 Model and Preliminaries

2.1 Capacitated Symmetric Cost Sharing Games

A capacitated, symmetric cost-sharing connection (CCS) game (also known as single commodity) is a tuple

$$\Delta = \langle n, G = (V, E), s, t, \{p_e\}_{e \in E}, \{c_e\}_{e \in E} \rangle,$$

where n is the number of agents and $G = (V, E)$ is an undirected graph, with $s, t \in V$ as its *source* and *sink* nodes, respectively. Every edge $e \in E$ is associated with a cost $p_e \in R^{\geq 0}$ and a capacity $c_e \in N$, where an edge capacity specifies the maximum number of agents that can use it. The set of agents $\{1, \ldots, n\}$ is also denoted by $[n]$. Every agent i wishes to construct an s-t path in G. The strategy space of an agent i, denoted Σ_i, is the set of $s - t$ paths in G, and a strategy of an agent i is denoted by $S_i \in \Sigma_i$. Since this is a symmetric game, all agents have the same strategy space. The joint action space is denoted by Σ.

We consider the *fair* cost-sharing game, where an edge's cost is shared equally by all the agents that use it in their path. Given a strategy profile $S = (S_1, \ldots, S_n)$, we denote by $x_e(S)$ the number of agents that use edge e in their path; i.e., $x_e(S) = |\{i : e \in S_i\}|$. A profile S is said to be *feasible* if for every $e \in E$, $x_e(S) \leq c_e$. The cost of agent i in a profile S is defined as

$$p_i(S) = \begin{cases} \sum_{e \in S_i} \frac{p_e}{x_e(S)} & \text{, if } S \text{ is feasible} \\ \infty & \text{, otherwise} \end{cases} \tag{1}$$

A profile S is said to be a *Nash equilibrium* if no agent can improve its cost by a unilateral deviation; i.e., for every $i, S'_i \in \Sigma_i, S_{-i} \in \Sigma_{-i}$, it holds that $p_i(S) \leq p_i(S'_i, S_{-i})$, where S_{-i} denotes the joint action of all agents except i.

Given a game Δ, let $\tau(\Delta)$ denote the set of all feasible profiles in Δ. A CCS game Δ is said to be feasible if it admits a feasible profile; i.e., $\tau(\Delta) \neq \emptyset$.

We consider two social cost functions. The *sum-cost* of a profile S is the total cost of the agents in S (and also equals the total cost of the purchased edges in S), and is given by

$$sc_\Delta(S) = \begin{cases} \sum_i p_i(S) & \text{, if } S \text{ is feasible} \\ \infty & \text{, otherwise} \end{cases}$$

The *max-cost* of a profile S is the maximum cost of any agent in S, and is given by

$$mc_\Delta(S) = \begin{cases} \max_{i \in [n]} p_i(S) & \text{, if } S \text{ is feasible} \\ \infty & \text{, otherwise} \end{cases}$$

We denote by $OPT_{sc}(\Delta)$ and $OPT_{mc}(\Delta)$ the optimal profiles with respect to the sum-cost and max-cost objectives, respectively. When clear in the context, we omit Δ, and also abuse notation and use $OPT_{sc}(\Delta)$ and $OPT_{mc}(\Delta)$ to denote the cost of the respective optimal solutions.

In the figures of the paper, every edge is associated with a tuple (c_e, p_e), denoting its capacity and cost, respectively.

2.2 Nash Equilibrium Existence

An uncapacitated fair cost sharing game is known to be a *potential game* [3]. Every potential game admits a pure NE [22]. Moreover, BRD (where agents sequentially apply their best-response moves) always converge to a pure NE. Capacitated versions are not guaranteed to admit a feasible solution; however, if a feasible solution exists, then so does a pure NE.

Observation 1. *[3] Let Δ be a CCS game s.t. $\tau(\Delta) \neq \emptyset$. Then, Δ admits a pure NE and every best response dynamics convergence to a NE.*

This proof relies on the existence of a potential function, $\Phi(S) = \sum_{e \in E} \sum_{i=1}^{x_e(S)} \frac{p_e}{x_e(S)}$, that emulates the cost of an agent when deviating from a feasible solution to another.

2.3 Efficiency Loss

To quantify the efficiency loss due to strategic behavior, we use the PoA and PoS measures. The PoA is the ratio of the worst Nash equilibrium and the social optimum, and is given by $PoA_{sc}(\Delta) = \frac{\max_{S \in NE(\Delta)} sc_{\Delta}(S)}{OPT_{sc}(\Delta)}$ and $PoA_{mc}(\Delta) = \frac{\max_{S \in NE(\Delta)} mc_{\Delta}(S)}{OPT_{mc}(\Delta)}$ with respect to the sum-cost and max-cost objectives, respectively, where $NE(\Delta)$ denotes the set of NE of Δ, and it is assumed that $NE(\Delta) \neq \emptyset$. Similarly, the PoS of sum-cost and max-cost are given by $PoS_{sc}(\Delta) = \frac{\min_{S \in NE(\Delta)} sc_{\Delta}(S)}{OPT_{sc}(\Delta)}$ and $PoS_{mc}(\Delta) = \frac{\min_{S \in NE(\Delta)} mc_{\Delta}(S)}{OPT_{mc}(\Delta)}$, respectively.

2.4 Graph Theoretic Preliminaries

In this section we provide some preliminaries regarding network topologies. A *symmetric* network is an undirected graph G along with two distinguished nodes, a source s and a sink t. When clear in the context, we refer to G as the symmetric network. A CCS game is symmetric (also called single-commodity) if its underlying network is symmetric with source s and sink t, and nodes s and t are the respective source and sink of all the agents.A symmetric network G is *embedded* in a symmetric network G' if G' is isomorphic to G or to a network derived from G by applying the following operations any number of times in any order: (i) *Subdivision* of an edge (i.e., its replacement by a path of edges), (ii) *Addition* of a new edge joining two existing nodes, (iii) *Extension* of the source or the sink (i.e., addition of a new edge joining s or t with a new node, which becomes the new source or sink, respectively).

Next, we define the following operations on symmetric networks:

Identification: The *identification* operation is the collapse of two nodes into one. More formally, given a graph $G = (V, E)$ we define the *identification* of nodes $v_1 \in V$ and $v_2 \in V$ forming a new edge $v \in V$ as creating a new graph $G' = (V', E')$ where $V' = V \setminus \{v_1, v_2\} \cup \{v\}$ and E' includes the edges of E where the edges of v_1 and v_2 are now connected to v.

Parallel Composition: Given two symmetric networks, $G_1 = (V_1, E_1)$ and $G_2 = (V_2, E_2)$, with sources $s_1 \in V_1$ and $s_2 \in V_2$ and sinks $t_1 \in V_1$ and $t_2 \in V_2$, respectively, we define a new symmetric network $G = G_1||G_2$ as follows. Let $G' = (V_1 \cup V_2, E_1 \cup E_2)$ be the union of network. To generate $G = G_1||G_2$ we identify the sources s_1 and s_2, forming a new source node s, and identify the the sinks t_1 and t_2, forming a new sink t.

Series Composition: Given two symmetric networks, $G_1 = (V_1, E_1)$ and $G_2 = (V_2, E_2)$, with sources $s_1 \in V_1$ and $s_2 \in V_2$ and sinks $t_1 \in V_1$ and $t_2 \in V_2$, respectively, we define a new symmetric network $G = G_1 \to G_2$ as follows. Let $G' = (V_1 \cup V_2, E_1 \cup E_2)$ be the union network. To generate $G = G_1 \to G_2$ from G' we identify the vertices t_1 and s_2, forming a new vertex u. The network G has a source $s = s_1$ and a sink $t = t_2$.

A Series-parallel (SP) network is a symmetric network that is constructed inductively from two SP networks by either a series composition or a parallel composition, where a single edge serves as the base of the induction. That is, a symmetric network consisting of a single edge is an SP network. In addition, given two SP networks, G_1 and G_2, the networks $G = G_1||G_2$ and $G = G_1 \to G_2$ are SP networks.

3 The Sum-Cost Objective Function

3.1 Price of Anarchy (PoA)

Throughout this section, we write PoA to denote PoA_{sc} for simplicity. In uncapacitated cost sharing games, the PoA is n (tightly). This is, however, not the case in capacitated games, as demonstrated by the following proposition.

Proposition 1. *The price of anarchy with respect to the sum-cost function in CCS games can be arbitrarily high.*

Proof. Consider a CCS game with two agents and an underlying graph as depicted in Fig. 1(a), and suppose that y is arbitrarily larger than x. The optimal profile is where one agent uses the path *s-a-t* and the other uses the path *s-b-t*, resulting in a total cost of $4x$. However, there is a NE in which one agent uses the path *s-a-b-t* and the other uses the path *s-b-a-t*, resulting in a total cost of $4x + y$. Therefore, $PoA_{sc}(\Delta) = \frac{4x+y}{4x}$, which can be arbitrarily high.

Our goal is to characterize network topologies in which such a "bad" example cannot occur; i.e., topologies in which the PoA is always bounded, independent of the specific edge costs and capacities. The lower bound of n for a network with two parallel links motivates the following definition.

Definition 1. *A symmetric network $G = (V, E)$ with source s and sink t is PoA bounded for a family of symmetric CCS games $\mathcal{F}$ if for every symmetric CCS game $\Delta \in \mathcal{F}$ on the symmetric network G, it holds that $PoA(\Delta) \leq n$.*

Our main result is a full characterization of PoA bounded network topologies.

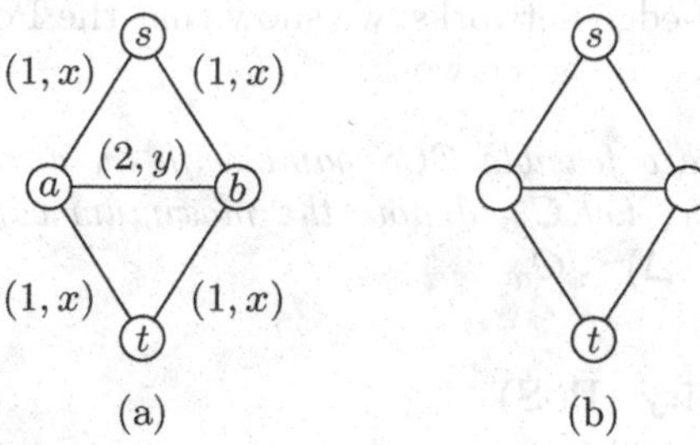

Fig. 1. (a) An example where the PoA can be arbitrarily high. (b) A *Braess* Graph.

Theorem 1. *For symmetric CCS games, a symmetric network topology G is PoA bounded w.r.t. sum-cost if and only if G is a series-parallel (SP) network.*

The proof of our characterization is composed of two parts. First, we show that for every symmetric CCS game that is played on an SP network $PoA_{sc} \leq n$. This is the content of Theorem 2. Second, we show that for every symmetric network topology G that is not an SP network, there exists a game that is played on G for which the PoA can be arbitrarily high. This part is the content of Theorem 2.

Theorem 2. *Let Δ be a feasible CCS game with an underlying graph G. If G is an SP graph then $PoA_{sc}(\Delta) \leq n$.*

In order to complete the characterization it remains to show that for every non-SP network G, there exists a symmetric CCS game on G that has an unbounded price of anarchy.

Theorem 3. *Let G be a non-SP symmetric network. Then, there exists a symmetric CCS game on G for which the price of anarchy is arbitrarily high.*

In order to prove the last theorem, we use the following result, established by Milchtaich [21].

Lemma 1. *[21] A symmetric network G is an SP network if and only if the symmetric network in Fig. 1(b) is not embedded in G.*

The network topology in the last lemma is precisely the network topology with the unbounded PoA that motivated our study. The last lemma asserts that this graph topology is embedded in every non-SP network. Thus, in order to establish the assertion of Theorem 1, it remains to show that the unbounded PoA given in Proposition 1 can be *extended* to every network topology that embeds it. This is established in the following lemma.

Lemma 2. *Let G be a symmetric network that is not PoA bounded with respect to sum-cost for a family of symmetric CCS games $\mathcal{F}$, and suppose G is embedded in a symmetric network G'. Then, G' is not PoA bounded with respect to sum-cost for the family $\mathcal{F}$ either.*

For the case of parallel-edge networks, we show that the PoA cannot exceed the maximum edge capacity in the network.

Theorem 4. *Let Δ be a feasible CCS game with an underlying graph G that consists of parallel edge. Let C_m denote the maximum capacity of any edge in G. It holds that $PoA_{sc}(\Delta) \leq C_m$.*

3.2 Price of Stability (PoS)

As mentioned above, for uncapacitated symmetric games, $PoS = 1$. In capacitated game, however, the PoS need not be optimal. Moreover, suboptimality is obtained already in parallel-link networks.

Theorem 5. *There exists a symmetric CCS game in which the PoS with respect to sum-cost is $H(n)$.*

Proof. Consider a CCS game with n agents played on a graph that consists of $n+1$ parallel links, $e_1, \ldots, e_{n+1}$, such that for $i \in [n]$, $p_i = 1/i$ and $c_i = 1$; and $p_{n+1} = 1 + \epsilon$ and $c_{n+1} = n$. It is easy to verify that the optimal solution is achieved when all the agents share edge e_{n+1}. However, this profile is not a NE since a single agent can benefit by deviating to edge e_n, incurring a cost of $1/n$ instead of $(1+\epsilon)/n$. Following similar reasonings, agents will continue to deviate, one by one, until reaching the profile in which for every agent $i \in [n]$, agent i uses edge e_i. The cost of this profile is $H(n)$; the assertion follows.

As established in [3], the potential function method can be used to show that the last bound is tight. The proof uses the potential function $\Phi(S) = \sum_{e \in E} \sum_{i=1}^{x_e(S)} \frac{p_e}{x_e(S)}$, and follows the same reasoning as in the uncapacitated case.

Theorem 6. *[3] For every feasible symmetric CCS game, it holds that $PoS_{sc} \leq H(n)$.*

4 The Max-cost Objective Function

In this section we study the max-cost objective function.

4.1 Price of Anarchy (PoA)

We first observe that the PoA can be arbitrarily high also with respect to the max-cost function.

Proposition 2. *The PoA with respect to max-cost in CCS games can be arbitrarily high.*

As in the sum-cost case, we wish to characterize network topologies in which the PoA cannot exceed n. Interestingly, we obtain the exact same characterization as in the sum-cost case.

Theorem 7. *A symmetric network topology G is PoA bounded w.r.t. max-cost if and only if G is an SP network.*

For the case of parallel-edge networks, we show that the PoA cannot exceed the maximum edge capacity in the network.

Theorem 8. *Let Δ be a feasible CCS game with an underlying graph G that consists of parallel edge. Let C_m denote the maximum cost of any edge in G. It holds that $PoA_{mc}(\Delta) \leq C_m$.*

4.2 Price of Stability (PoS)

For SP graphs, it follows directly from Theorem 7 that the PoS is bounded by n (since PoS is always bounded by PoA). This bound is tight, as follows from the example given in the proof of Theorem 5 . In this example, the unique NE is one in which every agent uses a distinct path, and the maximal cost incurred by any agent is 1, compared to $1/n$ in the optimal solution. For general networks, we establish the following bound.

Theorem 9. *For every CCS game Δ, it holds that $PoS_{mc}(\Delta)$ is bounded by $nH(n)$.*

Proof. Consider the function $\Phi(S) = \sum_{e \in E} \sum_{i=1}^{x_e(S)} \frac{p_e}{x_e(S)}$. It is shown by [3] that this is an exact potential function for the game; i.e., it emulates the change in the cost of a deviating agent. It is easy to verify that for every profile T,

$$sc(T) \leq \Phi(T) \leq H(n) \cdot sc(T). \tag{2}$$

Let S^* be an optimal solution with respect to max-cost, and consider a NE S that is obtained by running best-response dynamics with an initial profile S^*. We get that $mc(S) \leq sc(S) \leq \Phi(S) \leq \Phi(S^*) \leq H(n)sc(S^*) \leq nH(n)mc(S^*)$, where the second and fourth inequalities follow from Equation 2, the third inequality follows from the fact that Φ is a potential function and S is obtained from S^* through best-response steps, and the last inequality follows from the definition of max-cost. It follows that $mc(S)/mc(S^*) \leq nH(n)$, as promised.

5 Convergence Rate of BRD

In this section we study the convergence rate of best-response dynamics (BRD) to a NE. While BRD may in general take exponential number of steps depending on the number of agents to converge [3], the following proposition establishes that in the case of a symmetric, undirected graph, BRD converges to a pure NE within at most n steps, and this is tight. The intuition for this observation is that, in the uncapacitated version, after an agent deviates to some path P (as its best-response), the cost incurred by an agent using this path in the next iteration can only decrease; therefore, P remains a best-response move until all agents converge to the same path.

Observation 2. *For every uncapacitated cost-sharing game, every BRD converges to a NE within at most n steps, independent of the initial profile.*

In contrast, the following proposition shows that the convergence process of a capacitated game may be longer. In particular, we establish a lower bound of $\Omega(n^{3/2})$, even for parallel-link graphs.

Proposition 3. *There exists a symmetric CCS game and a best-response dynamics with convergence time of* $\Omega(n^{3/2})$.

6 Discussion

In this work we introduce a model of capacitated network design games, and study the implications of edge capacities on the existence and quality of Nash equilibria with respect to different objective functions, as well as on the convergence rate of best-response dynamics. We find that the consideration of edge capacities has a significant effect on all the above properties. Our main contribution is a full characterization of network topologies that have a bounded price of anarchy, independent of the edge capacities and costs. Our results suggest many avenues for future research. A few obvious directions include closing the gap of the PoS with respect to the max-cost objective for general networks, the consideration of non-symmetric networks and a better understanding of the convergence rate of best-response dynamics.

References

1. Albers, S.: On the value of coordination in network design. In: Proceedings of the Nineteenth Annual ACM-SIAM Symposium on Discrete Algorithms, SODA 2008, pp. 294–303. Society for Industrial and Applied Mathematics, Philadelphia (2008)
2. Andelman, N., Feldman, M., Mansour, Y.: Strong Price of Anarchy. In: SODA 2007 (2007)
3. Anshelevich, E., Dasgupta, A., Kleinberg, J., Tardos, E., Wexler, T., Roughgarden, T.: The price of stability for network design with fair cost allocation. In: Proceedings of the 45th Annual IEEE Symposium on Foundations of Computer Science, pp. 295–304. IEEE Computer Society, Washington, DC (2004)
4. Anshelevich, E., Dasgupta, A., Tardos, É., Wexler, T.: Near-optimal network design with selfish agents. In: STOC, pp. 511–520 (2003)
5. Bala, V., Goyal, S.: A noncooperative model of network formation. Econometrica 68(5), 1181–1230 (2000)
6. Corbo, J., Parkes, D.: The price of selfish behavior in bilateral network formation. In: Proceedings of the Twenty-Fourth Annual ACM Symposium on Principles of Distributed Computing, PODC 2005, pp. 99–107. ACM, New York (2005)
7. Devanur, N.R., Mihail, M., Vazirani, V.V.: Strategyproof cost-sharing mechanisms for set cover and facility location games. In: Proc. of ACM EC, pp. 108–114 (2003)
8. Epstein, A., Feldman, M., Mansour, Y.: Strong equilibrium in cost sharing connection games. In: Proceedings of the 8th ACM Conference on Electronic Commerce, EC 2007, pp. 84–92. ACM, New York (2007)

9. Epstein, A., Feldman, M., Mansour, Y.: Efficient graph topologies in network routing games. Games and Economic Behavior 66(1), 115–125 (2009)
10. Even-Dar, E., Kesselman, A., Mansour, Y.: Convergence Time to Nash Equilibria. In: Baeten, J.C.M., Lenstra, J.K., Parrow, J., Woeginger, G.J. (eds.) ICALP 2003. LNCS, vol. 2719, pp. 502–513. Springer, Heidelberg (2003)
11. Fabrikant, A., Papadimitriou, C., Talwar, K.: The complexity of pure nash equilibria. In: Proceedings of the Thirty-Sixth Annual ACM Symposium on Theory of Computing, STOC 2004, pp. 604–612. ACM, New York (2004)
12. Feldman, M., Tamir, T.: Convergence rate of best response dynamics in scheduling games with conflicting congestion effects. Working paper (2011)
13. Fotakis, D.: Congestion games with linearly independent paths: Convergence time and price of anarchy. Theory Comput. Syst. 47(1), 113–136 (2010)
14. Garey, M.R., Johnson, D.S.: Computers and Intractability; A Guide to the Theory of NP-Completeness. W. H. Freeman & Co., New York (1990)
15. Holzman, R., Law-Yone, N.: Strong equilibrium in congestion games. Games and Economic Behavior 21(1-2), 85–101 (1997)
16. Holzman, R., Law-Yone (Lev-tov), N.: Network structure and strong equilibrium in route selection games. Mathematical Social Sciences 46(2), 193–205 (2003)
17. Koutsoupias, E., Papadimitriou, C.: Worst-Case Equilibria. In: Meinel, C., Tison, S. (eds.) STACS 1999. LNCS, vol. 1563, pp. 404–413. Springer, Heidelberg (1999)
18. Chen, H.L., Roughgarden, T.: Network design with weighted players. In: Proceedings of the 18th ACM Symposium on Parallelism in Algorithms and Architextures (SPAA), pp. 29–38 (2006)
19. Milchtaich, I.: Topological conditions for uniqueness of equilibrium in networks. Mathematics of Operations Research 30, 225–244 (2005)
20. Milchtaich, I.: The Equilibrium Existence Problem in Finite Network Congestion Games. In: Spirakis, P.G., Mavronicolas, M., Kontogiannis, S.C. (eds.) WINE 2006. LNCS, vol. 4286, pp. 87–98. Springer, Heidelberg (2006)
21. Milchtaich, I.: Network topology and the efficiency of equilibrium. Games and Economic Behavior 57(2), 321–346 (2006)
22. Monderer, D.: Potential games. Games and Economic Behavior 14(1), 124–143 (1996)
23. Papadimitriou, C.: Algorithms, games, and the internet. In: Proceedings of the Thirty-Third Annual ACM Symposium on Theory of Computing, STOC 2001, pp. 749–753. ACM, New York (2001)
24. Rosenthal, R.W.: A class of games possessing pure-strategy nash equilibria. International Journal of Game Theory 2(1), 65–67 (1973)
25. Roughgarden, T., Tardos, É.: How bad is selfish routing? J. ACM 49(2), 236–259 (2002)

Decentralized Dynamics for Finite Opinion Games*

Diodato Ferraioli[1], Paul W. Goldberg[2], and Carmine Ventre[3]

[1] Dipartimento di Informatica, Università di Salerno, Fisciano, Italy
ferraioli@dia.unisa.it
[2] Department of Computer Science, University of Liverpool, Liverpool, UK
P.W.Goldberg@liverpool.ac.uk
[3] School of Computing, Teesside University, Middlesbrough, UK
C.Ventre@tees.ac.uk

Abstract. Game theory studies situations in which strategic players can modify the state of a given system, due to the absence of a central authority. Solution concepts, such as Nash equilibrium, are defined to predict the outcome of such situations. In the spirit of the field, we study the computation of solution concepts by means of decentralized dynamics. These are algorithms in which players move in turns to improve their own utility and the hope is that the system reaches an "equilibrium" quickly.

We study these dynamics for the class of opinion games, recently introduced by [1]. These are games, important in economics and sociology, that model the formation of an opinion in a social network. We study best-response dynamics and show that the convergence to Nash equilibria is polynomial in the number of players. We also study a noisy version of best-response dynamics, called logit dynamics, and prove a host of results about its convergence rate as the noise in the system varies. To get these results, we use a variety of techniques developed to bound the mixing time of Markov chains, including coupling, spectral characterizations and bottleneck ratio.

1 Introduction

Social networks are widespread in physical and digital worlds. The following scenario therefore becomes of interest. Consider a group of individuals, connected in a social network, who are members of a committee, and suppose that each individual has her own opinion on the matter at hand. How can this group of people reach *consensus*? This is a central question in economic theory, especially for processes in which people repeatedly average their own opinions. This line of work, see e.g. [2–5], is based on a model defined by DeGroot [6]. In this

* Work partially supported by EPSRC grant EP/G069239/1 "Efficient Decentralised Approaches in Algorithmic Game Theory" and by PRIN 2008 research project COGENT (COmputational and GamE-theoretic aspects of uncoordinated NeTworks), funded by the Italian Ministry of University and Research.

M. Serna (Ed.): SAGT 2012, LNCS 7615, pp. 144–155, 2012.

model, each person i holds an opinion given by a real number x_i, which might for example represent a position on a political spectrum. There is an undirected graph $G = (V, E)$ representing a social network, and node i is influenced by the opinions of her neighbors in G. In each time step, node i updates her opinion to be an average of her current opinion with the current opinions of her neighbors. A variation of this model of interest to our study is due to Friedkin and Johnsen [7]. In [7] it is additionally assumed that each node i maintains a persistent *internal belief* b_i, which remains constant even as node i updates her overall opinion x_i through averaging. (See Sect. 2 for the formal framework.)

However, as recently observed by Bindel et al. [1], consensus is hard to reach, the case of political opinions being a prominent example. The authors of [1] justify the absence of consensus by interpreting repeated averaging as a decentralized dynamics for selfish players. Consensus is not reached as players will not compromise further when this diminishes their *utility*. Therefore, these dynamics will converge to an equilibrium in which players might disagree; Bindel et al. study the cost of disagreement by bounding the price of anarchy in this setting.

In this paper, we continue the study of [1] and ask the question of how quickly equilibria are reached by decentralized dynamics in opinion games. We focus on the setting in which players have only a finite number of strategies available. This is motivated by the fact that in many cases although players have personal beliefs which may assume a continuum of values, they only have a limited number of strategies available. For example, in political elections, people have only a limited number of parties they can vote for and usually vote for the party which is *closer* to their own opinions. Motivated by several electoral systems around the world, we concentrate in this study on the case in which players only have two strategies available. This setting already encodes a number of interesting technical challenges as outlined below.

1.1 Our Contribution

For the finite version of the opinion games considered in [1], we firstly note that this is a potential game [8, 9] thus implying that these games admit pure Nash equilibria. The set of pure Nash equilibria is then characterized. We also notice the interesting fact that while the games in [1] have a price of anarchy of 9/8, our games have unbounded price of anarchy, thus implying that for finite games disagreeing has far deeper consequences on the social cost. These basic facts turn out to be useful in the study of decentralized dynamics for finite opinion games.

Given that the potential function is polynomial in the number of players, by proving that the potential decreases by a constant at each step of the best-response dynamics, we can prove that this dynamics quickly converges to pure Nash equilibria. This result is proved by "reducing" an opinion game to a version of it in which the internal beliefs can only take certain values. The reduced version is equivalent to the original one, as long as best-response dynamics is concerned. Note that the convergence rate for the version of the game considered in [1] is unknown.

In real life, however, there is some noise in the decision process of players. Arguably, people are not fully rational. On the other hand, even if they were, they might not exactly know what strategy represents the best response to a given strategy profile due to the incapacity to correctly determine their utility functions. To model this, we study *logit dynamics* [10] for opinion games. Logit dynamics features a *rationality level* $\beta \geq 0$ (equivalently, a noise level $1/\beta$) and each player is assumed to play a strategy with a probability which is proportional to the corresponding utility to the player and β. So the higher β is, the less noise there is and the more the dynamics is similar to best-response dynamics. Logit dynamics for potential games defines a Markov chain that has a nice structure. As in [11, 12] we exploit this structure to prove bounds on the convergence rate of logit dynamics to the so-called *logit equilibrium*. The logit equilibrium corresponds to the stationary distribution of the Markov chain. Intuitively, a logit equilibrium is a probability distribution over strategy profiles of the game; the distribution is concentrated around pure Nash equilibrium profiles.[1] It is observed in [12] how this notion enjoys a number of desiderata one would like solution concepts to have.

We prove a host of results on the convergence rate of logit dynamics that give a pretty much complete picture as β varies. We give an upper bound in terms of the cutwidth of the graph modeling the social network. The bound is exponential in β and the cutwidth of the graph, thus yielding an exponential guarantee for some topology of the social network. We complement this result by proving a polynomial upper bound when β takes a small value, namely, for β at most the inverse of the maximum degree of nodes of the graph. We complete the preceding upper bound in terms of the cutwidth with lower bounds. Firstly, we prove that in order to get an (essentially) matching lower bound it is necessary to evaluate the size of a certain subset of strategy profiles. For large enough β relative to this subset then we can prove that the upper bound is tight for any social network (specifically, we roughly need β bigger than $n \log n$ over the cutwidth of the graph). For smaller values of β, we are unable to prove a lower bound which holds for every graph. However, we prove that the lower bound holds in this case at both ends of the spectrum of possible social networks. In details, we look at two cases of graphs encoding social networks: cliques, which model monolithic, highly interconnected societies, and complete bipartite graphs, which model more sparse "antitransitive" societies. For these graphs, we firstly evaluate the cutwidth and then relate the latter to the size of the aforementioned set of states. This allows to prove a lower bound exponential in β and the cutwidth of the graph for (almost) any value of β. As far as we know, no previous result was known about the cutwidth of a complete bipartite graph; this might be of independent interest. The result on cliques is instead obtained by generalizing arguments in [13].

To prove the convergence rate of logit dynamics to logit equilibrium we adopt a variety of techniques developed to bound the mixing time of Markov chains.

---

[1] It is worth noting that the focus of best-response dynamics and logit dynamics is on two different solution concepts.

To prove the upper bounds we use some spectral properties of the transition matrix of the Markov chain defined by the logit dynamics, and coupling of Markov chains. To prove the lower bounds, we instead relay on the concept of bottleneck ratio and the relation between the latter and mixing time. (The interested reader might refer to [13] for a discussion of these concepts.)

Due to the lack of space some of the proofs are omitted or sketched.

1.2 Related Work

In addition to the papers mentioned above, our paper is related to the work on logit dynamics. This dynamics is introduced by Blume [10] and it is mainly adopted in the analysis of graphical coordination games [14–16], in which players are placed on vertices of a graph embedding social relations and each player wants to coordinate with neighbors: we highlight that an unique game is played on every edge, whereas, for opinion games, we need different games in order to encode beliefs (see below). Asadpour and Saberi [17] adopt the logit dynamics for analyzing a class of congestion games. However, none of these works evaluates the time the logit dynamics takes in order to reach the stationary distribution: this line of research is conducted in [11, 12].

A number of papers study the efficient computation of (approximate) pure Nash equilibria for 2-strategy games, such as, *party affiliation games* [18, 19] and *cut games* [20]. Similarly to these works, we focus on a class of 2-strategy games and study efficient computation of pure Nash equilibria; additionally we also study the convergence rate to logit equilibria.

Another related work is [21] by Dyer and Mohanaraj. They study graphical games, called *pairwise-interaction games*, and prove among other results, quick convergence of best-response dynamics for these games. However, our games do not fall in their class. The difference is that, in their case, there is a unique game being played on the edges of the graph; as noted above, we instead need a different game to encode the internal beliefs of the players.

2 The Game

Let $G = (V, E)$ be an undirected connected graph[2] with $|V| = n$. Every vertex of the graph represents a player. Each player i has an *internal belief* $b_i \in [0, 1]$ and only two strategies or *opinions* are available, namely 0 and 1. Motivated by the model in [1], we define the utility of player i in a strategy profile $\mathbf{x} \in \{0, 1\}^n$ as

$$u_i(\mathbf{x}) = -\left((x_i - b_i)^2 + \sum_{j:\ (i,j)\in E} (x_i - x_j)^2\right).$$

[2] A number of papers, including [1], assume that the graph is weighted to model neighbors' different levels of influence. Here we focus on the case in which all neighbors exert the same kind of "political" weight.

We call such a game an n-player opinion game on a graph G. Let $D_i(\mathbf{x}) = \{j : (i,j) \in E \wedge x_i \neq x_j\}$ be the set of neighbors of i that have an opinion different from i. Then $u_i(\mathbf{x}) = -(x_i - b_i)^2 - |D_i(\mathbf{x})|$.

Let $D(\mathbf{x}) = \{(u,v) \in E \colon x_u \neq x_v\}$ be the set of *discording edges* in the strategy profile $\mathbf{x}$, that is the set of all edges in G whose endpoints have different opinions. Then it is not hard to check that the function $\Phi(\mathbf{x}) = \sum_i (x_i - b_i)^2 + |D(\mathbf{x})|$ is an exact potential function for the opinion game described above. Interestingly, the potential function looks very similar to (but not the same as) the social cost $\mathsf{SC}(\mathbf{x}) = -\sum_{i=1}^n u_i(\mathbf{x}) = \sum_i (x_i - b_i)^2 + 2|D(\mathbf{x})|$.

Let B_i be the integer closer to the internal belief of the player i: that is, $B_i = 0$ if $b_i \leq 1/2$, $B_i = 1$ if $b_i > 1/2$. Moreover, let $N_i^s(\mathbf{x}) = |\{j \colon (i,j) \in E \text{ and } x_j = s\}|$ be the number of neighbors of i that play strategy s in the strategy profile $\mathbf{x}$.

It is not hard to verify that in Nash equilibria each player i selects B_i if and only if at least half his neighborhood has selected this opinion. The only special cases occur when players have beliefs in $\{0, 1/2, 1\}$: if $b_i = 1/2$ player i will be additionally indifferent when exactly half (assuming that Δ_i is even) of his neighbors are playing the same strategy and the other half are playing the other strategy; if $b_i = 0$ or $b_i = 1$ player i will also be indifferent when Δ_i is odd and only $\lfloor \Delta_i/2 \rfloor$ neighbors are playing B_i. Roughly speaking, in a Nash equilibrium players tend to form large coalitions, by preferring to play what the majority plays to their own beliefs.

It is easy to check that this game has infinite Price of Anarchy. Consider the opinion game on a clique where each player has internal belief 0: the profile where each player has opinion 0 has social cost 0. The profile where each player has opinion 1 is a Nash equilibrium and its social cost is $n > 0$. This is in sharp contrast with the bound 9/8 proved in [1].

3 Best-Response Dynamics

Given two games we say they are *best-response equivalent* if each player has identical best responses to every combination of opponents' strategies. For the opinion games the following observation is straightforward.

Observation 1. Let $\mathcal{G}$ be an opinion game where the player i has belief $b_i \in (0, 1/2)$: then $\mathcal{G}$ is best-response equivalent to the same game where the belief of i is set to $b_i = 1/4$. Similarly, if the player i has opinion $b_i \in (1/2, 1)$ the game is best-response equivalent to the same game where the belief of i is set to $b_i = 3/4$.

The following theorem shows that, for this class of games, the best-response dynamics quickly converges to a Nash equilibrium.

Theorem 2. *The best-response dynamics for an n-player opinion game $\mathcal{G}$ converges to a Nash equilibrium after a polynomial number of steps.*

Proof (Sketch). From Observation 1 we know that each opinion game is best-response equivalent to an opinion game where each player i has $b_i \in S = \{0, \frac{1}{4}, \frac{1}{2}, \frac{3}{4}, 1\}$. So, for a given opinion game $\mathcal{G}$ we construct a game $\mathcal{G}'$ with beliefs restricted to belong to S by "rounding" the beliefs of the original game and show that best-response dynamics converges quickly on $\mathcal{G}'$. We begin by observing that for every profile $\mathbf{x}$, we have $0 \leq \Phi(\mathbf{x}) \leq n^2 + n$. Thus, the theorem follows by showing that at each time step the cost of a player decreases by at least a constant value. □

4 Logit Dynamics for Opinion Games

Let $\mathcal{G}$ be an opinion game as from the above; moreover, let $S = \{0,1\}^n$ denote the set of all strategy profiles. For two vectors $\mathbf{x}, \mathbf{y} \in S$, we denote with $H(\mathbf{x}, \mathbf{y}) = |\{i \colon x_i \neq y_i\}|$ the Hamming distance between $\mathbf{x}$ and $\mathbf{y}$. The *Hamming graph* of the game $\mathcal{G}$ is defined as $\mathcal{H} = (S, \mathsf{E})$, where two profiles $\mathbf{x} = (x_1, \ldots, x_n), \mathbf{y} = (y_1, \ldots, y_n) \in S$ are adjacent in $\mathcal{H}$ if and only if $H(\mathbf{x}, \mathbf{y}) = 1$.

The *logit dynamics* for $\mathcal{G}$ runs as follows: at every time step (i) Select one player $i \in [n]$ uniformly at random; (ii) Update the strategy of player i according to the *Boltzmann distribution* with parameter β over the set $S_i = \{0,1\}$ of her strategies. That is, a strategy $s_i \in S_i$ will be selected with probability

$$\sigma_i(s_i \mid \mathbf{x}_{-i}) = \frac{1}{Z_i(\mathbf{x}_{-i})} e^{\beta u_i(\mathbf{x}_{-i}, s_i)}, \tag{1}$$

where $\mathbf{x}_{-i} \in \{0,1\}^{n-1}$ is the profile of strategies played at the current time step by players different from i, $Z_i(\mathbf{x}_{-i}) = \sum_{z_i \in S_i} e^{\beta u_i(\mathbf{x}_{-i}, z_i)}$ is the normalizing factor, and $\beta \geq 0$. As mentioned above, from (1), it is easy to see that for $\beta = 0$ player i selects her strategy uniformly at random, for $\beta > 0$ the probability is biased toward strategies promising higher payoffs, and for β that goes to ∞ player i chooses her best response strategy (if more than one best response is available, she chooses one of them uniformly at random).

The above dynamics defines a *Markov chain* $\{X_t\}_{t \in \mathbb{N}}$ with the set of strategy profiles as state space, and where the probability $P(\mathbf{x}, \mathbf{y})$ of a transition from profile $\mathbf{x} = (x_1, \ldots, x_n)$ to profile $\mathbf{y} = (y_1, \ldots, y_n)$ is zero if $H(\mathbf{x}, \mathbf{y}) \geq 2$ and it is $\frac{1}{n}\sigma_i(y_i \mid \mathbf{x}_{-i})$ if the two profiles differ exactly at player i. More formally, we can define the logit dynamics as follows.

Definition 3 (Logit dynamics [10]). *Let $\mathcal{G}$ be an opinion game as from the above and let $\beta \geq 0$. The* logit dynamics *for $\mathcal{G}$ is the Markov chain $\mathcal{M}_\beta = (\{X_t\}_{t \in \mathbb{N}}, S, P)$ where $S = \{0,1\}^n$ and*

$$P(\mathbf{x}, \mathbf{y}) = \frac{1}{n} \cdot \begin{cases} \sigma_i(y_i \mid \mathbf{x}_{-i}), & \text{if } \mathbf{y}_{-i} = \mathbf{x}_{-i} \text{ and } y_i \neq x_i; \\ \sum_{i=1}^n \sigma_i(y_i \mid \mathbf{x}_{-i}), & \text{if } \mathbf{y} = \mathbf{x}; \\ 0, & \text{otherwise;} \end{cases} \tag{2}$$

where $\sigma_i(y_i \mid \mathbf{x}_{-i})$ is defined in (1).

The Markov chain defined by (2) is ergodic. Hence, from every initial profile $\mathbf{x}$ the distribution $P^t(\mathbf{x}, \cdot)$ of chain X_t starting at $\mathbf{x}$ will eventually converge to a *stationary distribution* π as t tends to infinity.[3] As in [12], we call the stationary distribution π of the Markov chain defined by the logit dynamics on a game $\mathcal{G}$, the *logit equilibrium* of $\mathcal{G}$. In general, a Markov chain with transition matrix P and state space S is said to be *reversible* with respect to the distribution π if, for all $\mathbf{x}, \mathbf{y} \in S$, it holds that $\pi(\mathbf{x})P(\mathbf{x}, \mathbf{y}) = \pi(\mathbf{y})P(\mathbf{y}, \mathbf{x})$. If the chain is reversible with respect to π, then π is its stationary distribution. For the class of potential games the stationary distribution is the well-known *Gibbs measure.*

Theorem 4 ([10]). *If $\mathcal{G} = ([n], \mathcal{S}, \mathcal{U})$ is a potential game with potential function Φ, then the Markov chain given by (2) is reversible with respect to the Gibbs measure $\pi(\mathbf{x}) = \frac{1}{Z} e^{-\beta\Phi(\mathbf{x})}$, where $Z = \sum_{\mathbf{y} \in S} e^{-\beta\Phi(\mathbf{y})}$ is the normalizing constant.*

Mixing Time of Markov Chains. The most prominent measures of the rate of convergence of a Markov chain to its stationary distribution is the *mixing time.* For a Markov chain with transition matrix P and state space S, let us set $d(t) = \max_{\mathbf{x} \in S} \|P^t(\mathbf{x}, \cdot) - \pi\|_{\mathrm{TV}}$, where the *total variation distance* $\|\mu - \nu\|_{\mathrm{TV}}$ between two probability distributions μ and ν on the same state space S is defined as $\|\mu - \nu\|_{\mathrm{TV}} = \max_{A \subset S} |\mu(A) - \nu(A)|$. For $0 < \varepsilon < 1/2$, the mixing time is defined as $t_{\mathrm{mix}}(\varepsilon) = \min\{t \in \mathbb{N} \colon d(t) \leq \varepsilon\}$. It is usual to set $\varepsilon = 1/4$ or $\varepsilon = 1/2e$. If not explicitly specified, when we write t_{mix} we mean $t_{\mathrm{mix}}(1/4)$. Observe that $t_{\mathrm{mix}}(\varepsilon) \leq \lceil \log_2 \varepsilon^{-1} \rceil t_{\mathrm{mix}}$.

Bottleneck Ratio. An important concept to establish our lower bounds is represented by the *bottleneck ratio.* Consider an ergodic Markov chain with finite state space S, transition matrix P, and stationary distribution π. The probability distribution $Q(\mathbf{x}, \mathbf{y}) = \pi(\mathbf{x})P(\mathbf{x}, \mathbf{y})$ is of particular interest and is sometimes called the *edge stationary distribution.* Note that if the chain is reversible then $Q(\mathbf{x}, \mathbf{y}) = Q(\mathbf{y}, \mathbf{x})$. For any $L \subseteq S$, we let $Q(L, S \setminus L) = \sum_{\mathbf{x} \in L, \mathbf{y} \in S \setminus L} Q(\mathbf{x}, \mathbf{y})$. The bottleneck ratio of $L \subseteq S$, L non-empty, is $B(L) = \frac{Q(L, S \setminus R)}{\pi(L)}$.

The following theorem relates bottleneck ratio and mixing time.

Theorem 5 (Bottleneck ratio [13]). *Let $\mathcal{M} = \{X_t \colon t \in \mathbb{N}\}$ be an irreducible and aperiodic Markov chain with finite state space S, transition matrix P, and stationary distribution π. Then the mixing time is $t_{\mathrm{mix}} \geq \max_{L \colon \pi(L) \leq 1/2} \frac{1}{4 \cdot B(L)}$.*

4.1 Upper Bounds

For Every β. Consider the bijective function $\sigma \colon V \to \{1, \ldots, |V|\}$: it represents an ordering of vertices of G. Let $\mathcal{L}$ be the set of all orderings of vertices of G and set $V_i^\sigma = \{v \in V \colon \sigma(v) < i\}$. Then, the *cutwidth* of G is $\mathsf{CW}(G) = \min_{\sigma \in \mathcal{L}} \max_{1 < i \leq |V|} |E(V_i^\sigma, V \setminus V_i^\sigma)|$.

[3] The notation $P^t(\mathbf{x}, \cdot)$, standard in Markov chains literature [13], denotes the probability distribution over states of S after the chain has taken t steps starting from $\mathbf{x}$.

Theorem 6. *Let $\mathcal{G}$ be an n-player opinion game on a graph $G = (V, E)$. The mixing time of the logit dynamics for $\mathcal{G}$ is* $t_{\text{mix}} \leq (1 + \beta) \cdot \text{poly}(n) \cdot e^{\beta\Theta(\text{CW}(G))}$.

The proof is a generalization of a similar proof given by Berger et al. [22] based on spectral arguments.

For Small β. The following theorem shows that for small values of β the mixing time is polynomial. We remark that there are network topologies for which this theorem gives a bound higher than that guaranteed by Theorem 6 on the values of β for which the mixing time is polynomial.

Theorem 7. *Let $\mathcal{G}$ be an n-player opinion game on a connected graph G, with $n > 2$. Let $\Delta_{\max}$ be the maximum degree in the graph. If $\beta \leq 1/\Delta_{\max}$, then the mixing time of the logit dynamics for $\mathcal{G}$ is $\mathcal{O}(n \log n)$.*

Proof (Sketch). Consider two profiles $\mathbf{x}$ and $\mathbf{y}$ that differ only in the strategy played by player j and consider the coupling described in [11] for two chains X and Y starting respectively from $X_0 = \mathbf{x}$ and $Y_0 = \mathbf{y}$. We show the expected distance between X_1 and Y_1 after one step of the coupling is less then $e^{-1/(3n)}$. The bound on the mixing time follows from the well-known path coupling technique [23]. □

4.2 Lower Bounds

Recall that $\mathcal{H}$ is the Hamming graph on the set of profiles of an opinion games on a graph G. The following observation easily follows from the definition of cutwidth.

Observation 8. For every path on $\mathcal{H}$ between the profile $\mathbf{0} = (0, \ldots, 0)$ and the profile $\mathbf{1} = (1, \ldots, 1)$ there exists a profile for which there are at least $\text{CW}(G)$ discording edges.

From now on, let us write CW as a shorthand for $\text{CW}(G)$, when the reference to the graph is clear from the context. For sake of compactness, we set $\mathbf{b}(\mathbf{x}) = \sum_i (x_i - b_i)^2$. We denote as $\mathbf{b}^\star$ the minimum of $\mathbf{b}(\mathbf{x})$ over all profiles with CW discording edges.

Let $R_\mathbf{0}$ ($R_\mathbf{1}$) be the set of profiles $\mathbf{x}$ for which a path from $\mathbf{0}$ (resp., $\mathbf{1}$) to $\mathbf{x}$ exists on $\mathcal{H}$ such that every profile along the path has potential value less than $\mathbf{b}^\star + \text{CW}$. To establish the lower bound we use the technical result given by Theorem 5 which requires to compute the bottleneck ratio of a subset of profiles that is weighted at most a half by the stationary distribution. Accordingly, we set $R = R_\mathbf{0}$ if $\pi(R_\mathbf{0}) \leq 1/2$ and $R = R_\mathbf{1}$ if $\pi(R_\mathbf{1}) \leq 1/2$. (If both sets have stationary distribution less than one half, the best lower bound is achieved by setting R to $R_\mathbf{0}$ if and only if $\Phi(\mathbf{0}) \leq \Phi(\mathbf{1})$.) W.l.o.g., in the remaining of this section we assume $R = R_\mathbf{0}$.

For Large β. Let ∂R be the set of profiles in R that have at least a neighbor $\mathbf{y}$ in the Hamming graph $\mathcal{H}$ such that $\mathbf{y} \notin R$. Moreover let $\mathcal{E}(\partial R)$ the set of edges $(\mathbf{x}, \mathbf{y})$ in $\mathcal{H}$ such that $\mathbf{x} \in \partial R$ and $\mathbf{y} \notin R$: note that $|\mathcal{E}(\partial R)| \leq n|\partial R|$. The following lemma bounds the bottleneck ratio of R.

Lemma 9. *For the set of profiles R defined above, we have $B(R) \leq n \cdot |\partial R| \cdot e^{-\beta(\mathsf{CW}+\mathbf{b}^\star-\mathbf{b}(\mathbf{0}))}$.*

Proof. Since $\mathbf{0} \in R$, it holds $\pi(R) \geq \pi(\mathbf{0}) = \frac{e^{-\beta \mathbf{b}(\mathbf{0})}}{Z}$. Moreover, by (1) we have

$$\begin{aligned}
Q(R, \overline{R}) &= \sum_{\substack{(\mathbf{x},\mathbf{y}) \in \mathcal{E}(\partial R):\\ \mathbf{y}=(\mathbf{x}_{-i}, y_i)}} \frac{e^{-\beta\Phi(\mathbf{x})}}{Z} \frac{e^{\beta u_i(\mathbf{y})}}{e^{\beta u_i(\mathbf{x})} + e^{\beta u_i(\mathbf{y})}} \\
&= \sum_{\substack{(\mathbf{x},\mathbf{y}) \in \mathcal{E}(\partial R):\\ \mathbf{y}=(\mathbf{x}_{-i}, y_i)}} \frac{e^{-\beta\Phi(\mathbf{x})}}{Z} \frac{e^{-\beta\Phi(\mathbf{y})} e^{\beta(u_i(\mathbf{x})+\Phi(\mathbf{x}))}}{e^{-\beta\Phi(\mathbf{x})} e^{\beta(u_i(\mathbf{x})+\Phi(\mathbf{x}))} + e^{-\beta\Phi(\mathbf{y})} e^{\beta(u_i(\mathbf{x})+\Phi(\mathbf{x}))}} \\
&= \frac{1}{Z} \sum_{(\mathbf{x},\mathbf{y}) \in \mathcal{E}(\partial R)} \frac{e^{-\beta\Phi(\mathbf{x})} e^{-\beta\Phi(\mathbf{y})}}{e^{-\beta\Phi(\mathbf{x})} + e^{-\beta\Phi(\mathbf{y})}} = \frac{1}{Z} \sum_{(\mathbf{x},\mathbf{y}) \in \mathcal{E}(\partial R)} \frac{e^{-\beta\Phi(\mathbf{y})}}{1 + e^{\beta(\Phi(\mathbf{x})-\Phi(\mathbf{y}))}} \\
&\leq \frac{1}{Z} \sum_{(\mathbf{x},\mathbf{y}) \in \mathcal{E}(\partial R)} e^{-\beta\Phi(\mathbf{y})} \leq |\mathcal{E}(\partial R)| \cdot \frac{e^{-\beta(\mathbf{b}^\star + \mathsf{CW})}}{Z}.
\end{aligned}$$

The second equality follows from the definition of potential function which implies $\Phi(\mathbf{y}) - \Phi(\mathbf{x}) = -u_i(\mathbf{y}) + u_i(\mathbf{x})$ for $\mathbf{x}$ and $\mathbf{y}$ as above; last inequality holds because if by contradiction $\Phi(\mathbf{y}) < \mathbf{b}^\star + \mathsf{CW}$ then, by definition of R, it would be $\mathbf{y} \in R$, a contradiction. □

From Lemma 9 and Theorem 5 we obtain a lower bound to the mixing time of the opinion games that holds for every value of β, every social network G and every vector $(b_1, \ldots, b_n)$ of internal beliefs. However, it is not clear how close this bound is to the one given in Theorem 6. Nevertheless, by taking $b_i = 1/2$ for each player i and β high enough, we can state the following theorem.

Theorem 10. *Let $\mathcal{G}$ be an n-player opinion game on a graph G. Then, there exist a vector of internal beliefs such that for $\beta = \Omega\left(\frac{n \log n}{\mathsf{CW}}\right)$ it holds $t_{\text{mix}} \geq e^{\beta\Theta(\mathsf{CW})}$.*

Proof. If $b_i = 1/2$ for every player i, from Lemma 9 and Theorem 5, since $|\partial R| \leq 2^n$ then $t_{\text{mix}} \geq \frac{e^{\beta\mathsf{CW}}}{n2^n} = e^{\beta\mathsf{CW}-n\log(2n)} = e^{\beta\Theta(\mathsf{CW})}$. □

For Smaller β. Theorem 10 gives an almost tight lower bound for high values of β for each network topology. It would be interesting to prove a matching bound also for lower values of the rationality parameter: in this section we prove such a bound for specific classes of graphs: complete bipartite graphs and cliques.

We start by considering the class of complete bipartite graphs $K_{m,m}$.

Theorem 11. *Let $\mathcal{G}$ be an n-player opinion game on $K_{m,m}$. Then, there exist a vector of internal beliefs such that, for every $\beta = \Omega\left(\frac{1}{m}\right)$, we have $t_{\text{mix}} \geq \frac{e^{\beta\Theta(\mathsf{CW})}}{n}$.*

To prove the theorem above, we start by evaluating the cutwidth of $K_{m,m}$: in particular, we characterize the best ordering from which the cutwidth is obtained. We will denote with A and B the two sides of the bipartite graph. Then it is not hard to see that the ordering that obtains the cutwidth in $K_{m,m}$ is the one that selects alternatively a vertex from A and a vertex from B. Moreover, it turns out that the cutwidth of $K_{m,m}$ is $\lceil m^2/2 \rceil$. The following lemma gives a bound to the size of ∂R for this graph.

Lemma 12. *For the opinion game on the graph $K_{m,m}$ with $b_i = 1/2$ for every player i, there exists a constant c_1 such that $|\partial R| \leq e^{c_1\sqrt{\mathsf{CW}}}$.*

Proof (Sketch). Since $b_i = 1/2$ for every player i, we have that $\mathbf{b}(\mathbf{x}) = n/4$ for every profile $\mathbf{x}$. Therefore, by definition of R, all profiles in R (and therefore ∂R) have less then CW discording edges. Indeed, for $\mathbf{x} \in R$ we have $\mathbf{b}(\mathbf{x}) + |D(\mathbf{x})| = \Phi(\mathbf{x}) < \mathbf{b}^\star + \mathsf{CW}$. Moreover, if a profile $\mathbf{y}$ has less then $\mathsf{CW} - m$ discording edges, then $\mathbf{y}$ is not in ∂R as a state neighbor of $\mathbf{y}$ has at most $m-1$ additional discording edges.

Consequently, to bound the size of ∂R, we need to count the number of profiles in R that have potential between $\mathbf{b}^\star + \mathsf{CW} - m$ and $\mathbf{b}^\star + \mathsf{CW} - 1$ (i.e., the number of profiles with at least $\mathsf{CW} - m$ and at most $\mathsf{CW} - 1$ discording edges). By using the facts about the cut-width of bipartite graphs stated above, we have $|\partial R| \leq (5e)^m \leq e^{3m}$. The lemma follows since $m \leq \sqrt{2}\sqrt{\mathsf{CW}}$. □

Proof (of Theorem 11). If $b_i = 1/2$ for every player i, from Lemmata 9 and 12, we have $B(R) \leq n \cdot e^{c_1\sqrt{\mathsf{CW}}} \cdot e^{-\beta\mathsf{CW}} \leq n \cdot e^{-\beta\mathsf{CW}(1-c_2)}$, where $c_2 = \frac{c_1\sqrt{\mathsf{CW}}}{\beta\mathsf{CW}} < 1$ since by hypothesis $\beta > \frac{c_1}{\sqrt{\mathsf{CW}}} = \Omega(1/m)$; we also notice that c_2 goes to 0 as β increases. The theorem follows from Theorem 5. □

We remark that it is possible to prove a result similar to Theorem 11 also for the clique K_n: the proof follows from a simple generalization of Theorem 15.3 in [13] and by observing that the cutwidth of a clique is $\lfloor n^2/4 \rfloor$.

5 Conclusions and Open Problems

In this work we analyze two decentralized dynamics for binary opinion games: the best-response dynamics and the logit dynamics. For the best-response dynamics we show that it takes time polynomial in the number of players to reach a Nash equilibrium, the latter being characterized by the existence of clusters in which players have a common opinion. On the other hand, for the logit dynamics we show polynomial convergence when the level of noise is high enough and that it increases as β grows.

It is important to highlight, as noted above, that the convergence time of the two dynamics are computed with respect to two different equilibrium concepts, namely Nash equilibrium for the best-response dynamics and logit equilibrium for the logit dynamics. This explains why the convergence times of these two dynamics asymptotically diverge even though the logit dynamics becomes similar to the best response dynamics as β goes to infinity.

Theorem 6 and 10 which prove bounds to the convergence of logit dynamics can also be read in a positive fashion. Indeed, for social networks that have a bounded cutwidth, the convergence rate of the dynamics depends only on the value of β. (We highlight that checking if a graph has bounded cutwidth can be done in polynomial time [24].) In general, we have the following picture: as long as β is less than the maximum of (roughly) $\frac{\log n}{\mathsf{CW}}$ and $\frac{1}{\Delta}$ the convergence time to the logit equilibrium is polynomial. Moreover, Theorem 10 shows that for β lower bounded by (roughly) $\frac{n \log n}{\mathsf{CW}}$ the convergence time to the logit equilibrium is super-polynomial. Then for some network topology, there is a gap in our knowledge which is naturally interesting to close.

In [25] the concept of metastable distributions has been introduced in order to predict the outcome of games for which the logit dynamics takes too much time to reach the stationary distribution for some value of β. It would be interesting to investigate existence and structure of such distributions for our opinion games.

We also note that our proofs for logit dynamics can be extended to the case in which the social graph is weighted. In such a setting, however, we obtain non-matching bounds: it would be interesting to develop more sophisticated techniques in order to get tight bounds.

References

1. Bindel, D., Kleinberg, J., Oren, S.: How bad is forming your own opinion? In: 2011 IEEE 52nd Annual Symposium on Foundations of Computer Science (FOCS), pp. 57–66 (October 2011)
2. Acemoglu, D., Ozdaglar, A.: Opinion dynamics and learning in social networks. Dynamic Games and Applications 1, 3–49 (2011)
3. DeMarzo, P.M., Vayanos, D., Zwiebel, J.: Persuasion bias, social influence, and unidimensional opinions. The Quarterly Journal of Economics 118(3), 909–968 (2003)
4. Golub, B., Jackson, M.O.: Naïve learning in social networks and the wisdom of crowds. American Economic Journal: Microeconomics 2(1), 112–149 (2010)
5. Jackson, M.O.: Social and Economic Networks. Princeton University Press, Princeton (2008)
6. DeGroot, M.H.: Reaching a consensus. Journal of the American Statistical Association 69(345), 118–121 (1974)
7. Friedkin, N.E., Johnsen, E.C.: Social influence and opinions. The Journal of Mathematical Sociology 15(3-4), 193–206 (1990)
8. Rosenthal, R.: A class of games possessing pure-strategy nash equilibria. International Journal of Game Theory 2, 65–67 (1973)
9. Monderer, D., Shapley, L.S.: Potential games. Games and Economic Behavior 14(1), 124–143 (1996)
10. Blume, L.E.: The statistical mechanics of strategic interaction. Games and Economic Behavior 5(3), 387–424 (1993)

11. Auletta, V., Ferraioli, D., Pasquale, F., Persiano, G.: Mixing Time and Stationary Expected Social Welfare of Logit Dynamics. In: Kontogiannis, S., Koutsoupias, E., Spirakis, P.G. (eds.) SAGT 2010. LNCS, vol. 6386, pp. 54–65. Springer, Heidelberg (2010)
12. Auletta, V., Ferraioli, D., Pasquale, F., Penna, P., Persiano, G.: Convergence to equilibrium of logit dynamics for strategic games. In: Proceedings of the 23rd ACM Symposium on Parallelism in Algorithms and Architectures, SPAA 2011, pp. 197–206. ACM (2011)
13. Levin, D.A., Peres, Y., Wilmer, E.L.: Markov chains and mixing times. American Mathematical Society (2006)
14. Ellison, G.: Learning, local interaction, and coordination. Econometrica 61(5), 1047–1071 (1993)
15. Young, H.P.: The diffusion of innovations in social networks. In: Blume, B.L., Durlauf, S.N. (eds.) Economy as an Evolving Complex System. Proceedings volume in the Santa Fe Institute studies in the sciences of complexity, vol. 3, pp. 267–282. Oxford University Press, US (2006)
16. Montanari, A., Saberi, A.: Convergence to equilibrium in local interaction games. In: 50th Annual IEEE Symposium on Foundations of Computer Science, FOCS 2009, pp. 303–312 (October 2009)
17. Asadpour, A., Saberi, A.: On the Inefficiency Ratio of Stable Equilibria in Congestion Games. In: Leonardi, S. (ed.) WINE 2009. LNCS, vol. 5929, pp. 545–552. Springer, Heidelberg (2009)
18. Fabrikant, A., Papadimitriou, C., Talwar, K.: The complexity of pure nash equilibria. In: Proceedings of the Thirty-Sixth Annual ACM Symposium on Theory of Computing, STOC 2004, pp. 604–612. ACM, New York (2004)
19. Balcan, M.F., Blum, A., Mansour, Y.: Improved equilibria via public service advertising. In: Proceedings of the Twentieth Annual ACM-SIAM Symposium on Discrete Algorithms, SODA 2009, pp. 728–737. Society for Industrial and Applied Mathematics (2009)
20. Bhalgat, A., Chakraborty, T., Khanna, S.: Approximating pure nash equilibrium in cut, party affiliation, and satisfiability games. In: Proceedings of the 11th ACM Conference on Electronic Commerce, EC 2010, pp. 73–82. ACM (2010)
21. Dyer, M., Mohanaraj, V.: Pairwise-Interaction Games. In: Aceto, L., Henzinger, M., Sgall, J. (eds.) ICALP 2011, Part I. LNCS, vol. 6755, pp. 159–170. Springer, Heidelberg (2011)
22. Berger, N., Kenyon, C., Mossel, E., Peres, Y.: Glauber dynamics on trees and hyperbolic graphs. Probability Theory and Related Fields 131, 311–340 (2005)
23. Bubley, R., Dyer, M.: Path coupling: A technique for proving rapid mixing in markov chains, p. 223. IEEE Computer Society, Los Alamitos (1997)
24. Thilikos, D., Serna, M., Bodlaender, H.: Constructive Linear Time Algorithms for Small Cutwidth and Carving-Width. In: Lee, D.T., Teng, S.-H. (eds.) ISAAC 2000. LNCS, vol. 1969, pp. 192–203. Springer, Heidelberg (2000)
25. Auletta, V., Ferraioli, D., Pasquale, F., Persiano, G.: Metastability of logit dynamics for coordination games. In: Proceedings of the Twenty-Third Annual ACM-SIAM Symposium on Discrete Algorithms, SODA 2012, pp. 1006–1024. SIAM (2012)

On the Hardness of Network Design for Bottleneck Routing Games*

Dimitris Fotakis[1], Alexis C. Kaporis[2], Thanasis Lianeas[1], and Paul G. Spirakis[3,4]

[1] School of Electrical and Computer Engineering
National Technical University of Athens, 15780 Athens, Greece
[2] Department of Information and Communication Systems Engineering
University of the Aegean, 83200 Samos, Greece
[3] Department of Computer Engineering and Informatics
University of Patras, 26500 Patras, Greece
[4] Computer Technology Institute and Press - Diophantus
N. Kazantzaki Str., University Campus, 26500 Patras, Greece

fotakis@cs.ntua.gr, kaporisa@gmail.com,
tlianeas@mail.ntua.gr, spirakis@cti.gr

Abstract. In routing games, the network performance at equilibrium can be significantly improved if we remove some edges from the network. This counterintuitive fact, a.k.a. Braess's paradox, gives rise to the network design problem, where we seek to recognize routing games suffering from the paradox, and to improve the equilibrium performance by edge removal. In this work, we investigate the computational complexity and the approximability of network design for non-atomic bottleneck routing games, where the individual cost of each player is the bottleneck cost of her path, and the social cost is the bottleneck cost of the network. We first show that bottleneck routing games do not suffer from Braess's paradox either if the network is series-parallel, or if we consider only subpath-optimal Nash flows. On the negative side, we prove that even for games with strictly increasing linear latencies, it is NP-hard not only to recognize instances suffering from the paradox, but also to distinguish between instances for which the Price of Anarchy (PoA) can decrease to 1 and instances for which the PoA is $\Omega(n^{0.121})$ and cannot improve by edge removal. Thus, the network design problem for such games is NP-hard to approximate within a factor of $O(n^{0.121-\varepsilon})$, for any constant $\varepsilon > 0$. On the positive side, we show how to compute an almost optimal subnetwork w.r.t. the bottleneck cost of its worst Nash flow, when the worst Nash flow in the best subnetwork routes a non-negligible amount of flow on all edges. The running time is determined by the total number of paths, and is quasipolynomial if the number of paths is quasipolynomial.

* This work was supported by the project Algorithmic Game Theory, co-financed by the European Union (European Social Fund - ESF) and Greek national funds, through the Operational Program "Education and Lifelong Learning", under the research funding program Thales, by an NTUA Basic Research Grant (PEBE 2009), by the ERC project RIMACO, and by the EU FP7/2007-13 (DG INFSO G4-ICT for Transport) under Grant Agreement no. 288094 (Project eCompass).

M. Serna (Ed.): SAGT 2012, LNCS 7615, pp. 156–167, 2012.

1 Introduction

An typical instance of a non-atomic *bottleneck routing game* consists of a directed network, with origin s and destination t, where each edge has a non-decreasing function determining the edge's latency as a function of traffic. A traffic rate is controlled by an infinite population of players, each willing to route a negligible amount of traffic through an $s-t$ path. The players seek to minimize the maximum edge latency, a.k.a. the *bottleneck cost* of their path. Thus, the players reach a *Nash equilibrium flow*, or simply a *Nash flow*, where they all use paths with a common locally minimum bottleneck cost. Bottleneck routing games and their variants have received considerable attention due to their practical applications to communication networks (see e.g., [6,3] and the references therein).

Previous Work. Bottleneck routing games admit a Nash flow that is optimal for the network, in the sense that it minimizes the maximum latency on any used edge, a.k.a. the bottleneck cost of the network (see e.g., [3, Corollary 2]). However, bottleneck routing games usually admit many other Nash flows, some with a bottleneck cost quite far from the optimum. Hence, there has been a considerable interest in quantifying the performance degradation, due to the players' selfish behavior, in (several variants of) bottleneck routing games. This is measured by the *Price of Anarchy* (PoA) [13], that is the ratio of the bottleneck cost of the worst Nash flow to the optimal bottleneck cost of the network.

Simple examples (see e.g., [7, Fig. 2]) demonstrate that the PoA of bottleneck routing games with linear latencies can be $\Omega(n)$, where n is the number of nodes. For atomic splittable bottleneck routing games, where the population of players is finite, and each player has a non-negligible amount of traffic which can be split among different paths, Banner and Orda [3] observed that the PoA can be unbounded, even for very simple networks, if the players have different origins and destinations and the latency functions are exponential. On the other hand, Banner and Orda proved that if the players use paths that, as a secondary objective, minimize the number of bottleneck edges, then all Nash flows are optimal. For a variant of non-atomic bottleneck routing games, where the social cost is the average (instead of the maximum) bottleneck cost of the players, Cole, Dodis, and Roughgarden [7] proved that the PoA is 4/3, if the latency functions are affine and a subclass of Nash flows, called *subpath-optimal Nash flows*, is only considered. Subsequently, Mazalov et al. [16] studied the inefficiency of the best Nash flow under this notion of social cost.

For atomic unsplittable bottleneck routing games, where each player routes a unit of traffic through a single $s-t$ path, Banner and Orda [3] proved that for polynomial latencies of degree d, the PoA is $O(m^d)$, where m is the number of edges. On the other hand, Epstein, Feldman, and Mansour [8] proved that for series-parallel networks, all Nash flows are optimal. Busch and Magdon-Ismail [5] proved that the PoA of atomic unsplittable bottleneck routing games with identity latency functions can be bounded in terms of natural topological properties of the network. In particular, they proved that the PoA of such games is $O(l + \log n)$, where l is the length of the longest $s-t$ path, and $O(k^2 + \log^2 n)$, where k is length of the longest circuit.

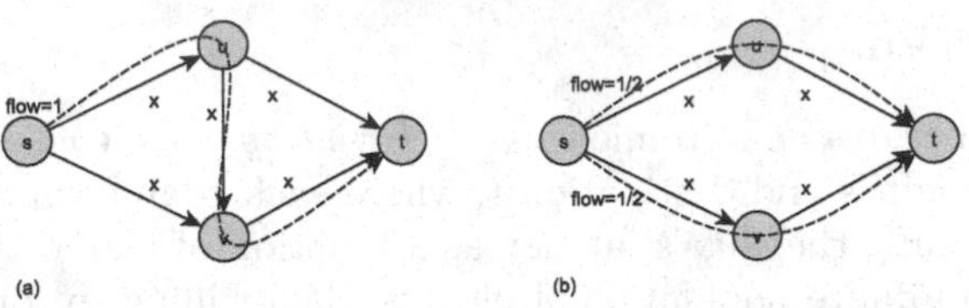

Fig. 1. Braess's paradox for bottleneck routing games. We consider identity latency functions and a unit of traffic to be routed from s to t. The worst Nash flow, in (a), has a bottleneck cost of 1. The optimal flow is the same as the flow in (b), and achieves a bottleneck cost of 1/2. Hence, PoA = 2. In the subnetwork (b), the Nash flow is unique and coincides with the optimal flow. Thus the PoA improves to 1. Hence the network on the left is *paradox-ridden*, and the network on the right is the *best subnetwork* of it.

With the PoA of bottleneck routing games so large and crucially depending on topological properties of the network, a natural approach to improving the equilibrium performance is to exploit Braess's paradox [4], namely that removing some edges may change the network topology (e.g., it may decrease the length of the longest path or cycle), and significantly improve the bottleneck cost of the worst Nash flow (see e.g., Fig. 1). This approach gives rise to the (selfish) *network design problem*, where we seek to recognize bottleneck routing games suffering from the paradox, and to improve the bottleneck cost of the worst Nash flow by edge removal. In particular, given a bottleneck routing game, we seek for the *best subnetwork*, namely, the subnetwork for which the bottleneck cost of the worst Nash flow is best possible. In this setting, one may distinguish two extreme cases: *paradox-free* instances, where edge removal cannot improve the bottleneck cost of the worst Nash flow, and *paradox-ridden* instances, where the bottleneck cost of the worst Nash flow in the best subnetwork is equal to the optimal bottleneck cost of the original network (see also [18,11]).

The approximability of selective network design, a generalization of network design where we cannot remove certain edges, was considered by Hou and Zhang [12]. For atomic unsplittable bottleneck routing games with a different traffic rate and a different origin and destination for each player, they proved that if the latency functions are polynomials of degree d, it is NP-hard to approximate selective network design within a factor of $O(m^{d-\varepsilon})$, for any $\varepsilon > 0$. Moreover, for atomic k-splittable bottleneck routing games with multiple origin-destination pairs, they proved that selective network design is NP-hard to approximate within any constant factor.

However, a careful look at the reduction of [12] reveals that their strong inapproximability results crucially depend on both (i) that we can only remove certain edges from the network, so that the subnetwork actually causing a large PoA cannot be destroyed, and (ii) that the players have different origins and destinations (and also are atomic and have different traffic rates). As for the importance of (ii), in a different setting, where the players' individual cost is the sum of edge latencies on their path and the social cost is the bottleneck cost of the network, it is known that Braess's paradox can be dramatically more severe for instances with multiple origin-destination pairs than for instances with a single origin-destination pair. More precisely, Lin et al. [14] proved that if the players have a common origin and destination, the removal of at most k edges from the network cannot improve the equilibrium bottleneck cost by a factor greater than $k+1$. On the other hand, Lin et al. [15] presented an instance with two origin-destination pairs where the removal of a single edge improves the equilibrium bottleneck cost by a factor of $2^{\Omega(n)}$. Therefore, both at the technical and at the conceptual level, the inapproximability results of [12] do not really shed light on the approximability of the (simple, non-selective) network design problem in the simplest, and most interesting, setting of non-atomic bottleneck routing games with a common origin and destination for all players.

Contribution. In this work, we investigate the approximability of the network design problem for the simplest, and seemingly easier to approximate, variant of non-atomic bottleneck routing games with a single origin-destination pair. Our main result is that network design is hard to approximate within reasonable factors, and holds even for strictly increasing linear latencies. To the best of our knowledge, this is the first work that investigates the approximability of the network design problem for the basic variant of bottleneck routing games.

In Section 3, we use techniques similar to those in [8,7], and show that bottleneck routing games do not suffer from Braess's paradox either if the network is series-parallel, or if we consider only subpath-optimal Nash flows.

On the negative side, we employ, in Section 4, a reduction from the 2-Directed Disjoint Paths problem, and show that for linear bottleneck routing games, it is NP-hard to recognize paradox-ridden instances (Lemma 1). In fact, the reduction shows that it is NP-hard to distinguish between paradox-ridden instances and paradox-free instances, even if their PoA is equal to 4/3, and thus, it is NP-hard to approximate the network design problem within a factor less than 4/3.

In Section 5, we apply essentially the same reduction, but in a recursive way, and obtain a much stronger inapproximability result. We assume the existence of a γ-gap instance, which establishes that network design is inapproximable within a factor less than γ, and show that the construction of Lemma 1, but with some edges replaced by copies of the gap instance, amplifies the inapproximability threshold by a factor of 4/3, while it increases the size of the network by roughly a factor of 8 (Lemma 2). Therefore, starting from the 4/3-gap instance of Lemma 1, and recursively applying this construction a logarithmic number times, we show that it is NP-hard to approximate the network design problem for linear bottleneck routing games within a factor of $O(n^{0.121-\varepsilon})$, for any constant $\varepsilon > 0$. An interesting technical point is that we manage to show this inapproximability result, even though we do not know how to efficiently compute the worst equilibrium bottleneck cost of a given subnetwork. Hence, our reduction uses a certain subnetwork structure to identify good approximations to the best subnetwork. To the best of our knowledge, this is the first rime that a similar recursive construction is used to amplify the inapproximability threshold of the network design problem, and of any other optimization problem related to selfish routing.

In Section 6, we consider general latency functions, and present an algorithm for finding a subnetwork that is almost optimal w.r.t. the bottleneck cost of its worst Nash flow, when the worst Nash flow in the best subnetwork routes a non-negligible amount of flow on all edges. The algorithm is based on Althöfer's Sparcification Lemma [1], and is motivated by its recent application to network design for additive routing games [11]. For any constant $\varepsilon > 0$, the algorithm computes a subnetwork and an $\varepsilon/2$-Nash flow whose bottleneck cost is within an additive term of $O(\varepsilon)$ from the worst equilibrium bottleneck cost in the best subnetwork. The running time is roughly $|\mathcal{P}|^{\mathrm{poly}(\log m)/\varepsilon^2}$, and is quasipolynomial, when the number $|\mathcal{P}|$ of paths is quasipolynomial.

Next, we present our results with as much technical justification as the space constraints permit. The interested reader may find the omitted proofs in [10].

Other Related Work. Considerable attention has been paid to the approximability of network design for *additive routing games*, where the players seek to minimize the sum of edge latencies on their path, and the social cost is the total latency incurred by the players. Roughgarden [18] introduced the selfish network design problem in this setting, and proved that it is NP-hard to recognize paradox-ridden instances. He also proved that it is NP-hard to approximate the network design problem for such games within a factor less than 4/3 for affine latencies, and less than $\lfloor n/2 \rfloor$ for general latencies. For atomic unsplittable additive routing games with weighted players, Azar and Epstein [2] proved that network design is NP-hard to approximate within a factor less than 2.618, for affine latencies, and less than $d^{\Theta(d)}$, for polynomial latencies of degree d.

On the positive side, Milchtaich [17] proved that non-atomic additive routing games on series-parallel networks do not suffer from Braess's paradox. Fotakis, Kaporis, and Spirakis [11] proved that we can efficiently recognize paradox-ridden instances when the latency functions are affine, and all, but possibly a constant number of them, are strictly increasing. Moreover, applying Althöfer's Sparsification Lemma [1], they gave an algorithm that approximates network design for affine additive routing games within an additive term of ε, for any constant $\varepsilon > 0$, in time that is subexponential if the total number of $s-t$ paths is polynomial and all paths are of polylogarithmic length.

2 Model, Definitions, and Preliminaries

Routing Instances. A *routing instance* is a tuple $\mathcal{G} = (G(V, E), (c_e)_{e \in E}, r)$, where $G(V, E)$ is a directed network with origin s and destination t, $c_e : [0, r] \mapsto \mathbb{R}_{\geq 0}$ is a continuous non-decreasing latency function associated with edge e, and $r > 0$ is the traffic rate entering at s and leaving at t. We consider a non-atomic model of selfish routing, where r is divided among an infinite population of players, each routing a negligible amount of traffic from s to t. We let $n \equiv |V|$ and $m \equiv |E|$, and let $\mathcal{P}$ denote the set of simple $s-t$ paths in G. A latency function $c_e(x)$ is *linear* if $c_e(x) = a_e x$, for some $a_e > 0$, and *affine* if $c_e(x) = a_e x + b_e$, for some $a_e, b_e \geq 0$. We say that a latency function $c_e(x)$ satisfies the *Lipschitz condition* with constant $\xi > 0$, if for all $x, y \in [0, r]$, $|c_e(x) - c_e(y)| \leq \xi |x - y|$.

Subnetworks and Subinstances. Given an instance $\mathcal{G} = (G(V,E),(c_e)_{e\in E},r)$, any subgraph $H(V,E')$, $E' \subseteq E$, obtained from G by edge deletions, is a *subnetwork* of G. H has the same origin s and destination t as G, and its edges have the same latency functions as in $\mathcal{G}$. Each instance $\mathcal{H} = (H(V,E'),(c_e)_{e\in E'},r)$, where $H(V,E')$ is a subnetwork of $G(V,E)$, is a *subinstance* of $\mathcal{G}$.

Flows. A ($\mathcal{G}$-feasible) *flow* f is a non-negative vector indexed by $\mathcal{P}$ so that $\sum_{p\in\mathcal{P}} f_p = r$. For a flow f and every edge e, we let $f_e = \sum_{p:e\in p} f_p$ denote the amount of flow that f routes through e. A path p (resp. edge e) is used by flow f if $f_p > 0$ (resp. $f_e > 0$). Given a flow f, the latency of each edge e is $c_e(f_e)$, and the *bottleneck cost* of each path p is $b_p(f) = \max_{e\in p} c_e(f_e)$. The *bottleneck cost* of a flow f, denoted $B(f)$, is $B(f) = \max_{p:f_p>0} b_p(f)$. An *optimal* flow of $\mathcal{G}$, denoted o, minimizes the bottleneck cost among all $\mathcal{G}$-feasible flows. We let $B^*(\mathcal{G}) = B(o)$. We note that for every subinstance $\mathcal{H}$ of $\mathcal{G}$, $B^*(\mathcal{H}) \geq B^*(\mathcal{G})$.

Nash Flows and their Properties. A flow f is at *Nash equilibrium*, or simply, is a *Nash flow*, if f routes all traffic on paths of a locally minimum bottleneck cost. Formally, f is a Nash flow if for all $p, p' \in \mathcal{P}$, if $f_p > 0$, then $b_p(f) \leq b_{p'}(f)$. Therefore, in a Nash flow f, all players incur a common bottleneck cost $B(f) = \min_p b_p(f)$, and for every $s-t$ path p', $B(f) \leq b'_p(f)$.

We observe that if a flow f is a Nash flow for an $s-t$ network $G(V,E)$, then the set of edges e with $c_e(f_e) \geq B(f)$ comprises an $s-t$ cut in G. For the converse, if for some flow f, there is an $s-t$ cut consisting of edges e either with $f_e > 0$ and $c_e(f_e) = B(f)$, or with $f_e = 0$ and $c_e(f_e) \geq B(f)$, then f is a Nash flow. Moreover, for all bottleneck routing games with linear latencies $a_e x$, a flow f is a Nash flow iff the set of edges e with $c_e(f_e) = B(f)$ comprises an $s-t$ cut.

It can be shown that every bottleneck routing game admits at least one Nash flow (see e.g., [7, Proposition 2]), and that there is an optimal flow that is also a Nash flow (see e.g., [3, Corollary 2]). In general, a bottleneck routing game admits many different Nash flows, each with a possibly different bottleneck cost. Given an instance $\mathcal{G}$, we let $B(\mathcal{G})$ denote the bottleneck cost of the players in the worst Nash flow of $\mathcal{G}$, i.e. the Nash flow f that maximizes $B(f)$ among all Nash flows. We refer to $B(\mathcal{G})$ as the worst equilibrium bottleneck cost of $\mathcal{G}$. For convenience, for an instance $\mathcal{G} = (G,c,r)$, we sometimes write $B(G,r)$, instead of $B(\mathcal{G})$, to denote the worst equilibrium bottleneck cost of $\mathcal{G}$. We note that for every subinstance $\mathcal{H}$ of $\mathcal{G}$, $B^*(\mathcal{G}) \leq B(\mathcal{H})$, and that there may be subinstances $\mathcal{H}$ with $B(\mathcal{H}) < B(\mathcal{G})$, which is the essence of Braess's paradox (see e.g., Fig. 1).

Subpath-Optimal Nash Flows. For a flow f and any vertex u, let $b_f(u)$ denote the minimum bottleneck cost of f among all $s-u$ paths. The flow f is a *subpath-optimal Nash flow* [7] if for any vertex u and any $s-t$ path p with $f_p > 0$ that includes u, the bottleneck cost of the $s-u$ part of p is $b_f(u)$. For example, the Nash flow f in Fig. 1.a is not subpath-optimal, because $b_f(v) = 0$, through the edge (s,v), while the bottleneck cost of the path (s,u,v) is 1. For this instance, the only subpath-optimal Nash flow is the optimal flow.

ε-Nash Flows. The definition of a Nash flow can be generalized to that of an "almost Nash" flow: For some constant $\varepsilon > 0$, a flow f is an ε-Nash flow if for all $s-t$ paths p, p', if $f_p > 0$, $b_p(f) \leq b_{p'}(f) + \varepsilon$.

Price of Anarchy. The *Price of Anarchy* (*PoA*) of an instance $\mathcal{G}$, denoted $\rho(\mathcal{G})$, is the ratio of the worst equilibrium bottleneck cost of $\mathcal{G}$ to the optimal bottleneck cost. Formally, $\rho(\mathcal{G}) = B(\mathcal{G})/B^*(\mathcal{G})$.

Paradox-Free and Paradox-Ridden Instances. A routing instance $\mathcal{G}$ is *paradox-free* if for every subinstance $\mathcal{H}$ of $\mathcal{G}$, $B(\mathcal{H}) \geq B(\mathcal{G})$. Paradox-free instances do not suffer from Braess's paradox and their PoA cannot be improved by edge removal. An instance $\mathcal{G}$ is *paradox-ridden* if there is a subinstance $\mathcal{H}$ of $\mathcal{G}$ such that $B(\mathcal{H}) = B^*(\mathcal{G}) = B(\mathcal{G})/\rho(\mathcal{G})$. Namely, the PoA of paradox-ridden instances can decrease to 1 by edge removal.

Best Subnetwork. Given an instance $\mathcal{G} = (G, c, r)$, the *best subnetwork* H^* of G minimizes the worst equilibrium bottleneck cost, i.e., for all subnetworks H of G, $B(H^*, r) \leq B(H, r)$.

Problem Definitions. Next, we study the complexity and the approximability of two basic selfish network design problems for bottleneck routing games:

- **Paradox-Ridden Recognition** (ParRidBC) : Given an instance $\mathcal{G}$, decide if $\mathcal{G}$ is paradox-ridden.
- **Best Subnetwork** (BSubNBC) : Given an instance $\mathcal{G}$, find the best subnetwork H^* of G.

The objective function of BSubNBC is the worst equilibrium bottleneck cost $B(H, r)$ of a subnetwork H. Thus, a (polynomial-time) algorithm A achieves an α-approximation for BSubNBC if for all instances $\mathcal{G}$, A returns a subnetwork H with $B(H, r) \leq \alpha B(H^*, r)$. A subtle point is that given a subnetwork H, we do not know how to efficiently compute the worst equilibrium bottleneck cost $B(H, r)$ (see also [2,12]). To deal with this delicate issue, our hardness results use a certain subnetwork structure to identify a good approximation to BSubNBC.

3 Paradox-Free Topologies and Paradox-Free Nash Flows

We start by discussing two interesting cases where Braess's paradox does not occur. We first observe that for any bottleneck routing game $\mathcal{G}$ defined on a series-parallel network, $\rho(\mathcal{G}) = 1$, and thus Braess's paradox does not occur. We recall that a directed $s - t$ network is series-parallel iff it does not contain a θ-graph with degree-2 terminals as a topological minor. Therefore, the example in Fig. 1 shows that series-parallel networks is the largest class of networks for which Braess's paradox does not occur (see also [17] for a similar result for additive routing games). The proof is conceptually similar to that of [8, Lemma 4.1].

Proposition 1. *Let $\mathcal{G}$ be a bottleneck routing game on an $s - t$ series-parallel network. Then, $\rho(\mathcal{G}) = 1$.*

Next, we observe that any subpath-optimal Nash flow achieves an optimal bottleneck cost. Thus, Braess's paradox does not occur if we only consider subpath-optimal Nash flows.

Proposition 2. *Let $\mathcal{G}$ be any bottleneck routing game, and let f be any subpath-optimal Nash flow of $\mathcal{G}$. Then, $B(f) = B^*(\mathcal{G})$.*

4 Recognizing Paradox-Ridden Instances Is Hard

Next, we show that given a linear bottleneck routing game $\mathcal{G}$, it is NP-hard not only to decide whether $\mathcal{G}$ is paradox-ridden, but also to approximate the best subnetwork within a factor less than 4/3. To this end, we employ a reduction from the 2-Directed Disjoint Paths problem (2-DDP), where we are given a directed network D and distinguished vertices s_1, s_2, t_1, t_2, and ask whether D contains a pair of vertex-disjoint paths connecting s_1 to t_1 and s_2 to t_2. 2-DDP is NP-complete, even if the network D is known to contain two edge-disjoint paths connecting s_1 to t_2 and s_2 to t_1 [9, Theorem 3]. In the following, we say that a subnetwork D' of D is *good* if D' contains (i) at least one path outgoing from each of s_1 and s_2 to either t_1 or t_2, (ii) at least one path incoming to each of t_1 and t_2 from either s_1 or s_2, and (iii) either no $s_1 - t_2$ paths or no $s_2 - t_1$ paths. We say that D' is *bad* if any of these conditions is violated by D'. We note that we can efficiently check whether a subnetwork D' of D is good, and that a good subnetwork D' serves as a certificate that D is a YES-instance of 2-DDP. The following lemma directly implies the hardness result of this section.

Lemma 1. *Let $\mathcal{I} = (D, s_1, s_2, t_1, t_2)$ be any 2-DDP instance. Then, we can construct, in polynomial time, an $s - t$ network $G(V, E)$ with a linear latency function $c_e(x) = a_e x$, $a_e > 0$, on each edge e, so that for any traffic rate $r > 0$, the bottleneck routing game $\mathcal{G} = (G, c, r)$ has $B^*(\mathcal{G}) = r/4$, and:*

1. *If $\mathcal{I}$ is a YES-instance of 2-DDP, there exists a subnetwork H of G with $B(H, r) = r/4$.*
2. *If $\mathcal{I}$ is a NO-instance of 2-DDP, for all subnetworks H' of G, $B(H', r) \geq r/3$.*
3. *For all subnetworks H' of G, either H' contains a good subnetwork of D, or $B(H', r) \geq r/3$.*

Proof sketch. We construct the network G by adding 4 vertices, s, t, v, u, to D and 9 "external" edges $e_1 = (s, u)$, $e_2 = (u, v)$, $e_3 = (v, t)$, $e_4 = (s, v)$, $e_5 = (v, s_1)$, $e_6 = (s, s_2)$, $e_7 = (t_1, u)$, $e_8 = (u, t)$, $e_9 = (t_2, t)$ (see also Fig. 2.a). The external edges e_1 and e_3 have latency $c_{e_1}(x) = c_{e_3}(x) = x/2$. The external edges $e_4, \ldots, e_9$ have latency $c_{e_i} = x$. The external edge e_2 and each edge e of D have latency $c_{e_2}(x) = c_e(x) = \varepsilon x$, for some $\varepsilon \in (0, 1/4)$.

We first observe that $B^*(\mathcal{G}) = r/4$. As for (1), by hypothesis, there are vertex-disjoint paths in D, p and q, connecting s_1 to t_1, and s_2 to t_2. Let H be the subnetwork of G that includes all external edges and only the edges of p and q from D (see also Fig. 2.b). We let $\mathcal{H} = (H, c, r)$ be the corresponding subinstance of $\mathcal{G}$. The flow routing $r/4$ units through each of the paths (e_4, e_5, p, e_7, e_8) and (e_6, q, e_9), and $r/2$ units through (e_1, e_2, e_3), is an $\mathcal{H}$-feasible Nash flow with a bottleneck cost of $r/4$.

We proceed to show that any Nash flow of $\mathcal{H}$ achieves a bottleneck cost of $r/4$. For sake of contradiction, let f be a Nash flow of $\mathcal{H}$ with $B(f) > r/4$. Since f is a Nash flow, the edges e with $c_e(f_e) \geq B(f)$ form an $s - t$ cut in H. Since the bottleneck cost of e_2 and of any edge in p and q is at most $r/4$, this cut includes either e_6 or e_9 (or both), either e_1 or e_3 (or both), and either e_4 or e_8

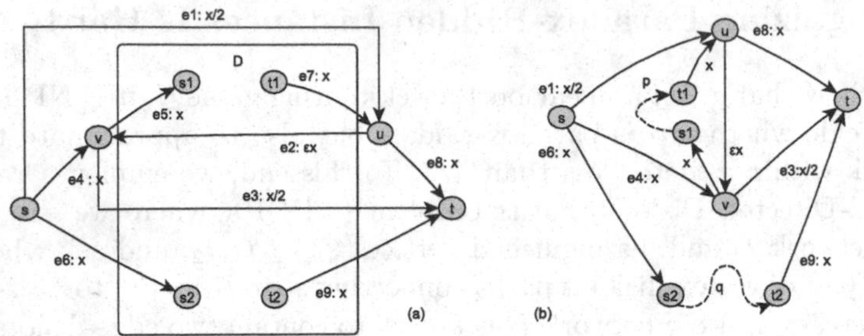

Fig. 2. (a) The network G constructed in the proof of Lemma 1. (b) The best subnetwork of G, with PoA = 1, for the case where D contains a pair of vertex-disjoint paths connecting s_1 to t_1 and s_2 to t_2.

(or e_5 or e_6, in certain combinations with other edges). Let us consider the case where this cut includes e_1, e_4, and e_6. Since the bottleneck cost of these edges is greater than $r/4$, we have more than $r/2$ units of flow through e_1 and more than $r/4$ units of flow through each of e_4 and e_6. Hence, we obtain that more than r units of flow leave s, a contradiction. All other cases are similar.

To conclude the proof, we first observe that (3) implies (2), because if $\mathcal{I}$ is a NO-instance, any two paths, p and q, connecting s_1 to t_1 and s_2 to t_2, have some vertex in common, and thus, D does not include any good subnetworks.

To sketch the proof of (3), we let H' be any subnetwork of G, and let $\mathcal{H}'$ be the corresponding subinstance of $\mathcal{G}$. We can show that either H' contains (i) all external edges, (ii) at least one path outgoing from each of s_1 and s_2 to either t_1 or t_2, and (iii) at least one path incoming to each of t_1 and t_2 from either s_1 or s_2, or H' includes a "small" $s-t$ cut, and any $\mathcal{H}'$-feasible flow f has $B(f) \geq r/3$.

Let us now consider a subnetwork H' of G that does not contain a good subnetwork of D, but it satisfies (i), (ii), and (iii) above. By (ii) and (iii), and the hypothesis that the subnetwork of D included in H' is bad, H' contains an $s_1 - t_2$ path p and an $s_2 - t_1$ path q. At the intuitive level, this corresponds to the case where no edges are removed from G. Then, routing $r/3$ units of flow on each of the $s - t$ paths (e_1, e_2, e_3), (e_1, e_2, e_5, p, e_9), and (e_6, q, e_7, e_2, e_3) has a bottleneck cost of $r/3$ and is a Nash flow, because the edges with bottleneck cost $r/3$ comprise an $s - t$ cut. □

The bottleneck routing game $\mathcal{G}$, in Lemma 1, has $\rho(\mathcal{G}) = 4/3$, and is paradox-ridden, if $\mathcal{I}$ is a YES instance of 2-DDP, and paradox-free, otherwise. Hence:

Theorem 1. *Deciding whether a bottleneck routing game with strictly increasing linear latencies is paradox-ridden is* NP*-hard.*

Moreover, Lemma 1 implies that it is NP-hard to approximate BSubNBC within a factor less than 4/3. A subtle point here is that given a subnetwork H, we do not know how to efficiently compute the worst equilibrium bottleneck cost $B(H, r)$. However, we can use the notion of a good subnetwork of D, and deal with this issue (see also the discussion before Theorem 2).

5 Approximating the Best Subnetwork Is Hard

Next, we recursively apply the construction of Lemma 1, and show that it is NP-hard to approximate BSubNBC within a factor of $O(n^{.121-\varepsilon})$, for any $\varepsilon > 0$.

We consider an $s-t$ network G that can be constructed in polynomial time from a 2-DDP instance $\mathcal{I}$, and includes (possibly many copies of) D. G has a linear latency $c_e(x) = a_e x$ on each edge e, and for any rate $r > 0$, the bottleneck routing game $\mathcal{G} = (G, c, r)$ has $B^*(\mathcal{G}) = r/\gamma_1$, for some $\gamma_1 > 0$. Moreover,

1. If $\mathcal{I}$ is a YES-instance, there exists a subnetwork H of G with $B(H, r) = r/\gamma_1$.
2. If $\mathcal{I}$ is a NO-instance, for all subnetworks H' of G, $B(H', r) \geq r/\gamma_2$, for some $\gamma_2 \in (0, \gamma_1)$.
3. For all subnetworks H' of G, either H' contains at least one copy of a good subnetwork of D, or $B(H', r) \geq r/\gamma_2$.

The existence of such a network G shows that it is NP-hard to approximate BSubNBC within a factor less than $\gamma = \gamma_1/\gamma_2$. Thus, we refer to G as a γ-gap instance. E.g., the network constructed in the proof of Lemma 1 has $\gamma_1 = 4$ and $\gamma_2 = 3$, and thus it is a 4/3-gap instance. We next show that given $\mathcal{I}$ and a γ_1/γ_2-gap instance G, we can construct a $4\gamma_1/(3\gamma_2)$-gap instance G'.

Lemma 2. *Let $\mathcal{I} = (D, s_1, s_2, t_1, t_2)$ be a* 2-DDP *instance, and let G be a γ_1/γ_2-gap instance with linear latencies, based on $\mathcal{I}$. Then, we can construct, in time polynomial in the size of $\mathcal{I}$ and G, an $s-t$ network G' with a linear latency function $c_e(x) = a_e x$, $a_e > 0$, on each edge e, so that for any traffic rate $r > 0$, the bottleneck routing game $\mathcal{G}' = (G', c, r)$ has $B^*(\mathcal{G}) = r/(4\gamma_1)$, and:*

1. *If $\mathcal{I}$ is a* YES*-instance, there is a subnetwork H of G' with $B(H, r) = r/(4\gamma_1)$.*
2. *If $\mathcal{I}$ is a* NO*-instance, for all subnetworks H', $B(H', r) \geq r/(3\gamma_2)$.*
3. *For all subnetworks H' of G', either H' contains at least one copy of a good subnetwork of D, or $B(H', r) \geq r/(3\gamma_2)$.*

The proof applies the construction of Lemma 1, but with all external edges, except for e_2, replaced by a copy of the gap-instance G. Hence, the number of vertices of G' is at most 8 times the number of vertices of G plus the number of vertices of D. If we start with an instance $\mathcal{I}$ of 2-DDP where D has k vertices, and apply Lemma 1 once, and subsequently apply Lemma 2 for $\lfloor \log_{4/3} k \rfloor$ times, we obtain a k-gap instance $\mathcal{G}'$ where G' has $n = O(k^{8.23})$ vertices. Suppose now that there is a polynomial-time algorithm A that approximates the best subnetwork of G' within a factor of $O(k^{1-\varepsilon}) = O(n^{0.121-\varepsilon})$, for a constant $\varepsilon > 0$. Then, if $\mathcal{I}$ is a YES-instance, algorithm A, applied to G', should return a best subnetwork H with at least one copy of a good subnetwork of D. Since H contains a polynomial number of copies of subnetworks of D, and we can check this in polynomial time, and efficiently recognize $\mathcal{I}$ as a YES-instance of 2-DDP. On the other hand, if $\mathcal{I}$ is a NO-instance, D includes no good subnetworks. Again, we can efficiently check that in the subnetwork returned by A, there are not any copies of a good subnetwork of D, and hence recognize $\mathcal{I}$ as a NO-instance of 2-DDP. Thus:

Theorem 2. *For bottleneck routing games with linear latencies, it is* NP*-hard to approximate* BSubNBC *within a factor of $O(n^{0.121-\varepsilon})$, for any constant $\varepsilon > 0$.*

6 Networks with Quasipolynomially Many Paths

In this section, we approximate, in quasipolynomial-time, the best subnetwork and its worst equilibrium bottleneck cost for instances $\mathcal{G} = (G, c, r)$ where the network G has quasipolynomially many $s-t$ paths, the latency functions satisfy a Lipschitz condition, and the worst Nash flow in the best subnetwork routes a non-negligible amount of flow on all edges.

The restriction to networks with quasipolynomially many $s-t$ paths is somehow necessary, in the sense that Theorem 2 shows that if the network has exponentially many $s-t$ paths, as it happens for the hard instances of 2-DDP, and thus for the networks G and G' in the proofs of Lemma 1 and Lemma 2, it is NP-hard to approximate BSubNBC within any reasonable factor. In addition, we assume here that there is a constant $\delta > 0$, such that the worst Nash flow in the best subnetwork H^* routes more than δ units of flow on all edges of H^*.

W.l.o.g., we normalize the traffic rate r to 1. Our algorithm is based on [11, Lemma 2], which applies Althöfer's Lemma [1], and shows that any flow can be approximated by a sparse flow using logarithmically many paths.

Lemma 3. *Let $\mathcal{G} = (G(V, E), c, 1)$ be an instance, and let f be a flow. Then, for any $\varepsilon > 0$, there exists a $\mathcal{G}$-feasible flow $\tilde{f}$ using at most $k(\varepsilon) = \lfloor \log(2m)/(2\varepsilon^2) \rfloor +$ 1 paths, such that for all edges e, $|\tilde{f}_e - f_e| \leq \varepsilon$, if $f_e > 0$, and $\tilde{f}_e = 0$, otherwise.*

By Lemma 3, there exists a sparse flow $\tilde{f}$ that approximates the worst Nash flow f on the best subnetwork H^* of G. Moreover, the proof of [11, Lemma 2] shows that the flow $\tilde{f}$ is determined by a multiset P of at most $k(\varepsilon)$ paths, selected among the paths used by f. Then, for every path $p \in \mathcal{P}$, $\tilde{f}_p = |P(p)|/|P|$, where $|P(p)|$ is number of times the path p is included in the multiset P. Therefore, if the total number $|\mathcal{P}|$ of $s-t$ paths in G is quasipolynomial, we can find, by exhaustive search, in quasipolynomial-time, a flow-subnetwork pair that approximates the optimal solution of BSubNBC. Based on this intuition, we can obtain an approximation algorithm for BSubNBC on networks with quasipolynomially many paths, under the technical assumption that the worst Nash flow in the best subnetwork routes a non-negligible amount of flow on all edges.

Theorem 3. *Let $\mathcal{G} = (G(V, E), c, 1)$ be a bottleneck routing game with latency functions that satisfy the Lipschitz condition with a constant $\xi > 0$, let H^* be the best subnetwork of G, and let f^* be the worst Nash flow in H^*. If for all edges e of H^*, $f^*_e > \delta$, for some constant $\delta > 0$, then for any constant $\varepsilon > 0$, we can compute in time $|\mathcal{P}|^{O(\log(2m)/\min\{\delta^2, \varepsilon^2/\xi^2\})}$ a flow f and a subnetwork H such that: (i) f is an $\varepsilon/2$-Nash flow in the subnetwork H, (ii) $B(f) \leq B(H^*, 1) + \varepsilon$, (iii) $B(H, 1) \leq B(f) + \varepsilon/4$, and (iv) $B(f) \leq B(H, 1) + \varepsilon/2$.*

The algorithm of Theorem 3 computes a flow-subnetwork pair (H, f) such that f is an $\varepsilon/2$-Nash flow in H, the worst equilibrium bottleneck cost of H approximates the worst equilibrium bottleneck cost of H^*, since $B(H^*, 1) \leq B(H, 1) \leq B(H^*, 1) + 5\varepsilon/4$, by (ii) and (iii), and the bottleneck cost of f approximates the worst equilibrium bottleneck cost of H, since $B(H, 1) - \varepsilon/4 \leq B(f) \leq B(H, 1) + \varepsilon/2$, by (iii) and (iv).

References

1. Althöfer, I.: On sparse approximations to randomized strategies and convex combinations. Linear Algebra and Applications 99, 339–355 (1994)
2. Azar, Y., Epstein, A.: The Hardness of Network Design for Unsplittable Flow with Selfish Users. In: Erlebach, T., Persinao, G. (eds.) WAOA 2005. LNCS, vol. 3879, pp. 41–54. Springer, Heidelberg (2006)
3. Banner, R., Orda, A.: Bottleneck routing games in communication networks. IEEE Journal on Selected Areas in Communications 25(6), 1173–1179 (2007)
4. Braess, D.: Über ein paradox aus der Verkehrsplanung. Unternehmensforschung 12, 258–268 (1968)
5. Busch, C., Magdon-Ismail, M.: Atomic routing games on maximum congestion. Theoretical Computer Science 410, 3337–3347 (2009)
6. Caragiannis, I., Galdi, C., Kaklamanis, C.: Network Load Games. In: Deng, X., Du, D.-Z. (eds.) ISAAC 2005. LNCS, vol. 3827, pp. 809–818. Springer, Heidelberg (2005)
7. Cole, R., Dodis, Y., Roughgarden, T.: Bottleneck links, variable demand, and the tragedy of the commons. In: Proc. of the 17th ACM-SIAM Symposium on Discrete Algorithms, SODA 2006, pp. 668–677 (2006)
8. Epstein, A., Feldman, M., Mansour, Y.: Efficient graph topologies in network routing games. Games and Economic Behaviour 66(1), 115–125 (2009)
9. Fortune, S., Hopcroft, J.E., Wyllie, J.: The directed subgraph homeomorphism problem. Theoretical Computer Science 10, 111–121 (1980)
10. Fotakis, D., Kaporis, A.C., Lianeas, T., Spirakis, P.: On the hardness of network design for bottleneck routing games. CoRR, abs/1207.5212 (2012)
11. Fotakis, D., Kaporis, A.C., Spirakis, P.G.: Efficient Methods for Selfish Network Design. In: Albers, S., Marchetti-Spaccamela, A., Matias, Y., Nikoletseas, S., Thomas, W. (eds.) ICALP 2009, Part II. LNCS, vol. 5556, pp. 459–471. Springer, Heidelberg (2009)
12. Hou, H., Zhang, G.: The Hardness of Selective Network Design for Bottleneck Routing Games. In: Cai, J.-Y., Cooper, S.B., Zhu, H. (eds.) TAMC 2007. LNCS, vol. 4484, pp. 58–66. Springer, Heidelberg (2007)
13. Koutsoupias, E., Papadimitriou, C.: Worst-Case Equilibria. In: Meinel, C., Tison, S. (eds.) STACS 1999. LNCS, vol. 1563, pp. 404–413. Springer, Heidelberg (1999)
14. Lin, H., Roughgarden, T., Tardos, É.: A stronger bound on Braess's paradox. In: Proc. of the 15th ACM-SIAM Symposium on Discrete Algorithms, SODA 2004, pp. 340–341 (2004)
15. Lin, H., Roughgarden, T., Tardos, É., Walkover, A.: Braess's Paradox, Fibonacci Numbers, and Exponential Inapproximability. In: Caires, L., Italiano, G.F., Monteiro, L., Palamidessi, C., Yung, M. (eds.) ICALP 2005. LNCS, vol. 3580, pp. 497–512. Springer, Heidelberg (2005)
16. Mazalov, V., Monien, B., Schoppmann, F., Tiemann, K.: Wardrop Equilibria and Price of Stability for Bottleneck Games with Splittable Traffic. In: Spirakis, P.G., Mavronicolas, M., Kontogiannis, S.C. (eds.) WINE 2006. LNCS, vol. 4286, pp. 331–342. Springer, Heidelberg (2006)
17. Milchtaich, I.: Network topology and the efficiency of equilibrium. Games and Economic Behavior 57, 321–346 (2006)
18. Roughgarden, T.: On the severity of Braess's paradox: Designing networks for selfish users is hard. Journal of Computer and System Sciences 72(5), 922–953 (2006)

Ad Auctions with Data*

Hu Fu[1], Patrick Jordan[2], Mohammad Mahdian[3], Uri Nadav[3], Inbal Talgam-Cohen[4], and Sergei Vassilvitskii[3]

[1] Cornell University, Ithaca, NY, USA
[2] Microsoft Inc., Mountain View, CA, USA
[3] Google Inc., Mountain View, CA, USA
[4] Stanford University, Stanford, CA, USA

Abstract. The holy grail of online advertising is to target users with ads matched to their needs with such precision that the users respond to the ads, thereby increasing both advertisers' and users' value. The current approach to this challenge utilizes information about the users: their gender, their location, the websites they have visited before, and so on. Incorporating this data in ad auctions poses an economic challenge: can this be done in a way that the auctioneer's revenue does not decrease (at least on average)? This is the problem we study in this paper. Our main result is that in Myerson's optimal mechanism, for a general model of data in auctions, additional data leads to additional expected revenue. In the context of ad auctions we show that for the simple and common mechanisms, namely second price auction with reserve prices, there are instances in which additional data decreases the expected revenue, but this decrease is by at most a small constant factor under a standard regularity assumption.

1 Introduction

When an item with latent characteristics is sold, information revealed by the seller plays a significant role in the value ascribed to the item by potential buyers. For example, when booking a hotel room on a website such as Priceline.com, every extra piece of information—including the hotel's star level or its location—affects the price a buyer is willing to pay. In a similar manner, in online advertising scenarios, any information revealed about the ad opportunity—including the description of the webpage's content or the type of user—plays a crucial role in determining the ad's value, in particular because this information is extremely useful in predicting the click and conversion rate of the user.

In online display advertising settings, the publisher auctions off opportunities to show an advertisement to its users in real time, often through online ad marketplaces operated by companies such as Yahoo!, Google or Microsoft. For example, every time a user visits The New York Times website, the opportunity to show an advertisement to the user is auctioned off. Both the publisher (in this case The New York Times) and the market operator have a great deal

* Work done while the authors were at Yahoo! Reseaerch

M. Serna (Ed.): SAGT 2012, LNCS 7615, pp. 168–179, 2012.

of information about the ad opportunity, including page specific features such as layout and content, as well as user specific features such as the user's age, gender, location, etc. *How much of this information should be revealed during the auction in order to maximize revenue?* This is the question we study in this work.

While concealing information can only decrease social efficiency, it may be advantageous in terms of revenue, since releasing information may decrease competition. As an example, suppose an advertiser values males at $2 and females at $8. In an incentive compatible auction, the advertiser bids his value when the user's gender is known, but will bid the expected value of $5 when the gender is not revealed (assuming each gender is equally likely). If there is a second advertiser who values males at $8 and females at $2, then revealing gender segments the buyers. As a result, when gender is revealed the auctioneer will face a bid of $8 and $2, and thus collect only $2 in a second price auction; if the gender is kept hidden, the auctioneer will have two bids of $5 and will collect $5.[1]

The example above may seem to suggest that it is never in the auctioneer's interest to release information about the item. Indeed, Board [5] has shown that revealing information can only decrease the expected revenue from a second price auction with two bidders. However, the auctioneer has additional tools to increase revenue at her disposal, namely she can set a reserve price for each bidder. The right reserve price may counter the potential loss in competition, allowing the auctioneer to preserve its revenue. In the example above, a reserve price of $8 for both advertisers would lead to a revenue of $8 precisely in the case where gender is revealed. On the other hand, it is not obvious that using reserve prices or even applying the optimal mechanism is sufficient to recover the lost revenue from revealing data; see Example 1 in Section 4.2 for a simple case in which this does not hold.

Our Contribution. In this work we study a general model of single-parameter auctions with data. We show that while revealing information can lead to a decrease in the expected revenue of second price auctions, using the revenue-optimal mechanism counteracts this trend. Our main result is that if Myerson [17]'s optimal auction mechanism is used, the expected revenue is guaranteed to (weakly) increase when more information is revealed. This result also applies to slot auctions and other settings.

We explore the assumptions of this result and show that they are necessary for revenue monotonicity to hold. In particular, if instead of Myerson's optimal mechanism, a simpler reserve price based mechanism is used, revealing information can lead to a decrease in expected revenue. However, we prove that in simple and practical second price auctions with reserve prices, fully revealing the auctioneer's information generates approximately the optimal revenue even compared with arbitrary intricate revealing schemes the auctioneer may adopt.

[1] Perturbing this example slightly shows withholding information can decrease social welfare.

1.1 Related Work

The following scenario has been extensively studied in auction theory: The auctioneer has access to a private source of data about the item; she wishes to maximize her expected revenue by pre-committing to a policy of revealing or concealing data. Two effects of revealing data have been identified: the *linkage principle* by Milgrom and Weber [15], and more recently the *allocation effect* by Board [5]. The linkage principle says that when bidders' valuations are positively correlated in a specific way to the auctioneer's data, the auctioneer can increase her revenue in first or second price auctions by revealing the data. However, in ad auctions, revealing information can increase the value to some advertisers and decrease it for the rest, and so the linkage principle does not apply. The allocation effect studies the effect of information revelation on revenue in second price auctions as the number of bidders changes.

A recent line of research [9, 16] considers the computational problem of finding the optimal information revelation scheme in second price auctions. In contrast, we study optimal auctions and their implications for auctions such as second price auctions with reserve prices, and never second price auctions per se. The valuation model of ours is also different. It is neither deterministic nor arbitrarily correlated, as studied by Emek et al. [9].

We briefly mention related work further afield. Levin and Milgrom [13] highlight disadvantages of information revelation from a market design point of view—too much information leads to thin markets that are hard to operate. Several proposed mechanisms address these issues [4, 7]. Dwork et al. [8] discuss fairness concerns arising from revealing user data. A separate body of work considers cases in which bidders, not the auctioneer, have private sources of information about the item, resulting in asymmetries among them; a recent example is Abraham et al. [1]. Ghosh et al. [10] study information revelation in ad auctions through the process of cookie-matching, and its impact on the revenue of the auction.

2 Preliminaries

We briefly describe Myerson [17]'s optimal truthful mechanism, under the interpretation of Bulow and Roberts [6]. Given a valuation distribution F, each probability quantile q corresponds to a value $v = F^{-1}(1-q)$. Each value, in turn, corresponds to an expected revenue $v(1-F(v))$ generated by setting a posted price of v. A revenue curve depicts such revenue $R(q) = qF^{-1}(1-q)$ as a function of the quantile q, and the ironed revenue curve $\tilde{R}(q)$ is the concave hull of this curve. The ironed virtual valuation of v under the distribution F is then $\tilde{\varphi}(v) = \frac{\mathrm{d}\tilde{R}(q)}{\mathrm{d}q}|_{q=1-F(v)}$.

Theorem 1 (Myerson 17). *In a revenue optimal truthful auction in which bidders' valuations are independently drawn from known distributions $D_1, \cdots, D_n$, the item is allocated to the bidder with the highest non-negative ironed virtual valuation, and the expected revenue is equal to* $\mathbf{E}[\max\{0, \tilde{\varphi}_1(v_1), \ldots, \tilde{\varphi}_n(v_n)\}]$.

3 Model

We describe a general model to which our main result applies, and show that ad auctions are captured in a natural way as an instantiation of this model.

General model. n bidders compete in an auction, in which the subsets of bidders who can win simultaneously are specified by $\mathcal{I} \subseteq 2^{[n]}$, the *feasible sets.* Every bidder $i \in [n]$ has a private, single-dimensional *signal* $s_i \in \mathbb{R}_+$, drawn independently from a publicly known distribution F_i with density f_i. In addition, the auctioneer also has a private signal $u \in U$, drawn from a publicly known discrete distribution F_U with density f_U independently of the bidder signals. We also call u an item type. Denote $|U|$ by m. Bidder i's value for winning the auction is a publicly known function v_i of his own signal s_i and the auctioneer's signal u: $v_{i,u} = v_i(s_i, u)$. For every u we assume that $v_i(\cdot, u)$ is non-negative and strictly increasing in its argument s_i. Note that, under these constraints, u can affect $v_{i,u}$ in a fairly general manner.

Signaling schemes. We adopt the framework for signaling schemes developed by Emek et al. [9] and Miltersen and Sheffet [16] which has its origins in Milgrom and Weber [15, see Theorem 9]. A signaling scheme is a set of m distributions over a signal set Σ of size k. On seeing type u, the auctioneer sends a signal $\sigma \in \Sigma$ with probability $\psi_{u,\sigma}$, and then bidders bid their expected value inferred from the posterior distribution on u given σ. It is important that the auctioneer commits to a signaling scheme before the auction starts. In the case of fully withdrawing the information, Σ has one element, which we call $\bar{u}$, and bidder i's posterior valuation in this case is denoted $v_{i,\bar{u}} = \mathbf{E}_{u \sim F_U}[v_{i,u}]$. In discussing this scenario, we often equivalently talk about a fictitious item type $\bar{u}$, for which each bidder i's valuation is $v_{i,\bar{u}}$.

Ad Auctions: An Instantiation. Ad auctions can be viewed as a special case of the above general model. In this case one opportunity of displaying an ad is auctioned to n bidders, and therefore the feasible sets $\mathcal{I}$ consist of single winning bidders. In this scenario, the item type u may refelect the auctioneer's information on the user to whom the ad is to be shown. (For this reason we also call u the *user type.*) A widely used model for ad auction is that a bidder i has a private value s_i for a user to click his ad, and for each user type u there is a particular probability $p_{i,u}$ with which the user does click. $p_{i,u}$ is the so-called *click-through rate.* Now a bidder's valuation is simply $v_{i,u} = v_i(s_i, u) = p_{i,u}s_i$.

4 Full Revelation in Myerson's Optimal Mechanism

In this section we present and prove our main result, which states that in Myerson's optimal mechanism, the expected revenue is monotone non-decreasing in revealed information, and so full revelation of the auctioneer's information maximizes the expected revenue.

Other results on information revelation in the private-value setting focus on the second price or English auctions [5, 14]. We first show, in the concrete context of ad auctions, why applying Myerson's mechanism gives different results. We then briefly discuss how to extend our results for ad auctions to slot auctions. In Section 4.2 we give a simple and general proof of the main result for the single-parameter model introduced in Section 3.

4.1 Second Price Auction vs. Myerson's Mechanism in Ad Auctions

Consider the two extreme signaling schemes of full revelation and no revelation. We show why in contrast to the result of Board [5] for the English auction, in Myerson's mechanism the former scheme is always preferable to the latter in terms of expected revenue (where expectations are taken over the random private signals and, where appropriate, over the random user type). The optimality of full revelation and monotonicity of expected revenue in information follow as corollaries.

Proposition 1. *In the ad auctions model, the expected revenue from Myerson's mechanism when the user's type u is revealed is at least as high as the expected revenue when u is not revealed.*

For completeness we include Board's result for 2 bidders (note that since $n = 2$, the second price and English auctions are the same).

Proposition 2 (Board [5]). *In a generalization of the ad auctions model with $n = 2$ bidders, the expected revenue from the second price or English auction when the user's type u is not revealed is at least as high as the expected revenue when u is revealed.*

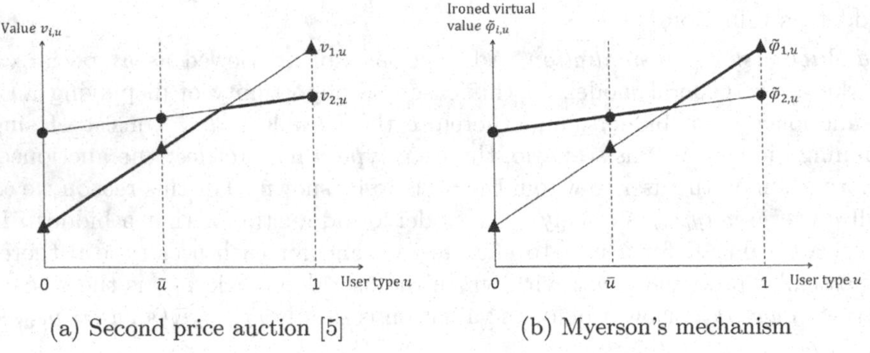

(a) Second price auction [5] (b) Myerson's mechanism

Fig. 1. The Effect of Information Revelation

Figure 1, adapted from Board [5], provides intuition for the difference between the above propositions (see also Palfrey [18], McAfee [14]). In the second price auction, for every signal profile of the bidders, the revenue is the minimum of their values and so a concave function. Therefore, while revealing information produces the average of pointwise minimums, no revelation does at least as well by producing the minimum of averages. By contrast, recall that applying Myerson's mechanism in our setting means that given user type $u \in U \cup \{\bar{u}\}$, the auctioneer transforms the advertisers' values $\{v_{i,u}\}$ into the corresponding ironed virtual values $\{\tilde{\varphi}_{i,u}\}$, and then allocates the impression to the advertiser with highest non-negative ironed virtual value. The expected revenue is equal to the expected ironed virtual surplus, and for every signal profile the ironed virtual surplus is convex, so the effect of data revelation is reversed.

To formalize this intuition we need to show that the same relation that holds for values before the transformation to ironed virtual values, when a bidder's value under no revelation is equal to his expected value under full revelation, continues to hold after the transformation as well. This is established in Observations 1 and 2. The proof of Proposition 1 then applies convexity and Jensen's inequality to get the result.

Observation 1. *Let advertiser* i*'s value be* $v_{i,u} = p_{i,u}s_i$*, where* $u \in U \cup \{\bar{u}\}$ *is the user type and* $s_i \sim F_i$*. Then* $v_{i,u}$ *is distributed according to* $F_{i,u}(x) = F_i(x/p_{i,u})$*, and the corresponding ironed virtual value function is* $\tilde{\varphi}_{i,u}(x) = p_{i,u}\tilde{\varphi}_i(x/p_{i,u})$*.*

Proof. The derivation of $F_{i,u}$ is straightforward. The expression for the ironed virtual value follows by looking at the revenue curves R_i and $R_{i,u}$ corresponding to distributions F_i and $F_{i,u}$ respectively:

$$\begin{aligned} R_{i,u}(1 - F_{i,u}(x)) &= x(1 - F_{i,u}(x)) \\ &= p_{i,u} \cdot \frac{x}{p_{i,u}}(1 - F_i(x/p_{i,u})) \\ &= p_{i,u}R_i(1 - F_i(x/p_{i,u})). \end{aligned}$$

The ironed revenue curves are concave hulls of the revenue curves, and therefore preserve the same relationship $\tilde{R}_{i,u}(1 - F_{i,u}(x)) = p_{i,u}\tilde{R}_i(1 - F_i(x/p_{i,u}))$. The ironed virtual valuations, which are their derivatives, also satisfy the same linear relationship. □

We can now compare the ironed virtual values with and without information revelation $\tilde{\varphi}_{i,u}$ and $\tilde{\varphi}_{i,\bar{u}}$. We show the latter equals the former in expectation.

Observation 2. $\tilde{\varphi}_{i,\bar{u}}(v_{i,\bar{u}}) = \mathbf{E}_{u\sim F_U}[\tilde{\varphi}_{i,u}(v_{i,u})]$.

Proof. We have

$$\begin{aligned} \tilde{\varphi}_{i,\bar{u}}(p_{i,\bar{u}}s_i) = p_{i,\bar{u}}\tilde{\varphi}_i(s_i) &= \mathbf{E}_{u\sim F_U}\left[p_{i,u}\right]\tilde{\varphi}_i(s_i) \\ &= \mathbf{E}_{u\sim F_U}\left[p_{i,u}\tilde{\varphi}_i(s_i)\right] = \mathbf{E}_{u\sim F_U}\left[\tilde{\varphi}_{i,u}(p_{i,u}s_i)\right], \end{aligned}$$

where the first and last equalities are by Observation 1, the second is by definition of $p_{i,\bar{u}}$, and the third is by linearity of expectation. □

Proof of Proposition 1. The expected revenue of Myerson's mechanism is equal to its expected ironed virtual surplus [17] [see also 11, Theorem 13.10]. We use this result by Myerson to prove the proposition as follows. We show that pointwise for every fixed profile of values per click $(s_1, \ldots, s_n)$, the ironed virtual surplus of Myerson's mechanism when u is revealed is at least as high as when u is not revealed, in expectation over u. Taking expectation over profiles $(s_1, \ldots, s_n)$ and applying Myerson's result completes the proof.

Fix $(s_1, \ldots, s_n)$ and let $u \in U$ be the user's type. The ironed virtual surplus of Myerson's mechanism when u is revealed is $\max\{0, \tilde{\varphi}_{1,u}(p_{1,u}s_1), \ldots, \tilde{\varphi}_{n,u}(p_{n,u}s_n)\}$. We will omit the term 0 from this point on, since we can always add a dummy bidder whose valuation (and virtual valuation) is constantly 0. Taking expectation over u gives

$$\mathbf{E}_{u \sim F_U}\left[\max\{\tilde{\varphi}_{1,u}(p_{1,u}s_1), \ldots, \tilde{\varphi}_{n,u}(p_{n,u}s_n)\}\right]. \tag{1}$$

If u is not revealed, the ironed virtual surplus of Myerson's mechanism is

$$\max\{\tilde{\varphi}_{1,\bar{u}}(p_{1,\bar{u}}s_1), \ldots, \tilde{\varphi}_{n,\bar{u}}(p_{n,\bar{u}}s_n)\}.$$

By Observation 2, this is equal to

$$\max\left\{\mathbf{E}_{u \sim F_U}\left[\tilde{\varphi}_{i,u}(p_{i,u}s_i)\right]\right\}_{i=1}^{n}. \tag{2}$$

Since max is a convex function, by Jensen's inequality (1) $\geq$ (2). We conclude that in expectation over u, revealing the user's type u does not reduce the ironed virtual surplus. □

So far we have considered only two possible signaling schemes for the auctioneer: either to fully reveal the user's type or to conceal it. A direct corollary of Proposition 1 is that the full revelation strategy yields the highest expected revenue among *all* possible signaling schemes.

Corollary 1. *In the ad auctions model, the expected revenue from Myerson's mechanism when the user type is revealed is optimal among all signaling schemes.*

Proof. Consider a signaling scheme $\{\psi_{u,\sigma}\}_{u \in U, \sigma \in \Sigma}$. Condition on the revealed signal σ. Recall that together with the scheme $\{\psi_{u,\sigma}\}$ and the distribution F_U, it induces an ex post distribution $F_{U|\sigma}$ on the user types. We can now apply Proposition 1 to the setting in which $u \sim F_{U|\sigma}$, and conclude that the expected revenue from full revelation of u is at least as high as the expected revenue from revealing σ. Taking expectation over $\sigma \in \Sigma$ completes the proof. □

Furthermore, Proposition 1 implies *monotonicity* of optimal expected revenue in information release—adding any signaling scheme to Myerson's mechanism can only improve expected revenue.

Corollary 2. *In the ad auctions model, the expected revenue from Myerson's mechanism with a signaling scheme is at least the expected revenue from Myerson's mechanism with no signaling.*

Proof. Let f_U be the density of the user types in the original setting, and let $\{\psi_{u,\sigma}\}_{u \in U, \sigma \in \Sigma}$ be the signaling scheme. Now consider the following alternative setting: A user type σ is sampled from Σ with probability $\sum_{u \in U} f_U(u)\psi_{u,\sigma}$, and the bidders' values are $\{v_{i,\sigma}\}$. Observe that the expected revenue from Myerson's mechanism with signaling scheme $\{\psi_{u,\sigma}\}$ in the original setting equals the expected revenue from Myerson's mechanism with full revelation in the new setting. Similarly, the expected revenue from Myerson with no signaling is the same in both settings. Applying Proposition 1 to the new setting we get that the expected revenue from full revelation of σ is at least as high as the expected revenue from no revelation, completing the proof. □

Generalization to Slot Auctions. A particular case of practical interest is slot auctions, in which the auctioneer has k slots $\{1, \cdots, k\}$ to sell to the advertisers, and an advertiser's value for winning depends on the particular slot he gets. We extend our result for ad auctions to show that full information revelation is the optimal signaling scheme for optimal slot auctions. The main thing to show is that the ironed virtual surplus remains convex.

Formally, the slots have intrinsic click through rates $\alpha_1 \geq \alpha_2 \geq \cdots \geq \alpha_k$. An advertiser's valuation for a user of type u at slot j is $\alpha_j p_{i,u} s_i$. By the same argument as in Observation 1, his ironed virtual valuation is $\alpha_j p_{i,u} \tilde{\varphi}_i(s_i)$. The optimal auction ranks the k bidders with highest non-negative ironed virtual valuations and maps them to the k slots accordingly (if there are fewer than k bidders with non-negative ironed virtual valuations then the remaining slots are not sold). The auctioneer's expected revenue is then the expected sum of the k highest ironed virtual valuations. Just as the proof of Proposition 1 and its corollaries relies on the fact that taking maximum is a convex function, a similar full revelation statement for slot auctions follows from the next observation, whose proof is a consequence of the rearrangement inequality and appears in the appendix.

Observation 3. *The function* $M_k(v_1, \ldots, v_n) = \sum_{j=1}^{k} \alpha_j \max\text{-}j\{v_1, \ldots, v_n\}$ *is a convex function in* $(v_1, \ldots, v_n)$*, where* $\max\text{-}j\{v_1, \ldots, v_n\}$ *is the* j*-th largest element from the set* $\{v_1, \ldots, v_n\}$*.*

For space consideration we omit the proof of this, which is a simple application of the rearrangement inequality.

By the same argument as before using Jensen's inequality, we obtain

Corollary 3. *In slot auctions, the expected revenue from Myerson's mechanism when the user type is revealed is optimal among all signaling schemes.*

4.2 General Model

We extend the optimal revelation results in Section 4 to the general single-parameter model introduced in Section 3. The proof there uses the specific form of ironed virtual values found in this model (also used in Section 5.1). The specific ironed virtual values form is not necessary for the result to hold, and here we prove a general full revelation result for Myerson's mechanism based only on its optimality and monotonicity, and not on the details of its allocation rule.

Proposition 3. *In the general single-parameter model with values* $v_{i,u} = v_i(s_i, u)$*, where* v_i *is non-negative, strictly increasing in* s_i *and continuously differentiable for every* i*, the expected revenue from Myerson's mechanism when the auctioneer's information* u *is revealed is optimal among all signaling schemes.*

Proof. Similarly to ad auctions, it is sufficient to compare full revelation to no revelation. Assume full revelation, and fix the revealed signal of the auctioneer to be $u \in U$. We define the following auxiliary mechanism M. Mechanism M

receives reported values $\{v_{i,u}\}$ from the bidders. By the assumption that $v_{i,u} = v_i(s_i, u)$ is strictly increasing in bidder i's signal s_i, for every i and u there is a one-to-one relation between bidder i's signals and values. Therefore mechanism M may recover the bidders' signals from their reported values. It then finds $\{v_{i,\bar{u}}\}$, the set of values that would have been reported by the bidders if no data had been revealed. Finally, M runs Myerson's mechanism on these values, assuming they're drawn from distributions $\{F_{i,\bar{u}}\}$.

We first claim that the auxiliary mechanism M is truthful, i.e., that its allocation rule is monotone in the reported values $\{v_{i,u}\}$. Fix i and s_{-i}. We want to show that increasing $v_{i,u}$ can only cause M to allocate to bidder i more often. By truthfulness of Myerson's mechanism, we know that M is monotone in $v_{i,\bar{u}}$. Again by the assumption that $v_{i,u}$ is strictly increasing in s_i for every u, the expectation $v_{i,\bar{u}}$ is also strictly increasing in s_i. So M is monotone in s_i, and thus also in $v_{i,u}$, as required.

Consider the expected revenue of the auxiliary mechanism M. On one hand, in expectation over $u \sim F_U$ and the signal profile $\boldsymbol{s}$, its revenue equals that of Myerson's mechanism with no data revelation. On the other hand, for every fixed $u \in U$, Myerson's mechanism with full revelation does at least as well as M in terms of expected revenue over the signal profile, simply because it is optimal. We conclude that the expected revenue of Myerson with full revelation is at least as high as with no revelation, completing the proof. □

The following example shows that optimality of full revelation does not hold without the assumption that a bidder's value is strictly increasing in his signal.

Example 1. Assume u is distributed uniformly over $\{0, 1\}$, and there's a single bidder whose private signal s is distributed uniformly over a discrete support $\{1, 2, 3\}$. When $u = 0$, the bidder's value is just his signal, i.e., $v(s, 0) = s$. When $u = 1$, the values are $v(s, 1) = 4 - s$. Then with full revelation, the maximum expected revenue is $\frac{4}{3}$ by setting a reserve price of 2. When no information is revealed, the bidder's value is $v_{\bar{u}} = 2$ and so the expected revenue is 2.

5 Full Revelation in Simple Auctions with Reserve Prices

In this section we show several results relating to simple, commonly-used ad auctions, namely second price auctions with *anonymous reserves*, and second price auctions with *monopoly reserves*. In the former, a single reserve price is applied to all advertisers, and only those who bid above the reserve compete in a second price auction. In the latter, a distinct monopoly reserve price is applied to each advertiser, and advertisers who bid above their respective reserves enter the second price auction. The monopoly reserve price for a bidder with regular distribution is the optimal price for the auctioneer to set in an auction where only this bidder participates. Equivalently, it is equal to the value v whose corresponding virtual value $\varphi(v)$ is 0.

First, in Section 5.1, we complement our results for optimal ad auctions by showing that in second price auctions with reserves, fully revealing information is approximately optimal among all signaling schemes, provided that advertisers' distributions are regular. This is encouraging in light of previous results on signaling in second price auctions without reserves: Emek et al. [9] showed that finding the optimal signaling scheme is NP-hard, and no approximation algorithm is known yet. We note that in practice, second price auctions with reserve are more common than those without reserve.

In Sections 5.2 and 5.3, we demonstrate that full revelation in simple auctions can be sub-optimal. In fact, revealing no information at all can sometimes leave the auctioneer better off, even for distributions such as the uniform distribution, although by no more than a small constant factor, as we show in Section 5.1.

5.1 Approximation Guarantee in Simple Ad Auctions

We recall the following result of Hartline and Roughgarden [12] on the performance of second price auctions with reserves.

Theorem 2. *For every single-item setting with values drawn independently from regular distributions,*

1. *the expected revenue of the second price auction with the optimal anonymous reserve price is a 4-approximation to the optimal expected revenue; and*
2. *the expected revenue of the second price auction with monopoly reserves is a 2-approximation to the optimal expected revenue.*

Corollary 4. *In ad auctions, when bidders' valuations per click s_i are independently drawn from regular distributions, fully revealing the type in a second price auction with anonymous reserve (monopoly reserves, resp.) is a 4-approximation (2-approximation, resp.) to the expected revenue of the optimal signaling scheme.*

Proof. Consider an optimal signaling scheme in a second price auction with reserves. Under the same signaling scheme, running Myerson's optimal auction would extract at least the same expected revenue. By Corollary 1, fully revealing the user type is optimal among all signaling schemes in Myerson's auction. Then for every user type u, we apply Observation 1 and the regularity of the s_i's to establish regularity of the $v_{i,u}$'s, and so a second price auction with reserves extracts a 4 (or 2)-approximation by Theorem 2. We conclude that fully revealing the information u in a second price auction with reserves extracts a 4 (or 2)-approximation to the revenue obtained by Myerson's optimal auction with full information revelation. The corollary follows from this chain of bounds. □

5.2 Revenue Loss with Anonymous Reserve

This section gives an example in which announcing the item type decreases the revenue of the second price ad auction with the optimal anonymous reserve price.

The example has $n = 2$ bidders and $m = 2$ item types, with F_U being uniform between the two types. Bidder 1's valuation for a "high" type is uniformly drawn from $[0, 2]$, and for a "low" type is 0. Bidder 2 is not sensitive to the types and her valuation is drawn uniformly from $[0, 1]$ regardless of the type. When the type is not announced, the optimal auction is a second price auction with reserve price 1/2, and the optimal revenue is 5/12. When the type is revealed to be low, the optimal auction is a second price auction with a reserve price 1/2, and the revenue is 1/4. When the type is revealed to be high, if we set a reserve price of $x \in [0,1]$, the revenue is

$$x\left[x(1-\frac{x}{2})+\frac{x}{2}(1-x)\right]+\int_x^1 y(1-\frac{y}{2})+\frac{y}{2}(1-y)\,dy=\frac{3}{4}x^2-\frac{2}{3}x^3+\frac{5}{12}.$$

This is maximized at $x = 3/4$, yielding a revenue of $\frac{9}{64}+\frac{5}{12}$. Setting a reserve price in $[1,2]$ does no give a revenue better than 0.5. Therefore, for a high type, the revenue of an optimal second price auction with anonymous reserve is $\frac{9}{64}$ more than $\frac{5}{12}$, whereas for a low type the revenue is $\frac{1}{6}$ less. On average, if we reveal the type, the expected revenue is strictly less than $\frac{5}{12}$.

5.3 Revenue Loss with Monopoly Reserves

This section presents an example in which announcing the item type decreases the revenue of the second price ad auction with monopoly reserve prices.

As in the previous section, we assume 2 bidders and 2 types, with F_U being uniform. Bidder 1's valuation is uniformly drawn from $[0,8]$ for a "high" type, and uniformly from $[0,4]$ for a "low" type, whereas bidder 2 is not sensitive to the item type and her valuation is uniformly drawn from $[0,6]$ regardless of the type. When the type is not revealed, the optimal auction is a second price auction with reserve price 3, and the expected revenue is 2.5. When the type is revealed to be high, the monopoly reserves are 4 and 3, respectively. The expected revenue is:

$$\begin{aligned}&4\cdot\Pr(v_1\in[4,8],v_2\in[0,3])+3\cdot\Pr(v_1\in[0,4],v_2\in[3,6])+\\&4\cdot\Pr(v_1\in[4,8],v_2\in[3,4])+\frac{14}{3}\cdot\Pr(v_1,v_2\in[4,6])+\\&\qquad 5\cdot\Pr(v_1\in[6,8],v_2\in[4,6])=2.889.\end{aligned}$$

When the type is revealed to be low, the monopoly reserves are 2 and 3, respectively. The expected revenue is:

$$\begin{aligned}&2\cdot\Pr(v_1\in[2,4],v_2\in[0,3])+3\cdot\Pr(v_1\in[0,2],v_2\in[3,6])+\\&3\cdot\Pr(v_1\in[2,3],v_2\in[3,6])+\frac{7}{2}\cdot\Pr(v_1\in[3,4],v_2\in[4,6])+\\&\qquad\frac{10}{3}\cdot\Pr(v_1,v_2\in[3,4])=2.0556.\end{aligned}$$

Thus, the expected revenue when the type is revealed is $2.4722 < 2.5$.

6 Conclusion and Open Questions

Incorporating data into ad auctions raises many questions of practical importance to which our work may be applicable. We mention two open questions:

(1) In simple second price ad auctions, an intermediate revelation scheme may generate more revenue than both full revelation and no revelation. Can the auctioneer find such a scheme in a computationally efficient way? This question was studied by Emek et al. [9] in settings either more general or more restricted than ours, and remains open for the ad auctions model. (2) Can the auctioneer increase her revenue by *asymmetric* revelation of information to the bidders, perhaps charging them appropriate prices for the information? The answer will involve overcoming several challenges, some of which are studied in [1, 2, 3].

References

[1] Abraham, I., Athey, S., Babaioff, M., Grubb, M.: Peaches, lemons, and cookies: Designing auction markets with dispersed information (2011) (manuscript)

[2] Alon, N., Feldman, M., Gamzu, I., Tennenholtz, M.: The asymmetric matrix partition problem (manuscript, 2012)

[3] Babaioff, M., Kleinberg, R., Leme, R.P.: Optimal mechanisms for selling information. In: ACM Conference on Electronic Commerce, pp. 92–109 (2012)

[4] Beck, M., Milgrom, P.: Auctions, adverse selection and internet display advertising. Working paper (2011)

[5] Board, S.: Revealing information in auctions: the allocation effect. Econ. Theory 38, 125–135 (2009)

[6] Bulow, J., Roberts, J.: The simple economics of optimal auctions. The Journal of Political Economy 97(5), 1060–1090 (1989)

[7] Celis, L.E., Lewis, G., Mobius, M., Nazerzadeh, H.: Buy-it-now or take-a-chance: A simple sequential screening mechanism. In: WWW (2011)

[8] Dwork, C., Hardt, M., Pitassi, T., Reingold, O., Zemel, R.: Fairness through awareness. In: ITCS (2012)

[9] Emek, Y., Feldman, M., Gamzu, I., Leme, R.P., Tennenholtz, M.: Signaling schemes for revenue maximization. In: ACM Conf. on Electronic Commerce (2012)

[10] Ghosh, A., Mahdian, M., McAfee, R.P., Vassilvitskii, S.: To match or not to match: Economics of cookie matching in online advertising. In: ACM Conference on Electronic Commerce (2012)

[11] Hartline, J.D., Karlin, A.R.: Profit maximization in mechanism design. In: Nisan, N., Roughgarden, T., Tardos, É., Vazirani, V.V. (eds.) Algorithmic Game Theory, ch. 13. Cambridge University Press (2007)

[12] Hartline, J.D., Roughgarden, T.: Simple versus optimal mechanisms. In: ACM Conference on Electronic Commerce, pp. 225–234 (2009)

[13] Levin, J., Milgrom, P.: Online advertising: Heterogeneity and conflation in market design. American Economic Review 100(2), 603–607 (2010)

[14] McAfee, R.P.: When does improved targeting increase revenue? Working paper, private communication (2012)

[15] Milgrom, P., Weber, R.J.: A theory of auctions and competitive bidding. Econometrica 50(5), 1089–1122 (1982)

[16] Miltersen, P.B., Sheffet, O.: Send mixed signals — earn more, work less. In: ACM Conference on Electronic Commerce (2012)

[17] Myerson, R.: Optimal auction design. Mathematics of Operations Research 6(1), 58–73 (1981)

[18] Palfrey, T.R.: Bundling decisions by a multiproduct monopolist with incomplete information. Econometrica 51(2), 463–483 (1983)

Commodity Auctions and Frugality Ratios

Paul W. Goldberg* and Antony McCabe

Department of Computer Science, University of Liverpool, U.K.
{P.W.Goldberg,A.McCabe}@liverpool.ac.uk

Abstract. We study set-system auctions whereby a single buyer wants to purchase Q items of some commodity. There are multiple sellers, each of whom has some known number of items, and a private cost for supplying those items. Thus a "feasible set" of sellers (a set that is able to comprise the winning bidders) is any set of sellers whose total quantity sums to at least Q. We show that, even in a limited special case, VCG has a *frugality ratio* of at least $n-1$ (with respect to the NTUmin benchmark) and that this matches the upper bound for any set-system auction. We show a lower bound on the frugality of any truthful mechanism of $\sqrt{Q}$ in this setting and give a truthful mechanism with a frugality ratio of $2\sqrt{Q}$. However, we show that similar types of 'scaling' mechanism, in the general (integer) case, give a frugality ratio of at least $\frac{4Qe^{-2}}{\ln^2 Q}$.

1 Introduction

In this paper we examine a simple and natural type of procurement auction, whereby some central authority wishes to purchase some items from amongst a set $\mathcal{E}$ of possible sellers, or *agents*, by requesting quotes for their costs of supplying the items, then selecting and paying the winners so as to incentivise true bidding. We examine some alternative *mechanisms*, which consist of a set of rules that determine how the auction is run. We assume each seller $e \in \mathcal{E}$ provides a (sealed) bid b_e to the auction mechanism. The auctioneer then utilises a mechanism, $\mathcal{M}$, to choose a set S of winning agents (a *selection* rule) and a price p_e to pay each agent (a *payment* rule).

We focus on so-called *truthful* mechanisms. In such a mechanism each agent may maximise its profit simply by making a bid equal to the value that they have (privately) determined as their true cost — the cost the agent incurs as a result of participating in the winning set — for agent e we denote this cost by c_e. At first glance, this may appear to be somewhat restrictive, but truthful mechanisms turn out to be widespread. The first study of a truthful mechanism was by Vickrey in 1961 [11] showing how a sealed-bid second-price auction is truthful (an item is sold to the highest bidder, at a price equal to the second-highest bid). Furthermore, due to the *revelation principle* (see, e.g., [5,9]), it is possible to take any mechanism that has a *dominant strategy* and convert it into a truthful mechanism.

* Supported by EPSRC Grant EP/G069239/1 "Efficient Decentralised Approaches in Algorithmic Game Theory"

M. Serna (Ed.): SAGT 2012, LNCS 7615, pp. 180–191, 2012.

However, a truthful mechanism may not be optimal in terms of revenue. For example, if there are two sellers with very different prices, we must end up paying the larger of the prices. While accepting that some measure of overpayment is necessary, it seems reasonable to try and keep this as low as possible, particularly if we are looking for any real-world motivation. This overpayment is often described (see, e.g., [1,10,7]) in terms of a *frugality ratio*. The frugality ratio is defined as the worst-case ratio between the payments made by a given truthful mechanism and a benchmark figure for the same instance. It has been called "the price of truthfulness" [4]. When frugality was first studied [1,10], it was in the context of path auctions, and benchmark figures were described as properties of the paths. More recently, Karlin, Kempe and Tamir [7] described a benchmark figure that can be used to express a benchmark figure for any monopoly-free set-system auction (where the solutions deemed to be acceptable are described as sets of the agents). They also proposed a *scaling* mechanism for path auctions, and describe its frugality ratio. They give a lower-bound on the frugality ratio for any truthful mechanism, and show that their mechanism is within a constant factor of this lower bound. This constant factor was later improved by Yan [12] and Chen et al. [2].

Since then, Elkind, Goldberg and Goldberg [4] considered alternatives to the benchmark that was proposed in [7] (in [4] they are denoted TUmin, TUmax, NTUmin, NTUmax). Formal definitions of these are given in Definition 1. They also described a polynomial-time mechanism, based on an approximation algorithm, which gives a frugality ratio which is close to that of the well-known Vickrey-Clarke-Groves (VCG) [11,3,6] mechanism (the VCG mechanism must solve the vertex cover problem exactly, which is known to be NP-complete and hence cannot be solved in polynomial time unless P=NP). We give, in Section 2.1, a more general framework for determining the frugality ratios of similarly well-behaved approximation algorithms. (An approximation algorithm is well-behaved if it is monotonic in the bid values, i.e. an agent cannot go from being a loser to a winner by increasing its bid.) Most recently, two groups of researchers [8,2] independently proposed a more general framework of 'scaling' mechanisms that produce improved frugality ratios for a number of set-system auctions, including vertex-covers, flows and cuts. In common with the scaling mechanisms of Karlin et al. [7] they take advantage of the idea that the size of the winning set has a large influence on the overpayment made by a mechanism, and that improvements can be made when the mechanism biases the choice of winning set towards smaller winning sets (by scaling the bids). The frugality results that we present in Section 3 are slightly different, in that the feasible sets may be of similar sizes, yet the frugality ratio can still vary by a large degree.

Preliminaries

Denote a set system as a pair $(\mathcal{E}, \mathcal{F})$, where $\mathcal{E}$ is the ground set of n elements and $\mathcal{F} \subseteq 2^{\mathcal{E}}$ is a collection of feasible sets.

Each element $e \in \mathcal{E}$ has cost c_e; denote the cost vector $\mathbf{c} = (c_1, \dots, c_n)$.

Definition 1. *Let $(\mathcal{E}, \mathcal{F})$ be a set system, let* $\mathbf{c}$ *be a cost vector, and let* S *be the lowest-cost feasible set (with ties broken lexicographically)* $S \in \mathrm{argmin}_{T \in \mathcal{F}} \sum_{e \in T} c_e$. *Let* $\mathrm{NTUmin}(\mathbf{c})$ *be the solution to the problem: Minimize* $B = \sum_{e \in S} b_e$ *subject to the following conditions.*

(1) $b_e \geq c_e$ *for all* $e \in S$
(2) $\sum_{e \in S \setminus T} b_e \leq \sum_{e \in T \setminus S} c_e$ *for all* $T \in \mathcal{F}$
(3) for every $e \in S$*, there is* $T_e \in \mathcal{F}$ *such that* $e \notin T_e$ *and* $\sum_{e' \in S \setminus T_e} b_{e'} = \sum_{e' \in T_e \setminus S} c_{e'}$

As noted, a mechanism $\mathcal{M}$ takes a cost vector $\mathbf{c}$, selects a winning feasible set S, and pays S, incurring a price $p_{\mathcal{M}}(\mathbf{c})$. The *frugality ratio* for mechanism $\mathcal{M}$ is

$$\phi_{\mathrm{NTUmin}}(\mathcal{M}) = \sup_{\mathbf{c}} (p_{\mathcal{M}}(\mathbf{c}) / \mathrm{NTUmin}(\mathbf{c})).$$

We will also consider one of the alternative benchmarks of Elkind et al. [4]. Let $\mathrm{NTUmax}(\mathbf{c})$ be the solution to the problem: Maximize $B = \sum_{e \in S} b_e$ subject to conditions (1), (2), and (3). Let $\phi_{\mathrm{NTUmax}}(\mathcal{M}) = \sup_{\mathbf{c}} (p_{\mathcal{M}}(\mathbf{c}) / \mathrm{NTUmax}(\mathbf{c}))$.

To simplify notation, define the aggregates for a set $V \subseteq \mathcal{E}$; let $b_V = \sum_{e \in V} b_e$, $c_V = \sum_{e \in V} c_e$, and $p_V = \sum_{e \in V} p_e$.

2 Preliminary Results

Let $d(V)$ be the best feasible set (with the lowest sum of costs) using only agents in V where $V \subseteq \mathcal{E}$. We will now see a lower bound for $\mathrm{NTUmin}(\mathbf{c})$ which, informally, states that NTUmin must be at least as large as the worst-case cost of replacing one of the agents to make a feasible set without it. (The proof is omitted due to space constraints.)

Lemma 1. $\mathrm{NTUmin} \geq \max_e c_{d(\mathcal{E} \setminus \{e\})}$.

This lower bound for $\mathrm{NTUmin}(\mathbf{c})$ is a useful tool in analysing frugality ratios, and we will now see how it can be used to prove an upper bound on the frugality of mechanisms based on approximation algorithms.

2.1 Frugality of Approximation Mechanisms

Let $\mathcal{P}$ be some approximation algorithm, and let $S^{\mathcal{P}}$ be the feasible set returned by $\mathcal{P}$ (which uses the bids as an input parameter). We will assume that $\mathcal{P}$ is monotonic in the bids (that is, given fixed bids of the other agents, no agent can be chosen in the winning set when some smaller bid may result in that agent not being chosen). So if we use this algorithm as a selection rule, and use threshold payments as a payment rule, then it is well-known (e.g. [9]) that we have a resulting truthful mechanism $\mathcal{M}^{\mathcal{P}}$. (A threshold payment is the supremum of the amounts that the agent can bid and still be selected in the winning set, given the fixed bids of the other agents.) Let k be the approximation ratio of the algorithm; i.e. some k, such that for all instances of the problem $b_{S^{\mathcal{P}}} \leq k \cdot b_S$ holds. (Note that, as the mechanism is truthful, we can assume that $b_e = c_e$).

Lemma 2. *Let k be the approximation ratio of the algorithm $\mathcal{P}$. Then $\forall e \in S^{\mathcal{P}}, p_e \leq k \cdot \mathrm{NTUmin}(\mathbf{c})$.*

Proof. We have defined $d(\mathcal{E}\setminus\{e\})$ to be a (lowest cost) feasible set, not containing e. Assume, for contradiction, that e were to make a threshold bid, $b_e > k \cdot \mathrm{NTUmin}(\mathbf{c})$, and the winning set $S^{\mathcal{P}}$ (chosen by $\mathcal{P}$) includes e. From Lemma 1 we can observe that $b_{d(\mathcal{E}\setminus\{e\})} \leq \mathrm{NTUmin}(\mathbf{c})$. As we have assumed that $b_e \geq k \cdot \mathrm{NTUmin}(\mathbf{c})$, and as $e \in S^{\mathcal{P}}$ we have $b_{S^{\mathcal{P}}} > k \cdot \mathrm{NTUmin}(\mathbf{c})$ (this holds for all choices of $S^{\mathcal{P}}$ when $e \in S^{\mathcal{P}}$). Hence, by transitivity, we have $b_{S^{\mathcal{P}}} > k \cdot b_{d(\mathcal{E}\setminus\{e\})}$. As $d(\mathcal{E} \setminus \{e\})$ is a feasible set, the approximation ratio of $\mathcal{P}$ is at least $\frac{b_{S^{\mathcal{P}}}}{b_{d(\mathcal{E}\setminus\{e\})}}$. Hence when $b_{S^{\mathcal{P}}} > k \cdot b_{d(\mathcal{E}\setminus\{e\})}$ we have $\frac{b_{S^{\mathcal{P}}}}{b_{d(\mathcal{E}\setminus\{e\})}} > k$, showing that $\mathcal{P}$ does not have an approximation ratio of k, giving a contradiction. Therefore for the threshold bid the inequality $b_e \leq k \cdot \mathrm{NTUmin}(\mathbf{c})$ holds, and hence the payment $p_e \leq k \cdot \mathrm{NTUmin}(\mathbf{c})$. □

Theorem 1. *Let $\mathcal{P}$ be a monotonic approximation algorithm with an approximation ratio of k. Then the resulting mechanism $\mathcal{M}^{\mathcal{P}}$ (with selection rule $\mathcal{P}$ and threshold payments) has $\phi_{\mathrm{NTUmin}(\mathbf{c})}(\mathcal{M}^{\mathcal{P}}) \leq k(n-1)$.*

Proof. In a monopoly-free setting we have a winning set $S^{\mathcal{P}}$ such that $|S| \leq n-1$. from Lemma 2, we have upper bounds on the payment for each $e \in S$, $p_e \leq k \cdot \mathrm{NTUmin}(\mathbf{c})$. Summing over $e \in S$ gives $p(S^{\mathcal{P}}) \leq (n-1)k \cdot \mathrm{NTUmin}(\mathbf{c})$. □

While the approximation result is not strictly relevant to the rest of this paper, it does imply, when $k = 1$, that $\phi_{\mathrm{NTUmin}}(VCG) \leq n - 1$. (This is more precise than the observation made by Karlin et al. [7] that the frugality ratio of VCG is $O(n)$.) We will also see, in Section 3.1, that even our most restricted commodity auction has a frugality ratio that is exactly as high as this upper bound.

3 The Single-Commodity Auction

We consider a *single-commodity auction* where we have some number of identical items for sale, and a quantity Q, the number of these items the auctioneer requires. Each agent $e \in \mathcal{E}$ can provide a fixed, indivisible, quantity of these items, denoted by q_e. The private cost value of e is denoted by c_e, while the bid made to the mechanism is denoted by b_e. Again, since we focus on truthful mechanisms, we can assume $b_e = c_e$.

One could regard this more abstractly as modelling a setting where each seller has some level of capacity to assist with a task, and the buyer wants the task done, and the total capacity to be at least some amount. However, for our results to apply we would need these capacities to be small integers.

The feasible sets $\mathcal{F}$, are defined based on these quantity parameters as follows:

$$\mathcal{F} = \{T \in 2^{\mathcal{E}} : \left(\sum_{e \in T} q_e\right) \geq Q\}. \tag{4}$$

Initially in Section 3.1 we focus on the special case where each agent e only has at most 2 items for sale. We call this the {1,2} single-commodity auction. In Section 3.3 we move to the more general *integer single-commodity auctions*, where a seller's capacity may be any positive integer, not just 1 or 2.

3.1 The $\{1, 2\}$ Single-Commodity Auction

The $\{1, 2\}$ Single-Commodity Auction is a single-commodity auction with the additional restriction, that $\forall e \in \mathcal{E}, q_e \in \{1, 2\}$. While we could simply use VCG to run this auction (recall that VCG chooses the lowest-cost solution and pays each winning agent a threshold value), Table 1 shows that VCG performs poorly in terms of frugality (in fact, matching the upper bound given in Section 2.1). It is also interesting to note that this frugality ratio is as large as Q, the number of items to purchase. We can argue that measuring the frugality ratio in terms of Q seems to make sense for these types of commodity auctions, as it is more naturally a parameter of the auction than the number of agents is. Hence, we will generally consider the frugality ratio in terms of Q, although the results in terms of n are generally similar.

Table 1. In this example we see that VCG has poor frugality; we have a commodity auction for quantity Q items and observe that the number of agents $n = Q + 1$. For each agent $e \in \mathcal{E}$ the quantity q_e and cost c_e are given in the table. A value $b_e^{\min}$ for a NTUmin bid vector is also given, as is the payment made by the VCG mechanism p_e^{VCG}.

	Agent	q_e	c_e	$b_e^{\min}$	p_e^{VCG}
S	1	1	0	1	1
S	2	1	0	0	1
S	$\vdots$	$\vdots$	$\vdots$	$\vdots$	$\vdots$
S	$n-1$	1	0	0	1
	n	2	1		
Total				1	$n-1$

In an attempt to improve frugality, we will now look at a class of (truthful) mechanisms that choose a winning set a little more intelligently.

3.2 The $\mathcal{M}^\alpha$ Mechanism

Here we analyse a class of mechanisms, $\mathcal{M}^\alpha$, each of which is uniquely defined by its 'scaling' value $\alpha \in \mathbb{R}$; a definition for this mechanism follows. $\mathcal{M}^\alpha$ will calculate 'virtual' bids v_e for each agent e by using a scaling factor as follows:

$$v_e = \begin{cases} \alpha b_e, & \text{if } q_e = 1 \\ b_e, & \text{otherwise.} \end{cases}$$

For ease of notation, let the aggregate be $v_V = \sum_{e \in V} v_e$. Let $S^\alpha \in \text{argmin}_{T \in \mathcal{F}} v_T$ be the winning set (the lexicographically first of the feasible sets that have the lowest sum of virtual bids). The payment rule is threshold payments. It is easy to observe that this selection rule is monotonic in the bids, and recall that these are sufficient conditions for a mechanism to be truthful.

Frugality Ratio for $\mathcal{M}^\alpha$. Recall that S is the lowest-cost feasible set, and partition S into two sets, S_1 having agents with quantity 1, and S_2 for those agents having quantity 2.

As choosing both S and S^α requires that ties are broken lexicographically, there is no agent in $S^\alpha \setminus S$ that has the same quantity as an agent in $S \setminus S^\alpha$ (if it is chosen in S^α then it would have been chosen in S). For any $\alpha > 1$, then where S contains some agent e having $q_e = 2$, then S^α must also contain agent e. (If there existed $i, j \notin S$ such that $v_i + v_j \leq v_e$, then $c_i + c_j \leq c_e/\alpha$ contradicting e being chosen in S in preference to $\{i, j\}$). Therefore, where S and S^α are different, $S^\alpha \setminus S$ contains only agents with quantity 2 and $S \setminus S^\alpha$ contains only agents with quantity 1.

We now partition the winning set S^α into three sets, $S^\alpha \cap S_1$, $S^\alpha \cap S_2$, and $S^\alpha \setminus S$ then consider the payments to members of each set separately.

Lemma 3. *For every instance of $\mathcal{M}^\alpha$ when $\alpha = \sqrt{Q}$ then $p_{S^\alpha \cap S_1} \leq \sqrt{Q} \cdot$ NTUmin.*

Proof. We will examine this as two cases. Case 1. Suppose that for every $e \in S^\alpha \cap S_1$ there exists a T_e set satisfying (3) when $(T_e \setminus S) \cap \mathcal{E}_1$ is not empty. Let j be some agent in $T_e \setminus S$ with $q_j = 1$. Assume, for contradiction, that $p_e > c_j$. Hence, agent e's threshold bid $b_e = p_e > c_j$. As j would bid c_j in a truthful mechanism, but $\mathcal{M}^\alpha$ chose e then $c_j \geq p_e$ giving a contradiction. W.l.o.g., we can assume that $T_e = S \setminus \{e\} \cup \{j\}$. Observe that $T'_e \setminus \{j\} \cup \{e\}$ is also a feasible set, hence it must satisfy condition (2), giving $b^{\min}_{S \setminus (T'_e \cup \{e\})} \leq c_{T'_e \setminus (S \cup \{j\})}$, and hence $b^{\min}_e \geq c_j$ or T'_e does not satisfy condition (3), showing that $T_e = S \setminus \{e\} \cup \{j\}$ satisfies condition (3). Using $b^{\min}_e = c_j$ we have $p_{S^\alpha \cap S_1} \leq b^{\min}_{S^\alpha \cap S_1}$ and hence $p_{S^\alpha \cap S_1} \leq$ NTUmin.

Case 2. Suppose that for some $e \in S^\alpha \cap S_1$ there is some T_e set satisfying (3) when $(T_e \setminus S) \cap \mathcal{E}_1$ is empty. There is some $j \in (T_e \setminus S) \cap \mathcal{E}_2$ such that $b^{\min}_{S \setminus T_e} = c_{T_e \setminus S}$. W.l.o.g. assume that $q_{S \setminus T_e} \leq 2$. For each $e \in S^\alpha \cap S_1$ agent e's threshold bid must be $b_e \leq c_j/\alpha$. Hence $p_e = b_e \leq c_j/\alpha$. As $\alpha = \sqrt{Q}$ and Q is trivially an upper bound on the size of S_1, $p_{S^\alpha \cap S_1} \leq \sqrt{Q} \cdot c_j$, with $b^{\min}_{S^\alpha \cap S_1} \geq c_j$ (from $S \setminus T_e \subseteq S^\alpha \cap S_1$), this gives $p_{S^\alpha \cap S_1} \leq \sqrt{Q} \cdot b^{\min}_{S^\alpha \cap S_1} \leq \sqrt{Q} \cdot$NTUmin.

Similar proofs for the other two sets are omitted due to space constraints.

Lemma 4. *For every instance of $\mathcal{M}^\alpha$ when $\alpha = \sqrt{Q}$ then $p_{S^\alpha \cap S_2} \leq \sqrt{Q} \cdot b^{\min}_{S^\alpha \cap S_2}$.*

Lemma 5. *For every instance of $\mathcal{M}^\alpha$ having $\alpha = \sqrt{Q}$ then $p_{S^\alpha \setminus S} \leq \sqrt{Q} \cdot b^{\min}_{S \setminus S^\alpha}$.*

Theorem 2. *For $\{1, 2\}$ Single-Commodity Auctions with quantity Q, the $\mathcal{M}^\alpha$ scaling mechanism when $\alpha = \sqrt{Q}$, gives $\phi_{\text{NTUmin}}(\alpha\mathcal{M}) \leq 2\sqrt{Q}$.*

Proof. From Lemmas 3,4, and 5, the inequalities $p_{S^\alpha \cap S_2} \leq \sqrt{Q} \cdot b^{\min}_{S^\alpha \cap S_2}$, $p_{S^\alpha \setminus S} \leq \sqrt{Q} \cdot c_{S \setminus S^\alpha}$, and $p_{S^\alpha \cap S_1} \leq \sqrt{Q} \cdot \text{NTUmin}$. hold. As $S \setminus S^\alpha$ and $S^\alpha \cap S_2$ are disjoint sets within S, then $b^{\min}_{S^\alpha \cap S_2} + c_{S \setminus S^\alpha} \leq b^{\min}_S \leq \text{NTUmin}$. Therefore, we have $p_{S^\alpha \cap S_2} + p_{S^\alpha \setminus S} \leq \sqrt{Q} \cdot \text{NTUmin}$, and add to give $p_S \leq 2\sqrt{Q} \cdot \text{NTUmin}$ and hence $\phi_{\text{NTUmin}}(\mathcal{M}^\alpha) \leq 2\sqrt{Q}$. □

A Lower Bound on Frugality. Here, we see that any truthful mechanism must pay at least $\sqrt{Q}$·NTUmin, showing that the $\mathcal{M}^\alpha$ mechanism with $\alpha = \sqrt{Q}$ is within at most a factor of two of optimal.

Theorem 3. *There exists a* $\{1, 2\}$ *single-commodity auction for* Q *items such that any truthful mechanism* $\mathcal{M}$, *must pay at least* $\sqrt{Q}$·NTUmin.

Proof. For any quantity Q, let I be an instance of a set-system auction having $\mathcal{E} = \{1, \ldots, Q + 1\}$ and $\mathbf{q} = \{1, \ldots, 1, 2\}$. Suppose that $\mathcal{M}$ is some truthful mechanism. Consider each $e \in \{1, \ldots, Q\}$ and suppose an instance such that $b_e = 1$, $b_{Q+1} = \sqrt{Q}$ and all other agents bid 0. We are interested in two cases, either every $e \in \{1, \ldots, Q\}$ would be chosen in the winning set by $\mathcal{M}$, or else there is some such e for which $Q + 1$ would be chosen instead.

Case 1. Suppose that every $e \in \{1, \ldots, Q\}$ is chosen in preference to $Q + 1$. Let $\mathbf{b} = (0, \ldots, 0, \sqrt{Q})$ be a bid vector. Observe that $S = \{1, \ldots, Q\}$ and that $\mathbf{b^{min}} = (\sqrt{Q}, 0, \ldots, 0)$ denotes a bid vector satisfying conditions (1),(2) and (3), hence NTUmin $\leq \sqrt{Q}$. As every agent in S, would have been chosen by $\mathcal{M}$ with a bid of 1 then $S^{\mathcal{M}} = S$ and each threshold bid must be at least 1, hence $p_{\mathcal{E}} \geq Q$ and $p_{\mathcal{E}}/\text{NTUmin} \geq \sqrt{Q}$.

Case 2. Suppose (w.l.o.g) that agent $Q + 1$ is chosen in preference to agent 1. Let $\mathbf{b} = (1, 0, \ldots, 0, 1)$ be a bid vector. Observe that $S = \{1, \ldots, Q\}$ (with the tie broken lexicographically) and that $\mathbf{b^{min}} = (1, 0, \ldots, 0)$ denotes a bid vector satisfying conditions (1),(2) and (3), hence NTUmin ≤ 1. As mechanism $\mathcal{M}$ will choose agent $Q + 1$ with bid $\sqrt{Q}$ in preference to 1, being truthful implies that $\mathcal{M}$ will still choose $Q + 1$ with a lower bid of 1, hence $Q + 1 \in S^{\mathcal{M}}$. As agent $Q + 1$ would still have been chosen had it bid $\sqrt{Q}$, its threshold bid is at least $\sqrt{Q}$, and hence $p_{Q+1} \geq \sqrt{Q}$. This gives $p_{\mathcal{E}} \geq \sqrt{Q}$ and hence $p_{\mathcal{E}}/\text{NTUmin} \geq \sqrt{Q}$.

For every truthful mechanism $\mathcal{M}$, either Case 1 or Case 2 applies, hence the frugality ratio $\phi_{\text{NTUmin}}(\mathcal{M}) \geq \sqrt{Q}$. □

3.3 Integer Single-Commodity Auctions

We consider improvements to frugality bounds in the more general setting, where the restriction on the quantity of each agent to 1 or 2 is relaxed. We have a lower bound on frugality of $\sqrt{Q}$ from the $\{1, 2\}$ single commodity auction, but we may believe that there is a stronger lower bound in the integer case. While we do not have a result for all truthful mechanisms, we obtain an asymptotically stronger lower bound on frugality that applies to a natural class of scaling mechanisms, of at least $\frac{4Qe^{-2}}{\ln^2 Q}$, for all mechanisms in this class.

Preliminaries. Let k be a 'maximum quantity' parameter such that $\forall e \in \mathcal{E}, q_e \leq k$ holds and assume that $k \leq \sqrt{Q}$. Let β be a scaling function, returning a linear scaling vector, $\mathbf{a} = \beta(Q, k)$ (with $a_e \in \mathbb{R}$). Let $\mathcal{M}^\beta$ be the mechanism that uses the scaling vector $\mathbf{a} = (a_1, \ldots, a_k)$ returned by β, as follows. Compute a 'virtual' bid v_e for each agent e as $v_e = b_e a_{q_e}$. Let $S \in \text{argmin}_{T \in \mathcal{F}} v_T$ be the winning set. Each agent e will be paid its threshold value, p_e. If we consider every scaling function β, and the resulting class of mechanisms, then we can think of $\mathcal{M}^\beta$ as the class of all 'blind-scaling' mechanisms; where the mechanism must choose a scaling factor for each possible quantity, based only on the quantity required Q and the maximum quantity parameter k.

A Lower Bound for Blind-Scaling Mechanisms. The proof will examine a series of example instances given, and show that at least one of them must cause a payment ratio that satisfies the lower bound. We can generalise the example given in Table 1, and will show this in Table 2. For each $j \in \{1, \ldots, k-1\}$ let Table 2 describe instance I_j. Observe the assumption that $j < k \leq \sqrt{Q}$ implies that $m \geq j$ which is required by the structure of the example (m is defined in the example as $m = \lceil \frac{Q}{j} \rceil$).

We can see that there are j agents in S that can have a (NTUmin) bid value $b_e^{\min} = 1$. We can show that there can be no more than j agents that can each bid 1 as follows; $j+1$ agents, each with quantity j, could be 'replaced' by the j agents outside S, each with quantity $j+1$, so no set of $j+1$ agents in S can bid a sum of more than j.

More formally, $\forall e \in S$, let $T_e = S \setminus \{1, \ldots, j, e\} \cup \{(m+1), \ldots, (m+j+1)\}$. Observe that $\left(\sum_{i=1}^{j} q_i\right) + q_e = j(j+1)$ and $\left(\sum_{i=1}^{j+1} q_{m+i}\right) = j(j+1)$ hence $q_S =$

Table 2. Instance I_j: In this example we have a $\{j, j+1\}$ commodity auction for quantity Q items. Let $m = \lceil \frac{Q}{j} \rceil$ and observe that the winning set is given by $S = \{1, \ldots, m\}$. For each agent $e \in \mathcal{E}$ the quantity q_e and cost c_e are given in the table. A value $b_e^{\min}$ for a NTUmin bid vector is also given, giving NTUmin$\leq j$. The payment made by the $\mathcal{M}^\beta$ mechanism is also given in Table 2 as p_e.

	Agent	q_e	c_e	$b_e^{\min}$	p_e
	1	j	0	1	a_{j+1}/a_j
	$\vdots$	$\vdots$	$\vdots$	$\vdots$	
	j	j	0	1	a_{j+1}/a_j
S	$j+1$	j	0	0	a_{j+1}/a_j
	$\vdots$	$\vdots$	$\vdots$	$\vdots$	
	m	j	0	0	a_{j+1}/a_j
	$m+1$	$j+1$	1		
	$\vdots$	$\vdots$	$\vdots$		
	$m+j+1$	$j+1$	1		
Total				j	ma_{j+1}/a_j

q_T and T_e is a feasible set. Using this T_e in condition (3) for all $e \in \{j+1, \ldots, m\}$ gives $\left(\sum_{i=1}^{j} b_i^{\min}\right) + b_e^{\min} = \sum_{i=1}^{j+1} c_{m+1}^{j}$. As $\sum_{i=1}^{j} b_i^{\min} = j$ and $\sum_{i=1}^{j+1} c_{m+i} = j$ then we have $b_e^{\min} = 0$, which shows that for all $e \in \{j+1, \ldots, m\}$ then vector $\mathbf{b}^{\min}$ has some T_e satisfying condition (3) of Definition 1. For all $e \in \{1, \ldots, j\}$, let $T_e = S \setminus \{e\} \cup \{m+1\}$ which gives $b_e^{\min} = 1$, showing that the bid vector $\mathbf{b}^{\min}$ has, for all $e \in S$, some T_e satisfying condition (3) and as we can observe $\mathbf{b}^{\min}$ satisfies conditions (1) and (2) then this shows NTUmin $\leq b_S^{\min}$ and hence NTUmin $\leq j$.

We can also generalise the payment to each $e \in S$. For each agent $e \in S$, if $v_e > v_{m+1}$ then agent e would not be chosen, as the winning set could become $S \setminus \{e\} \cup \{m+1\}$. Where $v_e = v_{m+1}$, then agent e may still be chosen, hence when agent e can submit a threshold bid b_e such that $v_e = v_{m+1}$ and this gives the threshold payment.

If we assume for all $e \in S$, that $b_e = \frac{a_{j+1}}{a_j}$ then as $v_e = b_e a_j$ we have $v_e = \frac{a_{j+1}}{a_j} a_j = a_{j+1} = v_{m+1}$. This shows that $b_e = \frac{a_{j+1}}{a_j}$ is a threshold bid for all $e \in S$, hence the payment is given by $p_e = \frac{a_{j+1}}{a_j}$.

Let $\mathbf{c}$ be a cost vector for instance I_j and let $p_\mathcal{E}$ be the sum of payments. We examine the payment ratio $\frac{p_\mathcal{E}}{\text{NTUmin}}$ as follows. There are at least $\frac{Q}{j}$ agents in S, each is paid $\frac{a_{j+1}}{a_j}$, and NTUmin $\leq j$; hence the payment ratio satisfies the inequality $\frac{p_\mathcal{E}}{\text{NTUmin}} \geq \frac{Q a_{j+1}}{j^2 a_j}$. We can then use this as we move onto the first part of the proof. We will use the 'maximum quantity' parameter, k, and will examine a series of instances where all agents have quantity at most k. We give a certain ratio, $\frac{Q^{\frac{k-1}{k}}}{k^2}$, and we will show (from these instances) that a minimum separation is needed between any consecutive scaling values (a_j, a_{j+1}) (where $j < k$) in order to satisfy this ratio. We will then show how having this minimum separation between consecutive scaling values implies a large separation between the first and k-th value, and give a further instance where a large separation will result in a frugality ratio larger than $\frac{Q^{\frac{k-1}{k}}}{k^2}$.

Finally we will show how to compute a value for k that gives a lower-bound for any given Q.

Proposition 1. *For instance I_j of $\mathcal{M}^\beta$ with $j \leq k-1$ and $\frac{a_j}{a_{j+1}} \leq Q^{\frac{1}{k}}$ the inequality $\frac{p_\mathcal{E}}{\text{NTUmin}} \geq \frac{Q^{\frac{k-1}{k}}}{k^2}$ holds.*

Proof. As $j \leq k$ implies $\frac{1}{j^2} \geq \frac{1}{k^2}$, then $\frac{Q a_{j+1}}{j^2 a_j} \geq \frac{Q a_{j+1}}{k^2 a_j}$. It follows, due to transitivity with $\frac{p_\mathcal{E}}{\text{NTUmin}} \geq \frac{Q a_{j+1}}{j^2 a_j}$ that $\frac{p_\mathcal{E}}{\text{NTUmin}} \geq \frac{Q a_{j+1}}{k^2 a_j}$. Also $\frac{a_j}{a_{j+1}} \leq Q^{\frac{1}{k}}$ can be be expressed as $\frac{a_{j+1}}{a_j} \geq Q^{\frac{-1}{k}}$ therefore, by transitivity $\frac{p_\mathcal{E}}{\text{NTUmin}} \geq \frac{Q}{k^2} \frac{a_{j+1}}{a_j} \geq \frac{Q Q^{\frac{-1}{k}}}{k^2}$. This can be simplified to state $\frac{p_\mathcal{E}}{\text{NTUmin}} \geq \frac{Q^{\frac{k-1}{k}}}{k^2}$, completing the proof. □

This minimum separation required between every a_j and a_{j+1} implies that there is large separation between a_1 and a_k. We will see, in Table 3, that such a large separation then results in a similarly large frugality ratio.

Table 3. Instance I_k: In this example we have a commodity auction for quantity Q items with the parameter k. Let $m = \lceil \frac{Q}{k} \rceil$ and observe that the winning set is given by $S = \{1, \ldots, m\}$. For each agent $e \in \mathcal{E}$ the quantity q_e and cost c_e are given in the table. A value $b_e^{\min}$ for a NTUmin bid vector is also given, showing NTUmin $\leq mk$. The payment made by the $\mathcal{M}^\beta$ mechanism is also given in the table as p_e.

	Agent	q_e	c_e	$b_e^{\min}$	p_e
S	1	k	0	k	ka_1/a_k
	$\vdots$	$\vdots$	$\vdots$	$\vdots$	
	m	k	0	k	ka_1/a_k
	$m+1$	1	1		
	$\vdots$	$\vdots$	$\vdots$		
	$m+k$	1	1		
	Total			mk	mka_1/a_k

Proposition 2. *For instance I_k of $\mathcal{M}^\beta$ the inequality $\frac{p_\mathcal{E}}{\text{NTUmin}} \geq \frac{a_1}{a_k}$ holds.*

Proof. For each $e \in S$, there is exactly one feasible set not containing e — that is $\mathcal{E} \setminus \{e\}$. Therefore the only bid vector that could satisfy NTUmin must satisfy condition (3) of Definition 1 with $T_e = \mathcal{E} \setminus \{e\}$. Therefore the NTUmin bid for each $e \in S$ must be given by $b_e^{\min} = c_{T_e \setminus S} = c_{\{m+1,\ldots,m+k+1\}} = k$. As there are m agents in S, each having a bid $b_e^{\min} = k$, we have NTUmin $\leq mk$. Similarly, the threshold bid for e must be where $v_e = v_{\{m+1,\ldots,m+k\}}$. Assuming $b_e = \frac{ka_1}{a_k}$ multiplying by the scaling factor a_k gives $v_e = \frac{ka_1}{a_k} a_k = ka_1$. The virtual bids of the competing agents $i \in \{m+1, \ldots, m+k+1\}$ are $v_i = a_1$, hence $v_{\{m+1,\ldots,m+k\}} = ka_1$ showing that $b_e = \frac{ka_1}{a_k}$ is a threshold bid, and hence the payment $p_e = \frac{ka_1}{a_k}$.

Therefore, in Instance I_k, there are m agents in S; each is paid $\frac{ka_1}{a_k}$ giving a total payment of $\frac{mka_1}{a_k}$. As we have seen NTUmin $\leq mk$ hence $\frac{p_\mathcal{E}}{\text{NTUmin}} \geq \frac{a_1}{a_k}$. □

We now see there is always some instance which implies a lower bound on the payment ratio, for any possible scaling vector of the mechanism.

Proposition 3. *For any scaling vector* **a** *given by $\mathcal{M}^\beta$ there is either some Instance I_j for $j \in \{1, \ldots, k-1\}$ or Instance I_k such that the inequality $\frac{p_\mathcal{E}}{\text{NTUmin}} \geq \frac{Q^{\frac{k-1}{k}}}{k^2}$ holds.*

Proof. If there existed some $j \in \{1, \ldots, k-1\}$ such that $\frac{a_j}{a_{j+1}} \leq Q^{\frac{1}{k}}$ then Proposition 1 implies that $\frac{p_\mathcal{E}}{\text{NTUmin}} \geq \frac{Q^{\frac{k-1}{k}}}{k^2}$. So, suppose that the expression $\forall j \in \{1, \ldots, k-1\}, \frac{a_j}{a_{j+1}} > Q^{\frac{1}{k}}$ holds. We can see this implies that the consecutive scaling values must have a certain separation. By way of example, this gives $\frac{a_1}{a_2} > Q^{\frac{1}{k}}$, $\frac{a_2}{a_3} > Q^{\frac{1}{k}}$ etc. By transitivity we would have $\frac{a_1}{a_3} > Q^{\frac{2}{k}}, \frac{a_1}{a_4} > Q^{\frac{3}{k}}$ etc. This can then be generalised, for $j \in \{1, \ldots, k-1\}$ to give $\frac{a_1}{a_{j+1}} > Q^{\frac{j}{k}}$.

For $j = k-1$, then we have $\frac{a_1}{a_k} > Q^{\frac{k-1}{k}}$. Referring back to Proposition 2, Instance I_k gives $\frac{p_\mathcal{E}}{\text{NTUmin}} \geq \frac{a_1}{a_k}$ and, by transitivity, $\frac{p_\mathcal{E}}{\text{NTUmin}} > Q^{\frac{k-1}{k}}$.

Hence there is some instance, either I_j for $j \in \{1, \dots, k-1\}$ or I_k that satisfies the proposition. □

Now that we have seen that there is always some instance that gives at least this payment ratio in terms of k, we can use this to prove a lemma that shows a lower bound on the frugality ratio for all Integer Single-Commodity Auctions.

Lemma 6. *For all Integer Single-Commodity Auctions with quantity Q and maximum quantity parameter $k \leq \sqrt{Q}$, for every blind-scaling scaling mechanisms $\mathcal{M}^\beta$ the inequality $\phi_{\text{NTUmin}}(\mathcal{M}^\beta) \geq \frac{Q^{\frac{k-1}{k}}}{k^2}$ holds.*

Proof. The blind-scaling mechanism $\mathcal{M}^\beta$ must, by definition, calculate its scaling vector $\mathbf{a}$ for use on any instance that it may be given with these parameters. Once this scaling vector is fixed the mechanism may possibly be given either Instance I_k or Instance I_j for any $j \in \{1, \dots, k-1\}$. Proposition 3 shows that at least one of these instances gives $\frac{p_\mathcal{E}}{\text{NTUmin}} \geq \frac{Q^{\frac{k-1}{k}}}{k^2}$. The existence of such an instance proves $\phi_{\text{NTUmin}}(\mathcal{M}^\beta) \geq \frac{Q^{\frac{k-1}{k}}}{k^2}$. □

Now that we have shown a lower bound on frugality for values of Q in terms of the parameter k, we can specify a value of k such as to give a lower bound entirely in terms of Q. To that end, suppose $k = \frac{\ln Q}{2}$, and we will see this implies a lower bound of $\frac{4Qe^{-2}}{\ln^2 Q}$ for $\mathcal{M}^\beta$ mechanisms.

Theorem 4. *Given any Integer Single-Commodity Auction having quantity Q, for every blind-scaling mechanism $\mathcal{M}^\beta$ the inequality $\phi_{\text{NTUmin}}(\mathcal{M}^\beta) \geq \frac{4Qe^{-2}}{\ln^2 Q}$ holds.*

Proof. Considering the proof of Lemma 6, suppose $k = \ln Q/2$. The expression given in Lemma 6 implies $\frac{Q^{\frac{k-1}{k}}}{k^2} = \frac{4Qe^{-2}}{\ln^2 Q}$, and hence, $\phi_{\text{NTUmin}}(\mathcal{M}^\beta) \geq \frac{4Qe^{-2}}{\ln^2 Q}$. □

4 Conclusion

While single-commodity auctions are quite simple, they show surprisingly high frugality ratios. Particularly in the $\{1, 2\}$ case, a lower bound on the frugality ratio for every truthful mechanism of $\sqrt{Q}$ seems unreasonably high. This result could also seem to call into question the suitability of NTUmin as a reasonable benchmark. Our scaling mechanism is shown to be within a factor of 2 of optimal; it may be that this factor of 2 could be reduced with a stronger analysis.

While we have shown a fairly large lower bound on the frugality of 'blind-scaling' mechanisms in the more general case of integer single-commodity auctions, it is not known if some other form of mechanism would result in better frugality. Also, we have not presented any mechanism that would give a frugality ratio of better than Q in this case, although it seems that some form of scaling mechanisms should, at least, give some slightly better result. Choosing to measure frugality in terms of Q or n makes little difference in the $\{1, 2\}$ case, but the difference is more pronounced in the integer case, and showing good frugality results in terms of n may be an interesting goal.

We have only considered frugality in this setting with respect to NTUmin. More recently (see, e.g., [2,8]) we have seen frugality ratios analysed with respect to NTUmax. It is likely that we will get more satisfactory frugality ratios with respect to NTUmax, particularly in the $\{1, 2\}$ case. Although, in the integer case, we may still get reasonably large frugality ratios. Take, for example, Theorem 3 and amend the quantity vector to be $\mathbf{q} = (1, \ldots, 1, Q)$. This would give NTUmax $= 1$ (as $T_e = \{Q + 1\}$ is the only alternative feasible set, and so must satisfy condition (3)). The rest of the proof could then be applied, with the obvious minor changes, to show that $\phi_{\text{NTUmax}}(\mathcal{M}) \geq \sqrt{Q}$.

References

1. Archer, A., Tardos, E.: Frugal path mechanisms. In: Proceedings of the Thirteenth Annual ACM-SIAM Symposium on Discrete Algorithms, SODA 2002, pp. 991–999. Society for Industrial and Applied Mathematics, Philadelphia (2002)
2. Chen, N., Elkind, E., Gravin, N., Petrov, F.: Frugal mechanism design via spectral techniques. In: IEEE Symposium on Foundations of Computer Science, pp. 755–764 (2010)
3. Clarke, E.H.: Multipart pricing of public goods. Public Choice 11(1) (September 1971)
4. Elkind, E., Goldberg, L., Goldberg, P.: Frugality ratios and improved truthful mechanisms for vertex cover. In: Proceedings of the 8th ACM Conference on Electronic Commerce, pp. 336–345 (2007)
5. Gibbard, A.: Manipulation of voting schemes: A general result. Econometrica 41(4), 587–601 (1973)
6. Groves, T.: Incentives in teams. Econometrica 41(4), 617–631 (1973)
7. Karlin, A.R., Kempe, D., Tamir, T.: Beyond VCG: Frugality of truthful mechanisms. In: FOCS 2005: Proceedings of the 46th Annual IEEE Symposium on Foundations of Computer Science, pp. 615–626. IEEE Computer Society, Washington, DC (2005)
8. Kempe, D., Salek, M., Moore, C.: Frugal and truthful auctions for vertex covers, flows and cuts. In: IEEE Symposium on Foundations of Computer Science, pp. 745–754 (2010)
9. Nisan, N., Roughgarden, T., Tardos, E., Vazirani, V.V.: Algorithmic Game Theory. Cambridge University Press, New York (2007)
10. Talwar, K.: The Price of Truth: Frugality in Truthful Mechanisms. In: Alt, H., Habib, M. (eds.) STACS 2003. LNCS, vol. 2607, pp. 608–619. Springer, Heidelberg (2003)
11. Vickrey, W.: Counterspeculation, Auctions, and Competitive Sealed Tenders. The Journal of Finance 16(1), 8–37 (1961)
12. Yan, Q.: On the Price of Truthfulness in Path Auctions. In: Deng, X., Graham, F.C. (eds.) WINE 2007. LNCS, vol. 4858, pp. 584–589. Springer, Heidelberg (2007)

On the Communication Complexity of Approximate Nash Equilibria

Paul W. Goldberg[1,*] and Arnoud Pastink[2]

[1] Department of Computer Science, University of Liverpool, U.K.
P.W.Goldberg@liverpool.ac.uk

[2] Utrecht University, Department of Information and Computing Science,
P.O. Box 80089, 3508TB Utrecht, The Netherlands
arnoudpastink@gmail.com

Abstract. We study the problem of computing approximate Nash equilibria, in a setting where players initially know their own payoffs but not the payoffs of other players. In order for a solution of reasonable quality to be found, some amount of communication needs to take place between the players. We are interested in algorithms where the communication is substantially less than the contents of a payoff matrix, for example logarithmic in the size of the matrix. At one extreme is the case where the players do not communicate at all; for this case (with 2 players having $n \times n$ matrices) ϵ-Nash equilibria can be computed for $\epsilon = 3/4$, while there is a lower bound of slightly more than $1/2$ on the lowest ϵ achievable. When the communication is polylogarithmic in n, we show how to obtain $\epsilon = 0.438$. For one-way communication we show that $\epsilon = 1/2$ is the exact answer.

1 Introduction

Algorithmic game theory is concerned not just with properties of a solution concept, but also how that solution can be obtained. It is considered desirable that the outcome of a game should be "easy to compute", and in that respect the PPAD-completeness results of [6,2] are interpreted as a "complexity-theoretic critique" of Nash equilibrium. Following those results, a line of work addressed the problem of computing ϵ-Nash equilibrium, where $\epsilon > 0$ is a parameter that bounds a player's incentive to deviate, in a solution. Thus, ϵ-Nash equilibrium imposes a weaker constraint on how players are assumed to behave, and an exact Nash equilibrium is obtained for $\epsilon = 0$.

Besides the existence of a fast algorithm, it is also desirable that a solution should be obtained by a process that is simple and decentralised, since that is likely to be a better model for how players in a game may eventually reach a solution. In that respect, most of the known efficient algorithms for computing ϵ-Nash equilibria are not entirely satisfying. They take as input the payoff matrices

* Supported by EPSRC Grant EP/G069239/1 "Efficient Decentralised Approaches in Algorithmic Game Theory".

M. Serna (Ed.): SAGT 2012, LNCS 7615, pp. 192–203, 2012.

and output the approximate Nash equilibrium. If we try to translate such an algorithm into real life, it would correspond to a process where the players pass their payoffs to a central authority, which returns to them some mixed strategies that have the "low incentive to deviate" guarantee. In this paper we try to model a setting where players perform individual computations and exchange some limited information.

There are various ways in which one can try to model the notion of a decentralised algorithm; here we consider a general approach that has previously been studied in [4,9] in the context of computing exact Nash equilibria. The players begin with knowledge of their own payoffs but not the payoffs of the other players. An algorithm involves communication in addition to computation; to reach an approximate equilibrium, a player usually has to know something about the other players' matrices, but hopefully not all of that information. We study the computation of ϵ-Nash equilibria in this setting, and the general topic is the trade-off between the amount of communication that takes place, and the value of ϵ that can be obtained.

1.1 Definitions

We consider 2-player games, with a *row player* and a *column player*, who both have n *pure strategies.* The game (R, C) is defined by two $n \times n$ *payoff matrices*, R for the row player, and C for the column player. The pure strategies for the row player are his rows and the pure strategies of the column player are her columns. If the row player plays row i and the column player plays column j, the *payoff* for the row player is R_{ij}, and C_{ij} for the column player. For the row player a *mixed strategy* is a probability distribution $\mathbf{x}$ over the rows, and a mixed strategy for the column player is a probability distribution $\mathbf{y}$ over the columns, where $\mathbf{x}$ and $\mathbf{y}$ are column vectors and $(\mathbf{x}, \mathbf{y})$ is a *mixed strategy profile.* The payoffs resulting from these mixed strategies $\mathbf{x}$ and $\mathbf{y}$ are $\mathbf{x}^T R\mathbf{y}$ for the row player and $\mathbf{x}^T C\mathbf{y}$ for the column player.

A *Nash equilibrium* is a pair of mixed strategies $(\mathbf{x}^*, \mathbf{y}^*)$ where neither player can get a higher payoff by playing another strategy assuming the other player does not change his strategy. Because of the linearity of a mixed strategy, the largest gain can be achieved by defecting to a pure strategy. Let $\mathbf{e}_i$ be the vector with a 1 at the ith position and a 0 at every other position. Thus a Nash equilibrium $(\mathbf{x}^*, \mathbf{y}^*)$ satisfies

$$\forall i = 1 \cdots n \quad \mathbf{e}_i^T R\mathbf{y}^* \leq (\mathbf{x}^*)^T R\mathbf{y}^* \text{ and } (\mathbf{x}^*)^T C\mathbf{e}_i \leq (\mathbf{x}^*)^T C\mathbf{y}^*$$

We assume that the payoffs of R and C are between 0 and 1, which can be achieved by rescaling. An ϵ*-approximate Nash equilibrium* (or, ϵ-Nash equilibrium) is a strategy pair $(\mathbf{x}^*, \mathbf{y}^*)$ such that each player can gain at most ϵ by unilaterally deviating to a different strategy. Thus, it is $(\mathbf{x}^*, \mathbf{y}^*)$ satisfying

$$\forall i = 1 \cdots n \quad \mathbf{e}_i^T R\mathbf{y}^* \leq (\mathbf{x}^*)^T R\mathbf{y}^* + \epsilon \text{ and } (\mathbf{x}^*)^T C\mathbf{e}_i \leq (\mathbf{x}^*)^T C\mathbf{y}^* + \epsilon$$

We say that the *regret* of a player is the difference between his payoff and the payoff of his best response.

The *support* of a mixed strategy $\mathbf{x}$, denoted by Supp($\mathbf{x}$), is the set of pure strategies that are played with non-zero probability by $\mathbf{x}$.

The communication model: Each player $p \in \{r, c\}$ has an algorithm $\mathcal{A}_p$ whose initial input data is p's $n \times n$ payoff matrix. Communication proceeds in a number of rounds, where in each round, each player may send a single bit of information to the other player. During each round, each player may also carry out a polynomial (in n) amount of computation. (One could alternatively omit the restriction to polynomial computation. Our lower bounds on communication requirement do not depend on computational limits.) At the end, each player p outputs a mixed strategy $\mathbf{x}_p$. We aim to design (pairs of) algorithms $(\mathcal{A}_r, \mathcal{A}_c)$ that output ϵ-Nash strategy profiles $(\mathbf{x}_r, \mathbf{x}_c)$, and are economical with the number of rounds of communication.

Notice that given $\Theta(n^2)$ rounds of communication, we can apply any centralised algorithm $\mathcal{A}$ by getting (say) the row player to pass additive approximations of all his payoffs to the column player, who applies $\mathcal{A}$ and passes to the row player the mixed strategy obtained by $\mathcal{A}$ for the row player. (The quality of the ϵ-Nash equilibrium is proportional to the quality of of the additive approximations used.) For this reason we focus on algorithms with many fewer rounds, and we obtain results for logarithmic or polylogarithmic (in n) rounds.

We also consider a restriction to *one-way communication*, where one player may send but not receive information.

1.2 Related Work

Algorithms for Approximate Equilibria. In recent years a number of algorithms have been developed that compute (in polynomial time) ϵ-Nash equilibria for various values of ϵ. This is not a complete overview of all existing algorithms. The algorithm with the best approximation that is known, gives a 0.3393-approximate Nash equilibrium [17]. However, here we mainly use ideas from certain earlier algorithms.

DMP-algorithm: The DMP-algorithm [7] works as follows to achieve a 0.5-approximate Nash equilibrium. The algorithm picks a arbitrary row for the row player, say row i. Let $j \in \text{argmax}_{j'} C_{ij'}$. Let $k \in \text{argmax}_{k'} R_{k'j}$. So j is a pure-strategy best response for the column player to row i and k is a best response strategy for the row player to column j. The strategy pair $(\mathbf{x}^*, \mathbf{y}^*)$ will now be $\mathbf{x}^* = \frac{1}{2}\mathbf{e}_i + \frac{1}{2}\mathbf{e}_k$ and $\mathbf{y}^* = \mathbf{e}_j$. With this strategy pair the row player plays a best response with probability $\frac{1}{2}$ to a pure strategy of the column player and the column player has a pure strategy that is with probability $\frac{1}{2}$ a best response.

The DMP-algorithm is well-adapted to the limited-communication setting. Suppose the row player uses $i = 1$ as his initial choice of row. The column player needs to tell the row player his value of j, a communication of $O(\log n)$ bits. No further communication is needed. Notice moreover that the communication is all one-way; the row player does not need to tell the column player anything.

Subsequent algorithms for computing ϵ-Nash equilibria cannot so easily be adapted to a limited-communication setting, but we can use some of the ideas they develop, to obtain values of ϵ below $\frac{1}{2}$ in this setting.

An algorithm of Bosse et al. [1]: The algorithm presented in [1] can be seen as a modification of the DMP-algorithm and achieves a 0.38197-approximate Nash equilibrium. Instead of a player playing a pure strategy with some positive probability, the algorithm starts with the row player allocating some probability to the row-player strategy $\mathbf{x}$ belonging to the Nash equilibrium of the zero-sum game $(R-C, C-R)$. In solving the zero-sum game efficiently we apply the connection of zero-sum games with linear programming [15,5,11]. If the (mixed) strategy profile $(\mathbf{x}, \mathbf{y})$ that is a Nash equilibrium of $(R-C, C-R)$ gives a 0.38197-approximate Nash equilibrium for (R, C), this solution is used. Otherwise, the column player plays a best response $\mathbf{e}_j$ to $\mathbf{x}$ and the row player plays a mixture of $\mathbf{x}$ and $\mathbf{e}_i$, where $\mathbf{e}_i$ is a best response to the strategy $\mathbf{e}_j$ of the column player. ([1] goes on to improve the worst-case performance to a 0.36395-approximate Nash equilibrium.)

Notice that this algorithm cannot be adapted in a straightforward way to our communication-bounded setup, since it requires a computation using knowledge of both matrices.

Communication Complexity. The "classical" setting of communication complexity is based on the model introduced by Yao in [18]. We will follow the representation in [12]. We have two agents[1], one holding an input $\mathbf{x} \in \{0,1\}^n$ and the other holding an input $\mathbf{y} \in \{0,1\}^n$. The objective is to compute $f(\mathbf{x}, \mathbf{y}) \in \{0,1\}$, a joint function of their inputs. The computation of $f(\mathbf{x}, \mathbf{y})$ is done via a communication protocol $\mathcal{P}$. During the execution of the protocol, the agents send messages to each other. While the protocol has not terminated, the protocol specifies what message the sender should send next, based on the input of the protocol and the communication so far. If the protocol terminates, it will output the value $f(\mathbf{x}, \mathbf{y})$. A communication protocol $\mathcal{P}$ computes f if for every input pair $(\mathbf{x}, \mathbf{y}) \in \{0,1\}^n \times \{0,1\}^n$, it terminates with the value $f(\mathbf{x}, \mathbf{y})$ as output.

The communication complexity of a communication protocol $\mathcal{P}$ for computing $f(\mathbf{x}, \mathbf{y})$ is the number of bits sent during the execution of $\mathcal{P}$, which we denote by $CC(\mathcal{P}, f, \mathbf{x}, \mathbf{y})$. The communication complexity of a protocol $\mathcal{P}$ for a function f is defined as the worst case communication complexity over all possible inputs for $(\mathbf{x}, \mathbf{y}) \in \{0,1\}^n \times \{0,1\}^n$, which we denote by $CC(\mathcal{P}, f)$:

$$CC(\mathcal{P}, f) = \max_{(\mathbf{x},\mathbf{y}) \in \{0,1\}^n \times \{0,1\}^n} CC(\mathcal{P}, f, \mathbf{x}, \mathbf{y})$$

The communication complexity of a function f is the minimum over all possible protocols:

$$CC(f) = \min_{\mathcal{P}} CC(\mathcal{P}, f)$$

[1] We use agents instead of players to avoid confusion, the communication does not have to be between the players of the game.

Existing Results on Communication Complexity of Nash Equilibria. There are a few results concerning the communication complexity of Nash equilibria. In [4] it is shown that a lower bound on the communication complexity for 2-player games of finding a pure Nash equilibrium is $\Omega(n^2)$, where n is the number of pure strategies for each player. They also show a simple algorithm that finds a pure Nash equilibrium (if it exists) in $O(n^2)$. They do not extend their analysis to mixed Nash equilibria; their method is about finding out whether there exists a pure Nash equilibrium, in contrast with the existence of a mixed Nash equilibrium, which is guaranteed [14].

In [9] the communication complexity of uncoupled equilibrium procedures is studied. They show that for reaching a pure Nash equilibrium, reaching a pure Nash equilibrium in a Bayesian setting and for reaching a mixed Nash equilibrium, a lower bound on the communication complexity is $\Omega(2^s)$, where s is the number of players. To show that reaching this equilibrium is not just due to the complexity of the input, they also show that you can reach a correlated equilibrium in a polynomial number of steps. The methods they use cannot be extended to analysing the communication complexity of ϵ-approximate Nash equilibria. For pure Nash equilibria, their analysis is based on games that might not have a Nash equilibrium and for mixed strategy Nash equilibrium the analysis is based on equilibria that require a large description. Approximate Nash equilibria always exist and can have small descriptions, so the developed techniques do not work for ϵ-approximate Nash equilibria.

1.3 Overview of Our Results

For general $n \times n$ games we show the following bounds on the approximate Nash equilibrium if we fix the amount of communication allowed. We start by considering a version where no communication is allowed. Theorem 1 gives a simple way to find a $\frac{3}{4}$-Nash equilibrium, in this setting. Theorem 3 identifies a contrasting lower bound of slightly more than $\frac{1}{2}$. For one-way communication we exhibit (Theorem 2) a lower bound of $0.5 - o(\frac{1}{\sqrt{n}})$. The DMP-algorithm can be implemented as a algorithm with one-way communication and gives a 0.5-approximate Nash equilibrium. Therefore the constant $\frac{1}{2}$ in the lower bound of Theorem 2 is tight, in this context. In Section 3 we show how to compute a 0.438-Nash equilibrium using polylogarithmic communication.

2 Computing Approximate Nash Equilibria with No Communication

The simplest version of our model is one where there is no communication between the players.[2] That means that for each player $p \in \{r, c\}$, we must find a

[2] This is to some extent inspired by earlier work of the first author [8] that studied an approach to pattern classification in which the set of observations of each class must be processed by an algorithm that proceeds independently of the corresponding algorithms that receive members of the other classes.

function f_p from p's payoff matrix to a mixed strategy, such that for all pairs of matrices (R, C), we have that $(f_r(R), f_c(C))$ is an ϵ-Nash equilibrium.

Theorem 1. *It is possible to guarantee a $\frac{3}{4}$-approximate Nash equilibrium, with no communication between the players.*

Proof. Each player allocates probability $\frac{1}{2}$ to his first pure strategy, and $\frac{1}{2}$ to his best response to the other player's first pure strategy. In detail, let $i \in \arg\max_{i'} R_{i'1}$ and let $j \in \arg\max_{j'} C_{1j'}$. The approximate Nash equilibrium will be $\mathbf{x}^* = \frac{1}{2}\mathbf{e}_1 + \frac{1}{2}\mathbf{e}_i$ and $\mathbf{y}^* = \frac{1}{2}\mathbf{e}_1 + \frac{1}{2}\mathbf{e}_j$.

Let i' be a best pure strategy response of the row player to $\mathbf{y}^*$. Then his incentive to deviate is

$$(\tfrac{1}{2}R_{i'1} + \tfrac{1}{2}R_{i'j}) - (\tfrac{1}{4}R_{11} + \tfrac{1}{4}R_{1j} + \tfrac{1}{4}R_{i1} + \tfrac{1}{4}R_{ij})$$
$$\leq (\tfrac{1}{4}R_{i'1} + \tfrac{1}{2}R_{i'j}) - (\tfrac{1}{4}R_{11} + \tfrac{1}{4}R_{1j} + \tfrac{1}{4}R_{ij}) \leq \tfrac{1}{4}R_{i'1} + \tfrac{1}{2}R_{i'j} \leq \tfrac{1}{4} + \tfrac{1}{2} = \tfrac{3}{4}$$

where the first inequality holds because i was a best response to column 1 (so $R_{i1} \geq R_{i'1}$) and the next inequalities hold because payoffs lie in $[0, 1]$. The same kind of argument holds for the column player. This proves the theorem. □

The following result gives a lower bound of $\frac{1}{2}$; in fact it provides a stronger result saying that $\frac{1}{2}$ is a lower bound for any amount of *one-way communication*, where one player (say, the row player) may send but not receive information about payoffs. Since the DMP-algorithm uses one-way communication, our result shows that it is optimal, in this context.

Theorem 2. *With one-way communication, it is impossibly to guarantee to find an ϵ-Nash equilibrium, for any constant $\epsilon < \frac{1}{2}$.*

Proof. We define a game $G = (R, C)$, where R and C are payoff matrices with dimensions $\binom{n}{k} \times n$, with $k \approx \sqrt{n}$. Consider the following set of column player payoff matrices $C^1, \ldots, C^n$, where C^ℓ has a payoff of 1 for every entry in the ℓth column and a 0 in every other place:

$$\forall i, j : \quad C^\ell_{ij} = 1 \text{ if } j = \ell;\ 0 \text{ otherwise}$$

The row player has matrix R with $\binom{n}{k}$ rows, where a row consists of k 1's and $(n-k)$ 0's. Every row is a different combination, so the $\binom{n}{k}$ rows are all distinct combinations of k 1's in a row of length n.

Let D^r be the strategy of the row player, resulting from matrix R. Let D^c_ℓ be the strategy of the column player resulting from matrices R and C^ℓ; note that with unlimited one-way communication we can assume that the row player sends all of R to the column player.

We will show that for this class of games, one cannot do better than a $(\frac{1}{2} - o(\frac{1}{\sqrt{n}}))$-approximate Nash equilibrium. This implies for large values of n approximately a $\frac{1}{2}$-approximate Nash equilibrium.

During the proof we will search for a lower bound of $\frac{1}{2} - z$, where the value of z is to be determined.

First observe that a best response for the column player having matrix C^ℓ is $\mathbf{e}_\ell$, the pure strategy of column ℓ. It has payoff 1 and other columns have payoff 0. So to reach a $(\frac{1}{2} - z)$-approximate Nash equilibrium, D^c_ℓ must allocate a probability at least $(\frac{1}{2} + z)$ to column ℓ.

The row player has one matrix R with all different combinations of k 1's in a row of length n. Now consider the columns of R. By construction each column of R consists of $\frac{k}{n} \cdot \binom{n}{k}$ 1's and $(1 - \frac{k}{n}) \cdot \binom{n}{k}$ 0's.

D^r assigns a probability to each row of R. Define an unnormalised probability distribution Φ over the columns as follows. Φ assigns to each column j a value $\Phi(j)$, which gives the probability that a 1 will be in this column given a row sampled from D^r. This value $\Phi(j)$ will be at most 1, when every row that is played with positive probability has a 1 in column j. Because every row contains k 1's, the sum of over all values will sum to k: $\sum_{j=1}^{n} \Phi(j) = k$.

We define column m to be one with a lowest value of Φ: $m \in \operatorname{argmin}_j \Phi(j)$. Suppose the column player has payoff matrix C^m. Note that the sum over all values $\Phi(j)$ is k and there are n columns. This means that $\Phi(m)$ is at most $\frac{k}{n}$. This means that column m, which is played at least $\frac{1}{2} + z$ of the time by the column player, gives a payoff of 0 with a probability of at least $1 - \frac{k}{n}$.

We now consider the row player's strategy D^r and construct an improved response D^* —that is supposed to be an improvement of at most $\frac{1}{2} - z$— as follows. D^* will differ from D^r in the following way. For every row i we see if there is a 1 on the mth entry. If this is the case, we do not change anything. If there is a 0 on the mth entry we do the following: look at the positions where there is a 1 in row i. Of all the entries where there is a 1, we select the entry to which the column player gives the lowest probability, say entry a. Now we move all the probability allocated by D^r this row, to the row of R that instead has a 0 on entry a and a 1 on entry m, and is otherwise the same as i.

The probability on entry a is defined as the smallest of all the entries where this row has a 1. We can bound the probability that was given to this entry by the column player. A probability at least $\frac{1}{2} + z$ is given to column m, so a probability of $\frac{1}{2} - z$ can be distributed over the remaining columns. The column belonging to entry a has the smallest probability of at least k columns, so the probability given to column a is at most $\frac{1/2-z}{k}$.

The result of this construction of D^* from D^r is that every row that is played with positive probability by D^* will have a 1 on the mth entry. There is a probability at least $(1 - \frac{k}{n})$ that a row sampled from D^r did not have a 1 on the mth entry. This means that the increase in payoff from replacing D^r with D^* is at least

$$\left(1 - \frac{k}{n}\right) \cdot \left(\frac{1}{2} + z\right) - \left(1 - \frac{k}{n}\right) \cdot \frac{1/2 - z}{k} = \left(1 - \frac{k}{n}\right) \cdot \left(\frac{1}{2} + z - \frac{\frac{1}{2} - z}{k}\right)$$

We will show that this increase in payoff is close to $\frac{1}{2}$ for well chosen k and z. Assume that z is chosen such that $z = \frac{(1/2)-z}{k}$. Equivalently, $z = 1/(2k+2)$.

This will make the difference in payoff between D^r and D^* at least

$$\left(1-\frac{k}{n}\right)\cdot\left(\frac{1}{2}+z-z\right)=\frac{1}{2}-\frac{k}{2n}.$$

So if the column player has a regret (as defined in Section 1.1) of $\leq \frac{1}{2}-z$, the row player has a regret of at least $\frac{1}{2}-\frac{k}{2n}$, and we put $z=\frac{1}{2k+2}$. We can use these two observations to find the value of k such that the regrets are the same for the row player and column player:

$$\begin{aligned}\tfrac{1}{2}-\tfrac{k}{2n} &= \tfrac{1}{2}-\tfrac{1}{2(k+1)}\\ \tfrac{k}{2n} &= \tfrac{1}{2(k+1)}\\ k=\tfrac{1}{2}(\sqrt{4n+1}-1) &\quad \vee\; k=\tfrac{1}{2}(-\sqrt{4n+1}-1)\end{aligned}$$

Since k should be greater than 0, only the first solution is feasible. So we have $k=\frac{1}{2}(\sqrt{4n+1}-1)$ and $z=\frac{\frac{1}{2}(\sqrt{4n+1}-1)}{2n}$, which is $o(\frac{1}{\sqrt{n}})$. We have proven now that for general games with one-way communication one cannot do better than a $(\frac{1}{2}-o(\frac{1}{\sqrt{n}}))$-approximate Nash equilibrium. □

Theorem 3. *It is impossible to guarantee a* 0.501*-Nash equilibrium, with no communication between the players.*

As we noted, the previous Theorem 2 already shows a lower bound of $\frac{1}{2}$ in this setting. Theorem 3 rules out the possibility that $\frac{1}{2}$ is the correct answer, as it was for one-way communication.

Proof. (sketch) For $p\in\{r,c\}$, let Ω^p be the set of (mixed) strategies p may use (the image of f_p). Let $\mathbf{c}^p$ be a distribution over $[n]$ that minimises the maximum variation distance $d_{\max}$ from $\mathbf{c}^p$ to elements of Ω^p; $\mathbf{c}^p$ is called the *centre strategy* for p, and p's *commitment* (denoted τ^p) is $1-d_{\max}$. Thus $\tau^p\in[0,1]$ and is high when p must choose a strategy close to some $\mathbf{c}^p$.

The proof is by case analysis on the values τ^r and τ^c. If either value (say τ^c) is ≥ 0.501, then c's matrix C is chosen to be C^ℓ as in the proof of Theorem 2 where column ℓ receives low probability from $\mathbf{c}^c$. c's high commitment prevents c from deviating sufficiently far from $\mathbf{c}^c$ to make a good enough response.

If either value (say τ^c) is ≤ 0.05 then c has 3 strategies s_1, s_2, s_3 that are all very far apart in variation distance. Design a matrix for r where row i is a very good response to s_i but a poor response to $s_j\neq s_i$. The row player has no strategy that is sure to fall short of optimal by ≤ 0.501.

If $\tau^r,\tau^c\in[0.05,0.501]$, assume $\tau^r\geq\tau^c$, and design a matrix R such that r's commitment forces him to allocate nearly 0.05 of his probability to rows that have zero payoff. The remaining rows $S\subset[n]$ have payoff 1 against "most" columns (w.r.t. measure $\mathbf{c}^c$). Each row in S is a good response to one of the remaining columns, associated with that row alone, but gets payoff 0 against others. The column player can be forced by matrix C to allocate probability ≥ 0.499 to one of those columns. r loses 0.05 due to having to allocate ≥ 0.05 to rows outside S, and a further ~ 0.49 due to not knowing which row in S is the best one to use, for a total regret > 0.501. □

3 A 0.438-Approximate Nash Equilibrium with Limited Communication

This section provides a 0.438-approximate Nash equilibrium where the amount of communication between the players is polylogarithmic in n. We present the algorithm as an α-approximate Nash equilibrium first and then optimize α. At various points the algorithm uses the operation of communicating a mixed strategy (a probability distribution over $[n]$) from one player to the other; the details of this operation are given in Section 3.1. The general idea is to send a sample of size $O(\log n)$ from the distribution and argue that the corresponding empirical distribution is a good enough estimate for our purposes.

First the row player finds a Nash equilibrium for the zero-sum game $(R, -R)$ and the column player computes a Nash equilibrium for the zero-sum game $(-C, C)$. Since both games are zero-sum, we know that the payoff values for their Nash equilibria will be unique. Both players compare this payoff value with α. We distinguish two cases, the Nash equilibrium of both players is lower than α (Case 1) or at least one of the players has a value equal to or higher than α for his Nash equilibrium (Case 2). With $O(1)$ communication, the case that holds can be identified.

Case 1:
Both players have a Nash equilibrium with value smaller than α. The row player finds a strategy pair $(\mathbf{x}_r^*, \mathbf{y}_r^*)$ and the column player a strategy pair $(\mathbf{x}_c^*, \mathbf{y}_c^*)$. The row player communicates $\mathbf{y}_r^*$ to the column player (as described in Section 3.1) and the column player sends $\mathbf{x}_c^*$ to the row player. They now play the game with the strategy pair $(\mathbf{x}_c^*, \mathbf{y}_r^*)$. Since $\mathbf{y}_r^*$ was a Nash equilibrium strategy in the zero-sum game $(R, -R)$ and the row player still plays with payoff matrix R, by definition of a Nash equilibrium, the row player has no strategy that can give him a payoff of α or higher. The row player has a best response with a value of at most α, so his regret is also at most α. This leads to an α-approximate Nash equilibrium for the row player. The strategy $\mathbf{x}_c^*$ was a Nash equilibrium strategy in the zero-sum game $(-C, C)$ and the column player still has payoff matrix C. So we can use the same argument for the column player to argue that when the row player has strategy $\mathbf{x}_c^*$, the column player has a α-approximate Nash equilibrium. This concludes Case 1.

Case 2:
If at least one of the players has a value of at least α for his zero-sum game, he can get a payoff of at least α if he plays this strategy, regardless the strategy of the other player. Assume w.l.o.g. that it is the row player who has a payoff of at least α in his zero-sum game. He communicates this strategy $\mathbf{x}_r^*$ to the column player (again, as described in Section 3.1). The column player identifies a pure strategy best response $\mathbf{e}_j$ to the strategy of the row player and communicates this strategy to the row player (using $\log n$ bits).

At this point in the algorithm we have the strategy pair $(\mathbf{x}_r^*, \mathbf{e}_j)$. The column player has a best response strategy, so at this point his strategy is a 0-approximate Nash equilibrium. The row player can guarantee a payoff of α.

Let $\beta \leq 1$ be the value of his best response to $\mathbf{e}_j$. So at this point the row player has a $\beta - \alpha$-approximate Nash equilibrium. We next deal with the possibility that $\beta - \alpha > \alpha$.

At this stage the column player has a 0-approximate Nash equilibrium while we are only looking for a α-approximate Nash equilibrium; meanwhile the row player has a strategy that might not be good enough for a α-approximate Nash equilibrium. To change this, we use a method used in [3] (Lemma 3.2), which allows the row player to shift some of his probability to his best response to $\mathbf{e}_j$. By shifting some of his probability, it could be that $\mathbf{e}_j$ no longer is a best response strategy for the column player. This is allowed, as long as the column player's regret while playing $\mathbf{e}_j$ is at most α. Suppose the row player shifts $\frac{1}{2}\alpha$ of his probability to a best response strategy. The payoff the column player gets could be $\frac{1}{2}\alpha$ lower because of this move. The payoff of some other strategy could go as much as $\frac{1}{2}\alpha$ higher because of this shift. The strategy $\mathbf{e}_j$ was a 0-approximate Nash equilibrium, so by the shift of $\frac{1}{2}\alpha$ of the row player's probability, the regret of the column player is at most $\frac{1}{2}\alpha + \frac{1}{2}\alpha = \alpha$, which constitutes an α-approximate Nash equilibrium, for the column player.

The row player is allowed to change the allocation of $\frac{1}{2}\alpha$ of his probability with the worst payoff. Since we rearrange the worst part of the row player, the remainder of his probability, $1 - \frac{1}{2}\alpha$ had already at least a payoff of α. The probability is shifted to his best response with a value of β, with $\alpha \leq \beta \leq 1$. This leads to the following inequality:

$$(1 - \frac{1}{2}\alpha)\alpha + \frac{1}{2}\alpha\beta \geq \beta - \alpha\,, \quad 0 \leq \alpha \leq \beta \leq 1$$

The solutions to this inequality are

$$\begin{array}{ll} 0 < \alpha \leq \frac{1}{2}(5 - \sqrt{17}) & \alpha \leq \beta \leq \frac{\alpha^2 - 4\alpha}{\alpha - 2} \\ \frac{1}{2}(5 - \sqrt{17}) < \alpha < 1 & \alpha \leq \beta \leq 1 \\ \alpha = 0 \quad \beta = 0 & \alpha = 1 \quad \beta = 1 \end{array}$$

where it holds that if $\alpha = \frac{1}{2}(5 - \sqrt{17})$ then $f(\alpha) = \frac{\alpha^2 - 4\alpha}{\alpha - 2} = 1$ and for $0 \leq \alpha \leq 1$ this function is monotone increasing. This procedure will give an α-approximate Nash equilibrium, so α should be as low as possible. Next to this it should also hold for every β with $\alpha \leq \beta \leq 1$. The lowest α such that this condition hold is when $f(\alpha) = 1$, thus $\alpha = \frac{1}{2}(5 - \sqrt{17}) \approx 0.438$.

So if the row player rearranges $\frac{1}{2} \cdot 0.438 = 0.219$ of his probability to his best response row, both players have a strategy that guarantees them a 0.438-approximate Nash equilibrium.

3.1 Communicating Mixed Strategies

We describe how to communicate an approximation of the mixed strategies that are computed, using $O(\log^2 n)$ bits. We ultimately obtain an ϵ of $0.438 + \delta$, for any $\delta > 0$.

We first look at the case where one of the players, assume w.l.o.g. the row player, has a payoff higher than α in the Nash equilibrium of his zero-sum game

$(R, -R)$. The column player plays a pure best response to the strategy of the row player, regardless of the support of the strategy of the row player. So we mainly consider the row player.

The zero-sum game $(R, -R)$ gives a strategy pair $(\mathbf{x}^*, \mathbf{y}^*)$. Fix $k = \frac{\ln n}{\delta^2}$ and form a multiset A by sampling k times from the set of pure strategies of the row player, independently at random according to the distribution $\mathbf{x}^*$. Let $\mathbf{x}'$ be the mixed strategy for the row player with a probability of $\frac{1}{k}$ for every member of A. We want the distribution $\mathbf{x}'$ to have a payoff close to the payoff of $\mathbf{x}^*$. This corresponds to the following event:

$$\phi = \{((\mathbf{x}')^T R\mathbf{y}^*) - ((\mathbf{x}^*)^T R\mathbf{y}^*) < -\delta\}$$

As noted in [13] the expression $((\mathbf{x}')^T R\mathbf{y}^*)$ is essentially a sum of k independent random variables each of expected value $((\mathbf{x}^*)^T R\mathbf{y}^*)$, where every random variable has a value between 0 and 1. This means we can bound the probability that ϕ does not hold, which we will call ϕ^c. When we apply a standard tail inequality [10] to bound the probability of ϕ^c, we get:

$$\Pr[\phi^c] \leq e^{-2k\delta^2}$$

With $k = \frac{\ln n}{\delta^2}$, this gives $\Pr[\phi^c] \leq \frac{1}{n^2}$ and $\Pr[\phi] \geq 1 - \frac{1}{n^2}$. If $\mathbf{x}'$ does not give payoffs close enough to $\mathbf{x}^*$, we sample again.

The strategy $\mathbf{x}'$ has a guaranteed payoff of $0.438 + \delta - \delta = 0.438$. This strategy is communicated to the column player. The support of this strategy is logarithmic and all probabilities are rational (multiples of $\frac{1}{k}$). Communication of one pure strategy has a communication complexity of $O(\log n)$. This will give a communication complexity for $\mathbf{x}'$ of $O(\log^2 n)$.

The column player computes a pure strategy best response to $\mathbf{x}'$ and communicates this strategy in $O(\log n)$ to the row player. The strategy of the row player might not yet lead to a 0.438-approximate Nash equilibrium, his payoff could be too low. As we have seen before, if the row player redistributes at most 0.219 of his probability, he is guaranteed to have a strategy that leads to a 0.438-approximate Nash equilibrium.

This change in strategy of the row player can decrease the payoff of the column player by as much as 0.219 and increase another pure strategy by as much as 0.219. His strategy was a best response, a 0-approximate Nash equilibrium, and the improvement to another pure strategy is maximal $0.219 + 0.219 = 0.438$, this leads to a 0.438-approximate Nash equilibrium.

In the alternative case, where both players have a low ($< \alpha$) payoff in their zero-sum games, the technique is essentially the same: each player samples k times from the opposing distribution, checks that it limits his own payoff to at most $\alpha + \delta$, re-samples as necessary, and communicates the k-sample.

4 Conclusions

The general topic of the communication complexity of approximate Nash equilibrium, seems to be a rich source of research questions. [16] considers some related ones, including the communication required for approximate *well-supported* equilibria, as well as games of fixed size. It may be that future work should address the issue of communication protocols where the players have an incentive to report their information truthfully.

Acknowledgements. The first author thanks Sergiu Hart for useful discussions during the iAGT workshop in May 2011.

References

1. Bosse, H., Byrka, J., Markakis, E.: New Algorithms for Approximate Nash Equilibria in Bimatrix Games. In: Deng, X., Graham, F.C. (eds.) WINE 2007. LNCS, vol. 4858, pp. 17–29. Springer, Heidelberg (2007)
2. Chen, X., Deng, X.: Settling the complexity of two-player Nash equilibrium. In: Procs. of the 47th FOCS Symposium, pp. 261–272. IEEE (2006)
3. Chen, X., Deng, X., Teng, S.H.: Settling the complexity of computing two-player Nash equilibria. J. ACM 56, 14:1–14:57 (2009)
4. Conitzer, V., Sandholm, T.: Communication complexity as a lower bound for learning in games. In: Proceedings of the 21st ICML, pp. 24–32 (2004)
5. Dantzig, G.B.: Linear Programming and Extensions. Princeton Univ. Press (1963)
6. Daskalakis, C., Goldberg, P.W., Papadimitriou, C.H.: The complexity of computing a Nash equilibrium. SIAM J. Comput. 39(1), 195–259 (2009)
7. Daskalakis, C., Mehta, A., Papadimitriou, C.: A Note on Approximate Nash Equilibria. In: Spirakis, P.G., Mavronicolas, M., Kontogiannis, S.C. (eds.) WINE 2006. LNCS, vol. 4286, pp. 297–306. Springer, Heidelberg (2006)
8. Goldberg, P.W.: Some discriminant-based PAC algorithms. Journal of Machine Learning Research 7, 283–306 (2006)
9. Hart, S., Mansour, Y.: How long to equilibrium? the communication complexity of uncoupled equilibrium procedures. GEB 69(1), 107–126 (2010)
10. Hoeffding, W.: Probability inequalities for sums of bounded random variables. Journal of the American Statistical Association 58(301), 13–30 (1963)
11. Karmarkar, N.: A new polynomial-time algorithm for linear programming. In: 16th STOC, pp. 302–311. ACM (1984)
12. Kushilevitz, E.: Communication complexity. Advances in Computers 44, 331–360 (1997)
13. Lipton, R.J., Markakis, E., Mehta, A.: Playing large games using simple strategies. In: Procs. of the 4th ACM-EC, EC 2003, pp. 36–41 (2003)
14. Nash, J.: Non-cooperative games. Ann. Math. 54(2), 286–295 (1951)
15. von Neumann, J.: Zur theorie der gesellschaftsspiele. Mathematische Annalen 100, 295–320 (1928)
16. Pastink, A.: Aspects of communication complexity for approximating Nash equilibria. MSc dissertation, Utrecht University (2012)
17. Tsaknakis, H., Spirakis, P.G.: An Optimization Approach for Approximate Nash Equilibria. In: Deng, X., Graham, F.C. (eds.) WINE 2007. LNCS, vol. 4858, pp. 42–56. Springer, Heidelberg (2007)
18. Yao, A.C.C.: Some complexity questions related to distributive computing (preliminary report). In: 11th STOC, pp. 209–213. ACM (1979)

Congestion Games with Capacitated Resources*

Laurent Gourvès[1], Jérôme Monnot[1], Stefano Moretti[1], and Nguyen Kim Thang[2]

[1] LAMSADE, CNRS UMR 7243, Université Paris Dauphine, France
[2] IBISC, Université d'Evry Val d'Essonne, France

Abstract. We extend congestion games to the setting where every resource is endowed with a capacity which possibly limits its number of users. From the negative side, we show that a pure Nash equilibrium is not guaranteed to exist in any case and we prove that deciding whether a game possesses a pure Nash equilibrium is **NP**-complete. Our positive results state that congestion games with capacities are potential games in the well studied singleton case. Polynomial algorithms that compute these equilibria are also provided.

1 Introduction

The players of a *congestion game* interact by allocating bundles of resources from a common pool [18]. This type of games leads to well studied models for analyzing strategic situations including routing [9], network design [3] and load balancing [8]. They are a prominent model for resource sharing among uncoordinated selfish users.

Significant interest has been addressed over the last years to the analysis of practical congestion problems in the Internet. Data delays and losses due to *data congestions*, or the network collapse as a consequence of exceeding the *data flow capacity* of some links or nodes, has long been a real problem for the Internet [4]. Several policies have been proposed to control congestion, in order to regulate and improve the availability of broadband access to the Internet. *Priority rules*, for instance, have been adopted to regulate the users who enter into the network, with the objective to prevent congestion and to obtain a *Quality of Service* (QoS) that otherwise would not be available to users [5]. A classical example of priorities of users is provided by the access categories of the IEEE 802.11e standard, that was developed in order to offer QoS capabilities to Wireless Local Area Networks (WLANs) [15].

Congestion games [18] can only partially model the practical situation described above. In order to catch other realistic factors like capacities of resources and the different priority of users on the network, a more sophisticated model is required.

* This work is supported by French National Agency (ANR), project COCA ANR-09-JCJC-0066-01.

M. Serna (Ed.): SAGT 2012, LNCS 7615, pp. 204–215, 2012.

For this purpose, we introduce the class of *congestion games with capacitated resources*, where each resource is associated both with a *capacity* level, representing the maximum number of users that such a resource may simultaneously *accommodate*, and with an *ordering* on the users, prescribing the *priority* of accommodation of the users. Given a certain profile of players' strategies, the cost of utilization of a resource for the players which have that resource in their strategy and which are accommodated on it, is a function of the number of players using it in that profile (as in the case of classical congestion games), whereas the cost of players having that resource in their strategy, but which are not accommodated, is prohibitive (supposed infinite).

In this paper we investigate the following questions: Do congestion games with capacitated resources always admit a pure strategy Nash equilibrium (NE in short) in any case as it holds for classical congestion games? If not, is it difficult to decide if an instance possesses a pure NE? Can we identify natural classes of instances admitting a pure NE? Are there polynomial (or more efficient) algorithms that build a pure NE for classes containing such an equilibrium?

2 Models and Notations

A *strategic (cost) game* is a tuple $\langle \mathcal{N}, (\Sigma_i)_{i\in\mathcal{N}}, (c_i)_{i\in\mathcal{N}}\rangle$, where $\mathcal{N} = \{1,\cdots,n\}$ is a finite set of *players*; Σ_i is a non-empty set of *pure strategies* for each player $i \in \mathcal{N}$; $c_i : \Sigma_1 \times \cdots \times \Sigma_n \to \mathbb{R}$ is an *individual cost function* specifying players i's cost $c_i(\sigma) \in \mathbb{R}$ for each strategy profile $\sigma = (\sigma_i)_{i\in\mathcal{N}} \in \Sigma_1 \times \cdots \times \Sigma_n$ and each $i \in \mathcal{N}$.

Using conventional notations, we denote by $\Sigma = \Sigma_1 \times \cdots \times \Sigma_n$ the *set of strategy profiles* or *strategy space* and we denote a *strategy profile* σ by (σ_i, σ_{-i}) if the choice of player i needs stressing. The *strategy space* Σ is *symmetric-strategy* if $\Sigma_1 = \Sigma_2 = \ldots = \Sigma_n$.

A *pure strategy Nash equilibrium* (or simply *pure Nash equilibrium*, NE in short) is a pure strategy profile $\sigma \in \Sigma$ such that, for all players $i \in N$, and all pure strategies $s_i \in \Sigma_i$, it holds that $c_i(\sigma) \leq c_i(s_i, \sigma_{-i})$. We only deal with pure strategies in this article so we often omit the word "pure".

For some given strategy profile, a *better move* of a player is a unilateral deviation such that his cost decreases strictly. If such a better move exists, we say that the corresponding player is *unhappy*, otherwise he is *happy*. In this setting a NE is a strategy profile where all players are happy. The *better-response dynamic* is the process of repeatedly choosing an arbitrary unhappy player and let him make an arbitrary better move. A *potential game* is a game in which, for any instance, the better-response dynamic always converges [17]. Such a property is typically shown by a potential function argument.

2.1 Congestion Models and Games

Rosenthal [18] defines a *congestion model* as a tuple $\langle \mathcal{N}, \mathcal{R}, (\Sigma_i)_{i\in\mathcal{N}}, (d_r)_{r\in\mathcal{R}}\rangle$ where $\mathcal{N} = \{1,\ldots,n\}$ is the set of players; $\mathcal{R}$ is a finite set of m *resources*; $\Sigma_i \subseteq 2^{\mathcal{R}}$ is the set of pure strategies of player i, for each $i \in \mathcal{N}$; $d_r : \{0,1,\ldots,n\} \to \mathbb{R}^+$

is a *delay function* associated with resource r, for each $r \in \mathcal{R}$. This function depends on the number of players using resource r, denoted by $n_r(\sigma)$ or simply n_r when the context is clear. The interpretation is that every player of a resource r incurs a cost of $d_r(n_r)$ (with the convention that $d_r(0) = 0$). Delay functions are sometimes supposed monotone (e.g. [9]) but we do not make this restriction in this paper.

Given a congestion model $\langle \mathcal{N}, \mathcal{R}, (\Sigma_i)_{i\in\mathcal{N}}, (d_r)_{r\in\mathcal{R}}\rangle$, an associated *congestion game* is defined as a strategic cost game $\langle \mathcal{N}, (\Sigma_i)_{i\in\mathcal{N}}, (c_i)_{i\in\mathcal{N}}\rangle$ where for each $\sigma \in \Sigma$ and $i \in \mathcal{N}$, $c_i(\sigma) = \sum_{r\in\sigma_i} d_r(n_r(\sigma))$. Better-response dynamic always converges in congestion games because every better move decreases Rosenthal's potential function $\sum_{r\in\mathcal{R}} \sum_{i=1}^{n_r} d_r(i)$ [18].

An important subclass of congestion games is the class of *singleton congestion games* (also known as *parallel-link games*) in which every player's strategy consists of a single resource [1, 8, 10–12, 14, 16].

2.2 Congestion Games with Capacitated Resources

This section describes the model introduced and studied in this paper. Given a congestion model $\langle \mathcal{N}, \mathcal{R}, (\Sigma_i)_{i\in\mathcal{N}}, (d_r)_{r\in\mathcal{R}}\rangle$, we also assume that every resource $r \in \mathcal{R}$ has a *capacity* κ_r – an integer between 1 and n – which is the maximal number of players that can use resource r. Moreover, every resource r is associated with a linear order $\mathsf{pos}_r : \mathcal{N} \to \{1, \ldots, n\}$, where $\mathsf{pos}_r(i) = t$ means that player i is in the t-th position of r (pos is strict total). We say that a player i has a *higher priority* than player j at resource r iff $\mathsf{pos}_r(i) < \mathsf{pos}_r(j)$. Notice that $\mathsf{pos}_r(i)$ is defined even if r does not appear in the strategy space of player i.

Let $N_r(\sigma)$ be the set of players using resource r in the strategy profile σ. A player $i \in N_r(\sigma)$ is *accommodated* by r iff the number of players in $N_r(\sigma)$ having a position lower than $\mathsf{pos}_r(i)$ is strictly smaller than the capacity of resource r, i.e., $|\{j \in N_r(\sigma) : \mathsf{pos}_r(j) < \mathsf{pos}_r(i)\}| < \kappa_r$. The *delay* $d_r(\sigma)$ of a resource r in profile σ is defined as $d_r(\min\{n_r(\sigma), \kappa_r\})$. The *delay* $d_r^i(\sigma)$ of player $i \in N_r(\sigma)$ on resource r is:

$$d_r^i(\sigma) = \begin{cases} d_r(\min\{n_r(\sigma), \kappa_r\}) & \text{if } i \text{ is accommodated,} \\ +\infty & \text{otherwise.} \end{cases} \tag{1}$$

A *congestion game with capacitated resources* (*capacitated congestion game* in short) is a strategic cost game where the *cost* of a player i in profile σ is defined as $c_i(\sigma) = \sum_{r\in\sigma_i} d_r^i(\sigma)$.

Note that capacitated congestion games follow the original congestion model of Rosenthal [18] when the resources are not overcrowded. When the capacity of a resource is exceeded, the game shares similarities with the player-specific model of Milchtaich [16] since we distinguish between accommodated and non accommodated players. However congestion games with capacitated resources are neither a refinement nor an extension of player-specific congestion games.

In congestion games with capacitated resources, a profile is a *Nash equilibrium* if the following conditions hold:

– no player, accommodated by *every* resource in his current strategy, can unilaterally deviate and decrease his cost;
– no player, not accommodated by at least one resource in his current strategy, can unilaterally deviate and incur a finite cost.

We say that a resource r is *saturated* if $n_r(\sigma) \geq \kappa_r$. We say that a player i is *displaced* by another player j in the following situation: i is accommodated by a resource r which is not used by j, j deviates so that r is in his new strategy and i is not accommodated by r anymore whereas j is (of course $\texttt{pos}_r(j) < \texttt{pos}_r(i)$).

3 Related Works

Various aspects of congestion games were investigated. The existence of pure NE, the convergence of better-response dynamic and the computation of equilibria are interleaved questions studied in [2, 6, 9, 14]. Computing a pure NE of a congestion game is a **PLS**-complete problem, even if strategies are symmetric. Nevertheless there are important subclasses for which a NE can be built in polynomial time, by the use of dedicated algorithms or simply via better response dynamic (see [19] for a survey).

Many extensions of the congestion model introduced in Rosenthal [18] have been studied in the literature of strategic games. *Player-specific congestion games*, have been introduced in [16] with the objective to model congestion situations where the delay of each resource in $\mathcal{R}$ depends not only on the number of players using that resource but also on the player's identity itself. The delay of a player $i \in \mathcal{N}$ on resource $r \in \mathcal{R}$ is a function $d_r^i : \mathbb{N} \mapsto \mathbb{R}^+$.

A generalization of this model are (player-specific) *congestion games with priorities*, which have been introduced in [1] with the objective to model situations where each resource can assign priorities to the players, and players with a higher priority can displace *all* players with a lower priority. Every resource $r \in \mathcal{R}$ is associated with a map (not necessarily a bijection) $\pi_r : \mathcal{N} \to \{1, .., |\mathcal{N}|\}$. Several players can *allocate* a resource r (those players form a set $N_r(\sigma)$) but only those with highest priority π_r are assigned to r. This latter subset of assigned players is denoted by $\hat{N}_r(\sigma)$.

Formally, for each strategy profile $\sigma \in \Sigma$ and each $r \in \mathcal{R}$ such that $N_r(\sigma) \neq \emptyset$, let $\hat{N}_r(\sigma) = \arg\max_{i \in N_r(\sigma)} \pi_r(i)$ be the set of players *assigned* to resource r. The delay incurred by an assigned player $i \in \hat{N}_r(\sigma)$ is $d_r^i(|\hat{N}_r(\sigma)|)$. Players in $N_r(\sigma) \setminus \hat{N}_r(\sigma)$, who are not assigned to resource r, incur an infinite delay.

Although there are some similarities between the congestion model with capacities introduced in this paper and the one with priorities introduced by [1] (e.g., the possibility to displace players with lower priority on a certain resource), in general, these two models generate well distinct strategic cost games.. Contrasting with the model discussed in this paper, Ackermann *et al* [1] suppose that there is no capacity on the resources, two players may have the same priority with respect to a given resource and two players with distinct priorities on a resource r can not be both assigned to r.

Finally, the notion of capacity in systems with congested resources has been considered in [7] (see also references therein). Nevertheless, capacitated congestion games and the model in [7] are different. In our setting, we consider a finite number of atomic players and resources have an order on the users, whereas in [7], players are non-atomic and resources are not endowed with an order.

4 Contribution and Organization

Our goal is two-fold: (i) characterize the existence of a NE in capacitated congestion games; and (ii) efficiently compute an equilibrium if it exists.

First, we consider capacitated congestion games in general. We prove that a capacitated congestion game always admits a NE if it consists of two resources; moreover, this equilibrium can be computed in linear time. Besides, a game with three resources (and more) does not necessarily possess a NE. This negative result holds even if the game is symmetric-strategy and all players' strategies except one are singleton. From a computational aspect, deciding whether a game, even symmetric-strategy and consisting of two players, has a NE is shown to be **NP**-complete. The results are presented in Section 5.

Next, we consider singleton capacitated congestion games. We show that the game is a potential game so it always admits a NE. The proof is based on a new geometrical approach of potential argument, which could be seen as a generalization of a dominant potential function in higher dimension. We believe that the approach would be useful in proving the existence of NE in other games and is of independent interest. In computational aspect, the better-response dynamic converges to a NE in at most $O(n^4 m)$ strategy changes (recall that n and m are the number of players and resources, respectively). Additionally, we give a more efficient algorithm to compute a NE when the game is symmetric-strategy. The results are presented in Section 6.

5 General Strategies

We begin with a simple symmetric-strategy game which does not admit a NE. There are two players, three resources x, y and z, and the priorities are the same for the three resources (priority is always given to the first player). The strategy space of the players is $\{\{x\}, \{y, z\}\}$. Resource x has capacity 1 and $d_x(1) = 2$. Resource y has capacity 2 and $d_y(1) = 3$ while $d_y(2) = 0$. Resource z has capacity 1 and $d_z(1) = 0$. The game is illustrated in Figures 1 and 2.

Notice that the example possesses some minimal characteristics for the existence of a NE: a game with one player obviously admits a NE and Theorem 1 states that capacitated congestion games defined on two resources always admit a NE. Moreover the instance falls into restricted cases which often make the existence of a NE likely: strategies are symmetric source-target paths of a directed network, delays are monotone and priorities on the resources are identical.

Theorem 1. *Every capacitated congestion game defined on two resources possesses a pure Nash equilibrium. Moreover, a NE can be computed in linear time.*

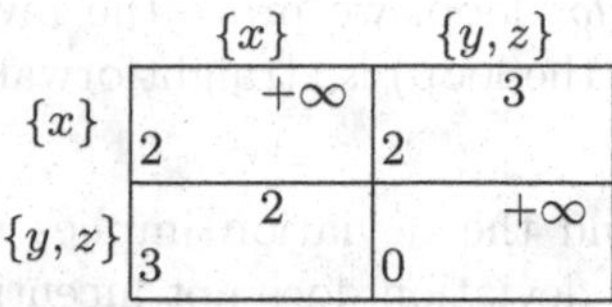

Fig. 1. A 3-resource 2-player symmetric-strategy capacitated congestion game without any pure Nash equilibrium

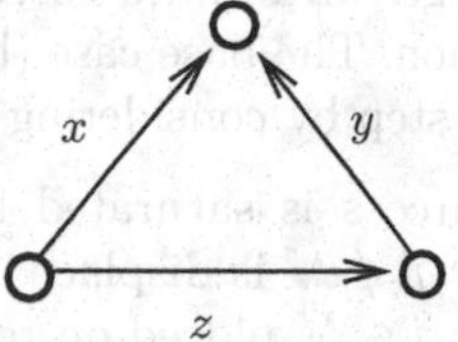

Fig. 2. The corresponding network where each arc is a resource

Proof (Sketch of proof). We prove that Algorithm 1 outputs an equilibrium σ. Denote by r and s the resources. Observe that players with strategy space $\{\{r\},\{r,s\}\}$, $\{\{s\},\{r,s\}\}$ and $\{\{r\},\{s\},\{r,s\}\}$ cannot prefer to play $\{r,s\}$ over $\{r\}$ or $\{s\}$, in any profile, as the delay of every resource is non-negative. Hence, we can reduce the strategy space of those players to be $\{\{r\}\}$, $\{\{s\}\}$ and $\{\{r\},\{s\}\}$, respectively. The action of the players having only one strategy in their (reduced) strategy space is obviously known. Denote by $\hat{\mathcal{N}}$ the players whose (reduced) strategy space is $\{\{r\},\{s\}\}$.

Algorithm 1. 2-resource

Input: a set $\mathcal{N}$ of players, two resources r, s
Output: A pure Nash equilibrium σ

$\hat{\mathcal{N}} \leftarrow \emptyset$
If a player i has only one strategy in his reduced strategy space then assign him to that strategy, else let $\sigma_i \leftarrow r$ and $\hat{\mathcal{N}} \leftarrow \hat{\mathcal{N}} \cup \{i\}$
Rename players in $\hat{\mathcal{N}}$ such that $\mathbf{pos}_s(1) < \mathbf{pos}_s(2) < \cdots < \mathbf{pos}_s(\hat{n})$ where $\hat{n} = |\hat{\mathcal{N}}|$
Let $\hat{\mathcal{N}}_\infty$ and $\hat{\mathcal{N}}_f$ be the set of players in $\hat{\mathcal{N}}$ with infinite cost and finite cost under the current profile σ, respectively
for $i = 1$ **to** $\hat{n}$ **do**
    If $i \in \hat{\mathcal{N}}_\infty$ and $c_i(s, \sigma_{-i}) < c_i(\sigma)$ then $\sigma_i \leftarrow s$
end for
for $i = 1$ **to** $\hat{n}$ **do**
    if $i \in \hat{\mathcal{N}}_f$ and $c_i(s, \sigma_{-i}) < c_i(\sigma)$ **then**
        $\sigma_i \leftarrow s$
        if i displaces a player $j \in \hat{\mathcal{N}}$ **then**
            $\sigma_j \leftarrow r$
        end if
    end if
end for
return profile σ

First, we show an invariant that at anytime, the algorithm maintains the property that no player of $\hat{\mathcal{N}}$ placed on s can or wants to move to r.

The property is clearly true before the first *for* loop. During the first *for* loop, no player who has moved from r to s has incentive to return back to r because he would get an infinite cost. For the second *for* loop, we prove the invariant by induction. The base case (before entering to the loop) is straightforward. We analyze a step by considering three subcases:

- Resource s is saturated before i moves and the deviation implies that a player $j' \notin \hat{\mathcal{N}}$ is displaced. In this case, the deviation does not incentivize a player $j \in \hat{\mathcal{N}}$ placed on resource s to move. Indeed j's cost is $d_s(\kappa_s)$ before and after i's deviation. After his deviation, i's cost is $d_s(\kappa_s)$ which is strictly smaller than his previous cost. Moving to r is not profitable to j.
- Resource s is saturated before i moves and the deviation implies that a player $j \in \hat{\mathcal{N}}$ is displaced. Observe that j cannot belong to $\hat{\mathcal{N}}_f$ because the loop follows the total order of priorities on s. The algorithm assigns j to r so that his cost is either equal to $+\infty$ or equal to the cost previously incurred by i. Then, the number of players on s remains unchanged. No player from $\hat{\mathcal{N}}$ placed on resource s has incentive to move, since otherwise the player can do it before the exchange of i and j, contradiction to the induction hypothesis.
- Resource s is not saturated before i moves and the deviation implies that at least one player $j \in \hat{\mathcal{N}}$ wants to unilaterally move to r. Players i and j have the same finite cost. By moving to r, player j would get either $+\infty$ or exactly the cost incurred by i before his deviation, contradiction.

The property holds at the end of the two phases. Now observe that a player $i \in \hat{\mathcal{N}}$ placed on r either has been displaced from s at some step or has had the opportunity to switch to s during the second loop but did not (could not) do so. Hence, those players are happy on resource r. The profile σ is then a pure Nash equilibrium. The algorithm is clearly linear in n.

When the number of resources is unbounded, the problem becomes much harder.

Proposition 1. *Deciding whether a symmetric-strategy capacitated congestion game has a NE is* **NP***-complete, even with two players.*

Proof (Sketch of proof). We reduce PARTITION — a **NP**-complete problem [13] — to the symmetric-strategy capacitated congestion game. In PARTITION, given n integers $\{a_1, \ldots, a_n\}$ such that $\sum_{j=1}^{n} a_j = 2B > 6$ and $0 < a_j < B$, one has to decide whether a subset $J \subseteq \{1, \ldots, n\}$ such that $\sum_{j \in J} a_j = B = \sum_{j \notin J} a_j$ exists.

Given an instance of PARTITION, we construct a capacitated congestion game with two players where the resources are the arcs of a network G and the players' strategies are all paths from a common source s to a common target t, see Figure 3. For arc e_0, $\kappa_{e_0} = 2$, $d_{e_0}(1) = B + 2$ and $d_{e_0}(2) = 0$. For arcs e_j and e'_j where $1 \le j \le n$, $\kappa_{e_j} = \kappa_{e'_j} = 2$, $d_{e_j}(1) = a_j$, $d_{e_j}(2) = B + 2$, and $d_{e'_j}(1) = 0, d_{e'_j}(2) = B + 2$. For arc e'_{n+2}, $\kappa_{e'_{n+2}} = 2$ and $d_{e'_{n+2}}(1) = 2$, $d_{e'_{n+2}}(2) = 0$. For arcs e_{n+1} and e'_{n+1}, their capacities are $\kappa_{e_{n+1}} = \kappa_{e'_{n+1}} = 1$ and player 1 has higher priority than player 2 in both arcs. Moreover, the delay functions are $d_{e_{n+1}}(1) = B$, $d_{e'_{n+1}}(1) = B - 1$.

One can show that the instance of PARTITION has a feasible solution iff the game defined on G admits a NE. □

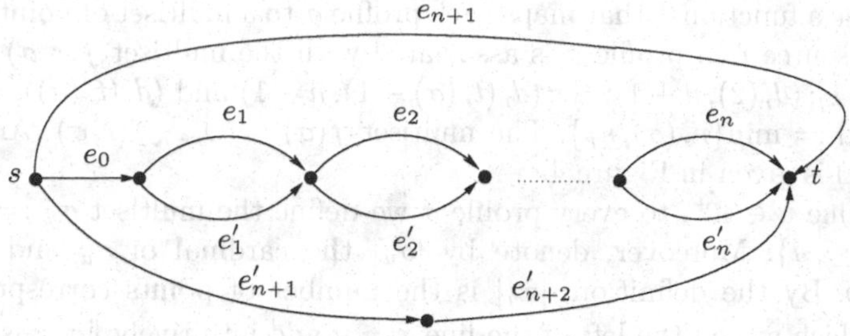

Fig. 3. The network associated with an instance of PARTITION

6 Singleton Strategies

In this section, we are interested in studying the existence of NE and efficient algorithms to compute a NE in singleton capacitated congestion games. First, we present intuitively our approach in proving the existence of a NE.

Starting Point. Consider the following dominant order $\prec'$. Let $A = \{a_1 \leq \ldots \leq a_k\}$ and $B = \{b_1 \leq \ldots \leq b_k\}$ be two sets of k real-value elements that are named in increasing order. We say that $A \prec' B$ if there exists an index $1 \leq \ell \leq k$ such that $a_i = b_i$ for all $1 \leq i < \ell$ and $a_\ell < b_\ell$. This order is well-defined and has been used in proving the existence of Nash equilibria (for example [8]). We interpret this order in a geometrical view. For each set A and B, map all elements to points on a real line where the coordinate of a point equals the value of its corresponding element. For $u \in \mathbb{R}$, let A_u and B_u be the number of points corresponding to elements in A and B with coordinate smaller than or equal to u, respectively. Then, the order $\prec'$ could be equivalently defined as follows: $A \prec' B$ if for the smallest $u \in \mathbb{R}$ such that $A_u \neq B_u$, it holds that $A_u < B_u$. In fact, the smallest $u \in \mathbb{R}$ such that $A_u \neq B_u$ is a_ℓ where ℓ is the index in the former definition.

As we have seen, the dominant order could be geometrically interpreted as a one-dimension order. Taking this geometrical approach, we prove the existence of NE by designing a two-dimension order. Intuitively, the two dimensions are due to the nature of the game where the cost of a player depends on the resource delay and the priority of the player on the resource.

Theorem 2. *Singleton capacitated congestion games are potential games. Moreover, the better-response dynamic necessarily converges in $O(n^4 m)$ strategy changes.*

Proof. First, we give some definitions which are useful in the proof.

For each profile σ, a function $\texttt{rank}_\sigma : \mathcal{R} \to \mathbb{N}$ is defined as follows. If resource r is saturated[1] then $\texttt{rank}_\sigma(r) = \max\{\texttt{pos}_r(j) : \sigma_j = r, j \text{ is accommodated}\}$. Otherwise, $\texttt{rank}_\sigma(r) := n + 1$.

[1] A resource r is saturated if $n_r(\sigma) \geq \kappa_r$.

We define a function f that maps each profile σ to a multiset of points in $\mathbb{R}^+ \times \mathbb{N}$. Each resource r in profile σ is associated with the multiset $f(r, \sigma)$ of points $(d_r(1), n+1); (d_r(2), n+1); \ldots; (d_r(t_r(\sigma)-1), n+1)$ and $(d_r(t_r(\sigma)), \mathtt{rank}_\sigma(r))$ where $t_r(\sigma) := \min\{n_r(\sigma), \kappa_r\}$. The multiset $f(\sigma) := \cup_{r \in \mathcal{R}} f(r, \sigma)$. An illustration of $f(\sigma)$ is given in Figure 4.

For a value $u \in \mathbb{R}^+$, to every profile σ we define the multiset $\sigma_u := \{(a, b) \in f(\sigma) : a \leq u\}$. Moreover, denote by $|\sigma_u|$ the cardinal of σ_u and $\|\sigma_u\| := \sum_{(a,b) \in \sigma_u} b$. By the definition, $|\sigma_u|$ is the number of points corresponding to profile σ which are on the left of the line $x = u$ and intuitively $\|\sigma_u\|$ is the total height of these points.

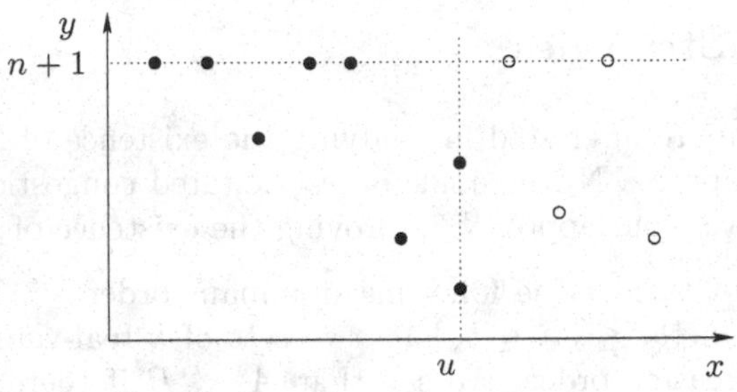

Fig. 4. An illustration of $f(\sigma)$, black filled dots if in σ_u

Now we define a partial order $\prec$ on profiles. Formally, two profiles ν and σ satisfy $\nu \prec \sigma$ if for the smallest $u > 0$ such that $(|\sigma_u|, \|\sigma_u\|) \neq (|\nu_u|, \|\nu_u\|)$ we have $|\sigma_u| < |\nu_u|$, or $|\sigma_u| = |\nu_u|$ but $\|\sigma_u\| > \|\nu_u\|$. Intuitively, we can interpret this order as follows. Two profiles ν and σ satisfy $\nu \prec \sigma$ if for the smallest $u > 0$ such that $(|\sigma_u|, \|\sigma_u\|) \neq (|\nu_u|, \|\nu_u\|)$, either (1) the half-space on the left of the line $x = u$ contains more points of ν than those of σ; or (2) if they are equal, the total height of such points in ν is smaller than that of σ.

Now we can prove that after a better move of some player i from resource r in profile σ to a resource s, resulting in profile ν, we get that $\nu \prec \sigma$. Note that $f(\sigma)$ and $f(\nu)$ only differ on some points corresponding to resources r and s. In the following, we consider only these points. Let u be the cost of player i after the move, which equals $d_s(t_s(\nu))$ — the delay of resource s in profile ν. (Note that player i is accommodated by resource s in profile ν as he has taken a better move.)

Consider the set of points corresponding to resource r in $f(\sigma)$ and $f(\nu)$. If i has unbounded cost in profile σ (meaning that i is not accommodated), then $f(r, \sigma) = f(r, \nu)$. If i is accommodated in profile σ then either $f(r, \sigma) = f(r, \nu) \cup (d_r(\sigma), \mathtt{rank}_\sigma(r))$ in case $n_r(\sigma) \leq \kappa_r$, or $f(r, \sigma) = f(r, \nu) \setminus (d_r(\kappa_r), \mathtt{rank}_\sigma(r)) \cup (d_r(\kappa_r), \mathtt{rank}_\nu(r))$ in case $n_r(\sigma) > \kappa_r$. However, as i has taken a better move, $d_r^i(\sigma) = d_r(\sigma) > u$. Hence, restricting to points with first coordinate smaller than or equal to u, $f(r, \sigma) = f(r, \nu)$.

Consider the set of point corresponding to resource s in $f(\sigma)$ and $f(\nu)$. If s is unsaturated before the move of i then $f(s,\nu) = f(s,\sigma) \cup (d_s(\nu), \mathbf{rank}_\nu(s))$ $= f(s,\sigma) \cup (u, \mathbf{rank}_\nu(s))$. If s is saturated before the move of i then $f(s,\nu) = f(s,\sigma) \cup (u, \mathbf{rank}_\nu(s)) \setminus (u, \mathbf{rank}_\sigma(s))$.

Therefore, for any $u' < u$, $(|\sigma_{u'}|, \|\sigma_{u'}\|) = (|\nu_{u'}|, \|\nu_{u'}\|)$. Moreover, if s is unsaturated before the move of i, $|\sigma_u| < |\nu_u|$. Otherwise, $|\sigma_u| = |\nu_u|$ but $\mathbf{rank}_\nu(s) < \mathbf{rank}_\sigma(s)$, so $\|\nu_u\| < \|\sigma_u\|$. Hence, $\nu \prec \sigma$, i.e., after each better move, a new profile is $\prec$-smaller than the previous one. In conclusion, the game is a potential game.

Now we bound the number of strategy changes to reach an NE from arbitrary profile in the better-response dynamic. Let σ be an arbitrary profile. By the definition of order $\prec$, there are at most nm values of u that we have to consider. Moreover, for each u, $0 \le |\sigma_u| \le n$ and $0 \le \|\sigma_u\| \le n(n+1)$. Hence, there are at most $O(n^4 m)$ couples $(|\sigma_u|, \|\sigma_u\|)$ (where σ is a profile) which are $\prec$-different. Thus, from an arbitrary profile, the better-response dynamic converges to a NE in at most $O(n^4 m)$ strategy changes. □

In the following, we consider singleton capacitated congestion games with additional property of symmetry on players' strategy sets. We give an algorithm to compute a NE that is more efficient than the better-response dynamic by exploiting that property.

Theorem 3. *A NE in a symmetric-strategy, singleton capacitated congestion game can be computed in* $\min\{n, \kappa\}$ *strategy changes and the overall time complexity of the algorithm is* $O(\min\{n^2 m, \kappa^2\})$, *where* $\kappa = \sum_{r\in\mathcal{R}} \kappa_r$.

Algorithm 2. Symmetric-strategy, singleton capacitated congestion games

---

Input: Set $\mathcal{N}$ of n players, $\mathbf{pos}_r$ and κ_r for all $r \in \mathcal{R}$
Output: An equilibrium σ

$n_r \leftarrow 0$ for all $r \in \mathcal{R}$
$\hat{n} \leftarrow \min\{n, \kappa\}$ where $\kappa = \sum_{r\in\mathcal{R}} \kappa_r$.
while $\hat{n} > 0$ **do**
  Find r^* and k_{r^*} such that $d_{r^*}(k_{r^*}) = \min\{d_r(k_r) : n_r < k_r \le \min\{n_r + \hat{n}, \kappa_r\}, r \in \mathcal{R}\}$.
  $\hat{n} \leftarrow \hat{n} - (k_{r^*} - n_{r^*})$
  $n_{r^*} \leftarrow k_{r^*}$
end while
Rename resources so that $d_{r_1}(n_1) \le d_{r_2}(n_2) \le \ldots \le d_{r_m}(n_m)$
for $j = 1$ **to** m **do**
  Assign to resource r_j the first n_j players $S \subset \mathcal{N}$ according to $\mathbf{pos}_{r_j}$
  $\mathcal{N} \leftarrow \mathcal{N} \setminus S$
end for
Assign all remaining players in $\mathcal{N}$ to an arbitrary resource, for example resource r_m.
output the current assignment σ.

Proof. We show that Algorithm 2 computes a NE.

First consider the case $n \geq \sum_{r \in \mathcal{R}} \kappa_r$. By the algorithm, at the end of the *while* loop, all resources become saturated with delays $d_{r_1}(\kappa_1) \leq \ldots \leq d_{r_m}(\kappa_m)$. Next, κ_{r_1} first players according to $\mathtt{pos}_{r_1}$ are assigned to resource r_1, then κ_{r_2} first players according to $\mathtt{pos}_{r_2}$ among the remaining players are assigned to resource r_2 then so on. Finally, assign all remaining players to resource r_m. The outcome is a NE because: (1) a player assigned to a resource r_j cannot displace other player assigned to a resource $r_{j'}$ where $j' < j$; (2) a player assigned to a resource r_j cannot decrease his cost by moving to other resource $r_{j'}$ where $j' > j$.

Now, consider the case $n < \sum_{r \in \mathcal{R}} \kappa_r$. In this case, every player is accommodated to some resource. Suppose a player i, assigned to resource r in profile σ, has incentive to deviate to resource s resulting in profile σ'.

If i's deviation displaces some player i' then we get a contradiction. Indeed, $d_r(n_r(\sigma)) = c_i(\sigma) > c_i(\sigma') = c_{i'}(\sigma) = d_s(n_s(\sigma))$ and $\mathtt{pos}_s(i) < \mathtt{pos}_s(i')$ hold. However, the algorithm fills resource s before resource r (steps 8 to 12 of the algorithm) and player i should have been assigned to s instead of player i'.

Assume i does not displace anyone when deviating. We have indeed $d_r(n_r(\sigma)) = c_i(\sigma) > c_i(\sigma') = d_s(n_s(\sigma)+1)$. Consider the moment at which n_r is modified for the last time (line 6 of the algorithm). Let k_r and k_s be the number of players already assigned to resource r and s at that time, respectively. By the algorithm, n_r is modified because $d_r(k_r) = d_r(n_r(\sigma))$ is minimum among other choices. Besides, observe that at that time, $\hat{n} \geq (n_s(\sigma) - k_s) + 1$ since later, the algorithm will set $n_s(\sigma)$ as the number of players (who are different to i) on resource s. Therefore, resource s and $n_s(\sigma)+1$ is a candidate for the choice of the algorithm in line 4. Thus, $d_r(n_r(\sigma)) \leq d_s(n_s(\sigma)+1)$ — contradiction. Hence, every player in σ is happy, meaning that it is a NE. By the algorithm, the number of strategy changes is obviously $\min\{n, \kappa\}$ and the time complexity is dominated by the *while* loop which needs at most $O(\min\{n^2 m, \kappa^2\})$ operations. □

7 Conclusion

In the paper, we have assumed that each capacitated resource r is endowed with a linear order $\mathtt{pos}_r$, indicating which players are accommodated when the resource is overcrowded. We believe that different and equally relevant ways to determine who is accommodated exist, and the existence of a NE should be investigated. For instance, an interesting open question is to know the computational complexity of symmetric-strategy capacitated congestion games with increasing delay functions. On a *dynamic* perspective, for instance, it would be interesting to study a model where the priorities of users depend on their timing of using resources (for routing problems, this could represent the arrival time to the starting node of an edge). On the other hand, in this perspective, dropping the assumption of priorities represented by linear orders could generate the technical problem of coordinating users asking for the same resource at the same time (on this issue, see the discussion about *timestamp games* in [10]).

References

1. Ackermann, H., Goldberg, P., Mirrokni, V., Röglin, H., Vöcking, B.: A unified approach to congestion games and two-sided markets. Internet Mathematics 5(4), 439–457 (2008)
2. Ackermann, H., Röglin, H., Vöcking, B.: On the impact of combinatorial structure on congestion games. J. ACM 55(6) (2008)
3. Anshelevich, E., Dasgupta, A., Kleinberg, J., Tardos, E., Wexler, T., Roughgarden, T.: The price of stability for network design with fair cost allocation. SIAM J. Comput. 38(4), 1602–1623 (2008)
4. Bauer, S., Clark, D., Lehr, W.: The evolution of internet congestion. In: 37th Research Conference on Communication, Information and Internet Policy, TPRC 2009 (2009)
5. Campbell, A., Aurrecoechea, C., Hauw, L.: A review of QoS architectures. ACM Multimedia Systems Journal 6, 138–151 (1996)
6. Chien, S., Sinclair, A.: Convergence to approximate nash equilibria in congestion games. In: SODA, pp. 169–178 (2007)
7. Correa, J., Schulz, A., Stier-Moses, N.: Selfish routing in capacitated networks. Math. Oper. Res. 29(4), 961–976 (2004)
8. Even-Dar, E., Kesselman, A., Mansour, Y.: Convergence time to nash equilibrium in load balancing. ACM Transactions on Algorithms 3(3) (2007)
9. Fabrikant, A., Papadimitriou, C., Talwar, K.: The complexity of pure nash equilibria. In: STOC, pp. 604–612 (2004)
10. Farzad, B., Olver, N., Vetta, A.: A priority-based model of routing. Chicago J. Theor. Comput. Sci. 1 (2008)
11. Fotakis, D., Kontogiannis, S., Koutsoupias, E., Mavronicolas, M., Spirakis, P.: The Structure and Complexity of Nash Equilibria for a Selfish Routing Game. In: Widmayer, P., Triguero, F., Morales, R., Hennessy, M., Eidenbenz, S., Conejo, R. (eds.) ICALP 2002. LNCS, vol. 2380, pp. 123–134. Springer, Heidelberg (2002)
12. Gairing, M., Lücking, T., Mavronicolas, M., Monien, B.: Computing nash equilibria for scheduling on restricted parallel links. In: STOC, pp. 613–622. ACM (2004)
13. Garey, M., Johnson, D.: Computers and Intractability: A Guide to the Theory of NP-Completeness. W. H. Freeman and Company, New York (1979)
14. Ieong, S., McGrew, R., Nudelman, E., Shoham, Y., Sun, Q.: Fast and compact: A simple class of congestion games. In: AAAI, pp. 489–494 (2005)
15. Mangold, S., Choi, S., May, P., Klein, O., Hiertz, G., Stibor, L.: IEEE 802.11e Wireless LAN for quality of service. In: Proc. European Wireless, vol. 18, pp. 32–39 (2002)
16. Milchtaich, I.: Congestion games with player-specific payoff functions. Games and Economic Behavior 13(1), 111–124 (1996)
17. Monderer, D., Shapley, L.: Potential games. Games and Economic Behavior 14, 124–143 (1996)
18. Rosenthal, R.: A class of games possessing pure-strategy Nash equilibria. International Journal of Game Theory 2, 65–67 (1973)
19. Vöcking, B.: Congestion games: Optimization in competition. In: ACiD Workshop. Texts in Algorithmics, vol. 7, pp. 9–20 (2006)

Network Bargaining: Using Approximate Blocking Sets to Stabilize Unstable Instances

Jochen Könemann[1], Kate Larson[2], and David Steiner[2]

[1] Department of Combinatorics and Optimization, University of Waterloo, Waterloo, Ontario N2L 3G1, Canada
jochen@uwaterloo.ca

[2] Cheriton School of Computer Science, University of Waterloo, Waterloo, Ontario N2L 3G1, Canada
{klarson,dasteine}@uwaterloo.ca

Abstract. We study a network extension to the Nash bargaining game, as introduced by Kleinberg and Tardos [6], where the set of players corresponds to vertices in a graph $G = (V, E)$ and each edge $ij \in E$ represents a possible deal between players i and j. We reformulate the problem as a cooperative game and study the following question: *Given a game with an empty core (i.e. an unstable game) is it possible, through minimal changes in the underlying network, to stabilize the game?* We show that by removing edges in the network that belong to a *blocking set* we can find a stable solution in polynomial time. This motivates the problem of finding small blocking sets. While it has been previously shown that finding the smallest blocking set is NP-hard [2], we show that it is possible to efficiently find approximate blocking sets in sparse graphs.

1 Introduction

In the classical *Nash bargaining* game [9], two players seek a mutually acceptable agreement on how to split a dollar. If no such agreement can be found, each player i receives her *alternative* α_i. Nash's solution postulates, that in an equilibrium, each player i receives her alternative α_i plus half of the surplus $1 - \alpha_1 - \alpha_2$ (if $\alpha_1 + \alpha_2 > 1$ then no mutually acceptable agreement can be reached, and both players settle for their alternatives).

In this paper, we consider a natural *network* extension of this game that was recently introduced by Kleinberg and Tardos [6]. Here, the set of players corresponds to the vertices of an undirected graph $G = (V, E)$; each edge $ij \in E$ represents a potential deal between players i and j of unit value. In Kleinberg and Tardos' model, players are restricted to bargain with at most one of their neighbours. Outcomes of the *network bargaining* game (NB) are therefore given by a matching $M \subseteq E$, and an *allocation* $x \in \mathbb{R}^V_+$ such that $x_i + x_j = 1$ for all $ij \in M$, and $x_i = 0$ if i is M-exposed; i.e., if it is not incident to an edge of M.

Unlike in the non-network bargaining game, the alternative α_i of player is not a given parameter but rather implicitly determined by the network neighbourhood of i. Specifically, in an outcome (M, x), player i's alternative is defined as

M. Serna (Ed.): SAGT 2012, LNCS 7615, pp. 216–226, 2012.

$$\alpha_i = \max\{1 - x_j : ij \in \delta(i) \setminus M\}, \tag{1}$$

where $\delta(i)$ is the set of edges incident to i. Intuitively, a neighbour j of i receives x_j in her current deal, and i may coerce her into a joint deal, yielding i a payoff of $1 - x_j$.

An outcome (M, x) of NB is called *stable* if $x_i + x_j \geq 1$ for all edges $ij \in E$, and it is *balanced* if in addition, the value of the edges in M is split according to Nash's bargaining solution; i.e., for a matching edge ij, $x_i - \alpha_i = x_j - \alpha_j$.

Kleinberg and Tardos gave an efficient algorithm to compute balanced outcomes in a graph (if these exist). Moreover, the authors characterize the class of graphs that admit such outcomes. In the following main theorem of [6], a vertex $i \in V$ is called *inessential* if there is a maximum matching in G that exposes i.

Theorem 1 ([6]). *An instance of NB has a balanced outcome iff it has a stable one. Moreover, it has a stable outcome iff no two inessential vertices are connected by an edge.*

The theory of *cooperative games* offers another useful angle for NB. In a cooperative game (with transferable utility) we are given a player set N, and a *valuation function* $v : 2^N \to \mathbb{R}_+$; $v(S)$ can be thought of as the value that the players in S can jointly create. The *matching game* [4,12] is a specific cooperative game that will be of interest for us. Here, the set of players is the set of vertices V of a given undirected graph. The matching game has valuation function ν where $\nu(S)$ is the size of a maximum matching in the graph $G[S]$ induced by the vertices in S.

One goal in a cooperative game is to allocate the value $v(N)$ of the so called *grand coalition* fairly among the players. The *core* is in some sense the gold-standard among the solution concepts that prescribe such a fair allocation: a vector $x \in \mathbb{R}_+^N$ is in the core if (a) $x(N) = v(N)$, and (b) $x(S) \geq v(S)$ for all $S \subseteq N$, where we use $x(S)$ as a short-hand for $\sum_{i \in S} x_i$. In the special case of the matching game, this is seen to be equivalent to the following:

$$\mathcal{C}(G) = \{x \in \mathbb{R}_+^V : x(V) = \nu(V) \text{ and } x_u + x_v \geq 1, \forall uv \in E\}. \tag{2}$$

Thus, the core of the matching game consists precisely of the set of stable outcomes of the corresponding NB game. This was recently also observed by Bateni et al. [1] who remarked that the set of balanced outcomes of an instance of NB corresponds to the elements in the intersection of core and *prekernel* (e.g., see [3,10] for a definition),of the associated matching game instance.

1.1 Dealing with Unstable Instances

Using the language of cooperative game theory and the work of Bateni et al. [1], we can rephrase the main results of [6] as follows: *Given an instance of NB, if the core of the underlying matching game is non-empty then there is an efficient algorithm to compute a point in the intersection of core and prekernel.* Such an algorithm had previously been given by Faigle et al. in [5]. It is not hard to see that the core of an instance of the matching game is non-empty if and only if the fractional matching LP for this instance has an integral optimum solution. We state this LP and its dual below; we let $\delta(i)$ denote the set of edges incident to vertex i in the underlying graph, and use $y(\delta(i))$ as a shorthand for the sum of y_e over all $e \in \delta(i)$.

$$\begin{array}{lll} \max & \sum_{e \in E} y_e & \text{(P)} \\ \text{s.t.} & y(\delta(i)) \leq 1 \quad \forall i \in V & \\ & y \geq \mathbb{0} & \end{array} \qquad \Bigg| \qquad \begin{array}{lll} \min & \sum_{i \in V} x_i & \text{(D)} \\ \text{s.t.} & x_i + x_j \geq 1 \quad \forall ij \in E & (3) \\ & x \geq \mathbb{0}, & \end{array}$$

LP (P) does of course typically have a fractional optimal solution, and in this case the core of the corresponding matching game instances is empty. Core assignments are highly desirable for their properties, but may simply not be available for many instances. For this reason, a number of more forgiving alternative solution concepts like *bargaining sets, kernel, nucleolus,* etc. have been proposed in the cooperative game theory literature (e.g., see [3,10]).

This paper addresses network bargaining instances that are *unstable*; i.e., for which the associated matching game has an empty core. From the above discussion, we know that there is no solution x to (D) that also satisfies $\mathbb{1}^T x \leq \nu(V)$. We therefore propose to find an allocation x of $\nu(V)$ that violates the stability condition in the smallest number of places. Formally, we call a set B of edges a *blocking set* if there is $x \in \mathbb{R}_+^V$ such that $\mathbb{1}^T x \leq \nu(V)$, and $x_i + x_j \geq 1$ for all $ij \in E \setminus B$.

Blocking sets were previously discussed by Biró et al. [2]. The authors showed that finding a smallest such set is NP-hard (via a reduction from *maximum independent set*). In this paper, we complement this result by showing that approximate blocking sets can be computed in *sparse* graphs. A graph $G = (V, E)$ is ω-*sparse* for some $\omega \geq 1$ if for all $S \subseteq V$, the number of edges in the induced graph $G[S]$ is bounded by $\omega\,|S|$. For example, if G is planar, then we may choose $\omega = 3$ by Euler's formula.

Theorem 2. *Given an ω-sparse graph $G = (V, E)$, there is an efficient algorithm for computing blocking sets of size at most $8\omega + 2$ times the optimum.*

The main idea in our algorithm is a natural one: formulate the blocking set problem as a linear program, and extract a blocking set from one of its optimal fractional solutions via an application of the powerful technique of *iterative rounding* (e.g., see [8]). We first show that the proposed LP has an unbounded integrality gap in general graphs, and is therefore not useful for the design of approximation algorithms for such instances. We turn to the class of sparse graphs, and observe that, even here, extreme points of the LP can be highly fractional, ruling out the direct use of standard techniques. We carefully characterize problem extreme-points, and develop a direct rounding method for them. Our approach exploits problem-specific structure as well as the sparsity of the underlying graph.

Given a blocking set B, let $E' = E \setminus B$ be the non-blocking set edges, and let $G' = (V, E')$ be the induced graph. Notice that the matching game induced by G' may *still* have an empty core, and that the maximum matching in G' may even be smaller than that in G. We are however able to show that we can find a balanced allocation of $\nu(V)$ as follows: let M' be a maximum matching in G', and define the alternative of player i as

$$\alpha'_i = \max\{1 - x_i : ij \in \delta_{G'}(i) \setminus M'\},$$

for all $i \in V$. Call an assignment x is balanced if it satisfies the stability condition (3) for all edges $ij \in M'$, and

$$x_i - \alpha'_i = x_j - \alpha'_j,$$

for all $ij \in M'$. A straight-forward application of an algorithm of Faigle et al. [5] yields a polynomial-time method to compute such an allocation. Details are omitted from this extended abstract.

2 Finding Small Blocking Sets in Sparse Graphs

We attack the problem of finding a small blocking set via iterative linear programming rounding. In order to do this, it is convenient to introduce a slight generalization of the blocking set problem. In an instance of the *generalized blocking set* problem (GBS), we are given a graph $G = (V, E)$, a partition $E_1 \cup E_2$ of E, and a parameter $\nu \geq 0$. The goal is to find a blocking set $B \subseteq E_1$, and an allocation $x \in \mathbb{R}^V_+$ such that $\mathbb{1}^T x \leq \nu$ and $x_u + x_v \geq 1$ for all $uv \in E \setminus B$, where $\mathbb{1}$ is a vector of 1s of appropriate dimension. The problem is readily formulated as an integer program. We give its relaxation below on the left.

$$\begin{aligned}
\min\quad & \mathbb{1}^T z && (\mathrm{P}_B)\\
\text{s.t.}\quad & x_u + x_v + z_{uv} \geq 1 \quad \forall uv \in E_1 && (4)\\
& x_u + x_v \geq 1 \quad \forall uv \in E_2 && (5)\\
& \mathbb{1}^T x \leq \nu && (6)\\
& x, z \geq \mathbb{0}
\end{aligned}$$

$$\begin{aligned}
\max\quad & \mathbb{1}^T a + \mathbb{1}^T b - \gamma\nu && (\mathrm{D}_B)\\
\text{s.t.}\quad & a(\delta_{E_1}(u)) + b(\delta_{E_2}(u)) \leq \gamma \quad \forall u \in V && (7)\\
& a \leq \mathbb{1}\\
& a, b \geq \mathbb{0}
\end{aligned}$$

The LP on the right is the dual of (P_B). It has a variable a_e for all $e \in E_1$, a variable b_e for all $e \in E_2$, and variable γ corresponds to the primal constraint limiting $\mathbb{1}^T x$. We can show the LP is weak and hence not useful for approximating the generalized blocking set problem in general graphs (for details, see [7]).

Lemma 1. *The integrality gap of* (P_B) *is* $\Omega(n)$*, where* n *is the number of vertices in the given instance of the blocking set problem.*

Given this negative result, we will focus on sparse instances (G, ν) and prove Theorem 2. We first characterize the extreme points of (P_B).

2.1 Extreme Points of (P_B)

In the following, we assume that the underlying graph G is bipartite; this assumption will greatly simplify our presentation, and will turn out to be w.l.o.g. Let (x, z) be a feasible solution of LP (P_B), and let $A^{=}(x, z)^T = b^{=}$ be the set of tight constraints of the LP. It is well known (e.g., see [11] and also [8]) that (x, z) is an extreme point of the feasible region if $A^{=}$ has full column-rank. In particular, (x, z) is uniquely determined by any full-rank sub-system $A'(x, z)^T = b'$ of $A^{=}(x, z)^T = b^{=}$. If constraint (6) is not part of this system of equations, then

$$A' = [A'', I],$$

where A'' is a submatrix of the edge-vertex incidence matrix of a bipartite graph, and I is an identity matrix of appropriate dimension. Such matrices A' are well-known to be *totally unimodular* (e.g., see [11]), and (x, z) is therefore integral in this case. From now on, we therefore assume that constraint (6) is tight, and that (x, z) is the unique solution to

$$\begin{bmatrix} A'' & I \\ \mathbb{1}^T & \mathbb{0}^T \end{bmatrix} \begin{pmatrix} \bar{x} \\ \bar{z} \end{pmatrix} = \begin{pmatrix} \mathbb{1} \\ \nu \end{pmatrix}, \tag{8}$$

where A'' is a submatrix of the edge, vertex incidence matrix of bipartite graph G, I is an identity matrix, and $\mathbb{1}^T$ and $\mathbb{0}^T$ are row vectors of 1's and 0's, respectively. We obtain the following useful lemma.

Lemma 2. *Let (x, z) be a non-integral extreme point solution to (P_B) satisfying (8). Then there is an $\alpha \in (0, 1)$ such that $x_u, z_{uv} \in \{0, \alpha, 1-\alpha, 1\}$ for all $u \in V$, and $uv \in E_1$.*

Proof. Standard linear algebra implies that the solution space to the the system $[A''\, I](\bar{x}, \bar{z})^T$ is a line; i.e., it has dimension 1. Hence, there are two extreme points (x^1, z^1) and (x^2, z^2) of the integral polyhedron defined by constraints (4), (5), and the non-negativity constraints, and some $\alpha \in [0, 1]$ such that

$$\begin{pmatrix} x \\ z \end{pmatrix} = \alpha \begin{pmatrix} x^1 \\ z^1 \end{pmatrix} + (1 - \alpha) \begin{pmatrix} x^2 \\ z^2 \end{pmatrix}.$$

In fact, α must be in $(0, 1)$ as (x, z) is assumed to be fractional. This implies the lemma. □

We call an extreme point *good* if there is a vertex u with $x_u = 1$, or an edge $uv \in E_1$ with $z_{uv} \in \{0\} \cup [1/3, 1]$. Let us call an extreme point *bad* otherwise. We will now characterize the structure of a bad extreme point (x, z). Let $G = (V, E_1 \cup E_2)$ be the bipartite graph for a given GBS instance. Let $\mathcal{T}_1 \subseteq E_1$ and $\mathcal{T}_2 \subseteq E_2$ be E_1 and E_2 edges corresponding to tight inequalities of (P_B) that are part of the defining system (8) for (x, z). Let α be as in Lemma 2. Since (x, z)

is bad, it must be that either α or $1-\alpha$ is larger than $2/3$; w.l.o.g., assume that $\alpha > 2/3$. We define the following useful sets:

$$X = \{u \in V : x_u = 1-\alpha\}$$
$$Y = \{u \in V : x_u = \alpha\}$$
$$O = \{u \in V : x_u = 0\}.$$

Lemma 3. *Let (x, z) be a bad extreme point. Using the notation defined above, we have*

(a) $z_{uv} = (1-\alpha)$ for all $uv \in E_1$,
(b) $O \cup X$ is an independent set in G
(c) Each $\mathcal{T}_1$ edge is incident to exactly one O and one Y vertex, and the edges of $\mathcal{T}_2$ form a tree spanning $X \cup Y$. Each edge in E is incident to exactly one Y vertex.

Proof. We know from Lemma 2 that $z_{uv} \in \{0, 1-\alpha, \alpha, 1\}$ for all $uv \in E_1$; (a) follows now directly from the fact that (x, z) is bad.

No two vertices $u, v \in O$ can be connected by an edge, as such an edge uv must then have $z_{uv} = 1$. Similarly, no two vertices $u, v \in X$ can be connected by an edge as otherwise $z_{uv} \geq 1 - 2(1-\alpha) > 1/3$. Finally, for an edge uv between O and X, we would have to have $z_{uv} \geq 1 - (1-\alpha) > 2/3$, which once again can not be the case. This shows (b).

To see (c), consider first an edge uv in $\mathcal{T}_1$; we must have $x_u + x_v = \alpha$, and this is only possible if uv is incident to one O and one Y vertex. Similarly, $x_u + x_v = 1$ for all $uv \in \mathcal{T}_2$, and therefore one of u and v must be in X, and one must be in Y. It remains to show that the edges in $\mathcal{T}_2$ induce a tree. Let us first show acyclicity: suppose for the sake of contradiction that $u_1v_1, \ldots, u_pv_p \in \mathcal{T}_2$ form a cycle (i.e., $u_1 = v_p$). Then since G is bipartite, this cycle contains an even number of edges. Let $\chi_1, \ldots, \chi_p$ be the $0,1$-coefficient vector of the left-hand sides of the constraints belonging to these edges. We see that

$$\sum_{i=1}^{p} (-1)^i \chi_i = \mathbb{0},$$

contradicting the fact that the system in (8) has full (row) rank. Note that the size of the support of (x, z) is

$$|\mathcal{T}_1| + |X| + |Y| \tag{9}$$

by definition. On the other hand, the rank of the system in (8) is

$$|\mathcal{T}_1| + |\mathcal{T}_2| + 1 \leq |\mathcal{T}_1| + (|X| + |Y| - k) + 1,$$

where k is the number of components formed by the edges in $\mathcal{T}_2$. The rank of (8) must be at least the size of the support, and this is only the case when $k = 1$; i.e., when $\mathcal{T}_2$ forms a tree spanning $X \cup Y$. Since G is bipartite, X must be fully contained in one side of the bipartition of V, and Y must be fully contained in the other. Since Y is a vertex cover in G by (b), every edge in E must have exactly one endpoint in Y. □

2.2 Blocking Sets in Sparse Graphs via Iterative Rounding

In this section we propose an iterative rounding (IR) type algorithm to compute a blocking set in a given sparse graph $G = (V, E)$. Recall that this means that there is a fixed parameter $\omega > 0$ such that the graph induced by any set S of vertices has at most $\omega|S|$ edges. Recall that we also initially assume that the underlying graph G is bipartite.

The algorithm we propose follows the standard IR paradigm (e.g., see [8]) in many ways: given some instance of the blocking set problem, we first solve LP (P_B) and obtain an extreme point solution (x, z). We now generate a *smaller* sub-instance of GBS such that (a) the *projection* of (x, z) onto the sub-instance is feasible, and (b) any integral solution to the sub-instance can cheaply be extended to a solution of the original GBS instance. In particular, the reader will see the standard steps familiar from other IR algorithms: if there is an edge $uv \in E_1$ with $z_{uv} = 0$ then we may simply drop the edge, if $z_{uv} \geq 1/3$ then we include the edge into the blocking set, and if $x_u = 1$ for some vertex, then we may install one unit of x-value at u permanently and delete u and all incident edges.

The problem is that the feasible region of (P_B) has bad extreme points, even if the underlying graph is sparse and bipartite. We will exploit the structural properties documented in Lemma 3 and show that a small number of edges can be added to our blocking set even in this case. Crucially, these edges will have to come from both E_1 and E_2.

In an iteration of the algorithm, we are given a sub-instance of GBS. We first solve (P_B) for this instance, and obtain an optimal basic solution (x, z). Inductively we maintain the following: The algorithm computes a set $\hat{B} \subseteq E$ of edges, and vector $\hat{x} \in \mathbb{R}^V$ such that

[I1] $\hat{x}_u + \hat{x}_v \geq 1$ for all $uv \in E \setminus \hat{B}$,
[I2] $\mathbb{1}^T \hat{x} \leq \nu$, and
[I3] $|\hat{B}| \leq (2\omega + 1) \cdot \mathbb{1}^T z$,

where ω is the sparsity parameter introduced above. Let us first assume that the extreme point solution (x, z) is good. In this case we proceed according to one of the following cases:

Case 1. ($\exists u \in V$ with $x_u = 1$) In this case, all edges incident to u are covered. We obtain a subinstance of GBS by removing u and all incident edges from G, and by reducing ν by 1.

Case 2. ($\exists uv \in E$ with $z_{uv} = 0$) In this case, obtain a new instance of GBS by removing uv from E_1, and adding it to E_2.

Case 3. ($\exists uv \in E_1$ with $z_{uv} \geq 1/3$) In this case add uv to the approximate blocking set B, and remove uv from E_1.

In each of these three cases, we inductively solve the generated sub-instance of GBS. If this subinstance is the empty graph, then we can clearly return the empty set.

Let us now consider the case where (x, z) is a bad extreme point. This case will constitute a leaf of the recursion tree, and we will show that we can directly find a small blocking set. In the following lemma, we define the sets $X, Y, O \subseteq V$ as in Lemma 3. Its proof is deferred to [7].

Lemma 4. *Let (x, z) be a bad extreme point, and let ν be the current bound on $\mathbb{1}^T x$. Then $(|X| + |Y|)/2 < \nu < |Y|$.*

We can use this bound on ν to prove that we can find small blocking sets given a bad extreme point for (P_B).

Lemma 5. *Given a bad extreme point (x, z) to (P_B), we can find a blocking set $\hat{B} \subseteq E$, and corresponding $\hat{x}$ such that $\mathbb{1}^T \hat{x} \leq \nu$, and $|\hat{B}| \leq (2\omega + 1) \cdot \mathbb{1}^T z$.*

Proof. We will construct a blocking set $\hat{B}$ as follows: let $\hat{x}_u = 1$ for a carefully chosen set $\hat{Y}$ of ν vertices from the set Y, and let $\hat{x}_u = 0$ for all other vertices in V. Recall once more from Lemma 3 (b) that Y is a vertex cover in G, and hence it suffices to choose

$$\hat{B} = \bigcup_{u \in Y \setminus \hat{Y}} \delta(u) = \bigcup_{u \in Y \setminus \hat{Y}} (\delta_{E_1}(u) + \delta_{E_2}(u)) \tag{10}$$

as our blocking set, where $\delta_{E_i}(u)$ denotes the set of E_i edges incident to vertex u. Let (a, b, γ) be the optimal dual solution of (D_B) corresponding to extreme point (x, z). Then note that complementary slackness together with the fact that $z_{uv} > 0$ for all $uv \in E_1$ implies that $a_{uv} = 1$ for these edges as well. Thus γ is an upper bound on the number E_1-edges incident to a vertex u by dual feasibility. With (10) we therefore obtain

$$|\hat{B}| \leq \sum_{u \in Y \setminus \hat{Y}} (\gamma + |\delta_{E_2}(u)|) \leq (|Y| - \nu)\gamma + \sum_{u \in Y \setminus \hat{Y}} |\delta_{E_2}(u)|. \tag{11}$$

Lemma 3 (c) shows that each E_2 edge is incident to one X, and one Y vertex. As the subgraph induced by X and Y is sparse, there therefore must be a vertex $u_1 \in Y$ of degree at most $\omega(|X| + |Y|)/|Y|$. Removing this vertex from G leaves a sparse graph, and we can therefore find a vertex u_2 of degree at most $\omega(|X| + |Y| - 1)/(|Y| - 1)$. Repeating this $|Y| - \nu$ times we pick a set $u_1, \ldots, u_{|Y|-\nu}$ of vertices such that

$$\sum_{i=1}^{|Y|-\nu} |\delta_{E_2}(u_i)| \leq \sum_{i=1}^{|Y|-\nu} \frac{\omega(|X| + |Y| - i)}{|Y| - i} \leq$$

$$(|Y| - \nu) \cdot \frac{\omega(|X| + |Y|)}{\nu} \leq 2\omega(|Y| - \nu), \tag{12}$$

where the last inequality follows from Lemma 4. We now let $\hat{Y} = Y \setminus \{u_1, \ldots, u_{|Y|-\nu}\}$, and hence let $\hat{x}_u = 1$ for $u \in \hat{Y}$, and $\hat{x}_u = 0$ for all other vertices $u \in V$; (11) and (12) together imply that

$$|\hat{B}| \leq (|Y| - \nu)(\gamma + 2\omega) \leq (2\omega + 1)\gamma(Y - \nu),$$

where the last inequality follows from the fact that $\gamma \geq 1$. Lemma 3(c) shows that each edge $e \in E$ has exactly one endpoint in Y. Applying complementary slackness together with the fact that $x_u > 0$ for all $u \in Y$, we can therefore rewrite the objective function of (D_B) as

$$\mathbb{1}^T a + \mathbb{1}^T b - \gamma\nu = \gamma(|Y| - \nu).$$

The lemma follows. □

We can now put things together.

Lemma 6. *Given an instance of GBS, the above procedure terminates with a set $\hat{B} \subseteq E$, and $\hat{x} \in \mathbb{R}^V$ such that $\mathbb{1}^T\hat{x} \leq \nu$, and $\hat{x}_u + \hat{x}_v \geq 1$ for all $uv \in E \setminus \hat{B}$. The set $\hat{B}$ has size at most $(2\omega + 1)\mathbb{1}^T z$, where (x, z) is an optimal solution to (P_B) for the given GBS instance.*

Proof. The proof uses the usual induction on the recursion depth. Let us first consider the case where the current instance is a leaf of the recursion tree. The lemma follows vacuously if the graph in the given GBS instance is empty. Otherwise it follows immediately from Lemma 5.

Any internal node of recursion tree corresponds to an instance of GBS where (x, z) is a good extreme point. We claim that, no matter which one of the above cases we are in, we have that (a) a suitable projection of (x, z) yields a feasible solution for the created GBS sub-instance, and (b) we can *augment* an approximate blocking set for this sub-instance to obtain a *good* blocking set for the instance given in this iteration. We proceed by looking at the three cases discussed above.

Case 1. Let (x', z') be the natural projection of (x, z) onto the GBS sub-instance; i.e., x'_v is set to x_v for all vertices in $V - u$, and $z'_{vw} = z_{vw}$ for the remaining edges $vw \in E_1 \setminus \delta(u)$. This solution is easily verified to be feasible. Inductively, we therefore know that we obtain a blocking set $\bar{B}$ and corresponding vector $\bar{x}$ such that $\bar{B}$ has no more than $(2\omega + 1)\,\mathbb{1}^T\bar{z} \leq (2\omega + 1)\,\mathbb{1}^T z$ elements, and $\mathbb{1}^T\bar{x} \leq \nu - 1$. Thus, letting $\hat{x}_v = \bar{x}_v$ for all $v \in V - u$, and $\hat{x}_u = 1$ together with $\hat{B} = \bar{B}$ gives a feasible solution for the original GBS instance.

Case 2. The argument for this case is virtually identical to that of Case 1, and we omit the details.

Case 3. Once again we project the current solution (x, z) onto the GBS subinstance; i.e., let $x' = x$, and $z'_{qr} = z_{qr}$ for all $qr \in E_1 - uv$. Clearly (x', z') is feasible for the GBS subinstance, and inductively we therefore obtain a vector $\bar{x}$ and corresponding feasible blocking set $\bar{B}$ of size at most $(2\omega + 1) \cdot \mathbb{1}^T z'$. Adding uv to $\bar{B}$ yields a feasible blocking set $\hat{B}$ for the original instance together with $\hat{x} = \bar{x}$. Its size is at most $(2\omega + 1)\,\mathbb{1}^T z' + 1 \leq (2\omega + 1)\,1^T z$ as $\omega \geq 1$. □

Suppose now that we are given a non-bipartite, sparse instance of the blocking set problem: $G = (V, E)$ is a general sparse graph, and $\nu > 0$ is a parameter. We create a bipartite graph H in the usual way: for each vertex $u \in V$ create two copies u_1 and u_2 and add them to H. For each edge $uv \in E$, add two edges u_1v_2 and u_2v_1 to H. The new blocking set instance is given by (H, ν') where $\nu' = 2\nu$.

Given a feasible solution (x, z) to (P_B) for the instance (G, ν) , we let $x'_{u_i} = x_u$ for all $u \in V$ and $i \in \{1, 2\}$, and $z'_{u_iv_j} = z_{uv}$ for all edges u_iv_j. For any edge u_iv_j in H, we now have

$$x'_{u_i} + x'_{v_j} + z_{u_iv_j} = x_u + x_v + z_{uv} \geq 1,$$

and $\mathbb{1}^T x' \leq 2\mathbb{1}^T x \leq 2\nu$. Thus, (x', z') is feasible to (P_B) for instance (H, ν'), and its value is at most twice that of $\mathbb{1}^T z$. Let $\hat{x}, \hat{B}$ be a feasible solution to the instance on graph H. Then let

$$B = \{uv \in E \,:\, u_1v_2 \text{ or } u_2v_1 \text{ are in } \hat{B}\},$$

and note that B has size at most that of $\hat{B}$. Also let $x_u = (\hat{x}_{u_1} + \hat{x}_{u_2})/2$ for all $u \in V$. Clearly, $\mathbb{1}^T x \leq \nu$, and for any edge $uv \in E$, we have

$$x_u + x_v \geq \frac{\hat{x}_{u_1} + \hat{x}_{u_2} + \hat{x}_{v_1} + \hat{x}_{v_2}}{2},$$

and the right-hand side is at least 1 if none of the two edges u_1v_2, u_2v_1 is in $\hat{B}$. This shows feasibility of the pair x, B. In order to prove Theorem 2 it now remains to show that graph H is sparse. Pick any set S of vertices in H, and let

$$S' = \{v \in V \,:\, \text{at least one of } v_1 \text{ and } v_2 \text{ are in } S\}.$$

Then $|S'| \leq |S|$, and the number of edges of $H[S]$ is at most twice the number of edges in $G[S']$, and hence bounded by $2\omega\,|S|$; we let $\omega' = 2\omega$ be the sparsity parameter of H. Let (x, z) and (x', z') be optimal basic solutions to (P_B) for instances (G, ν), and (H, ν'), respectively. The blocking set B for G has size no more than

$$(2\omega' + 1)\mathbb{1}^T z' \leq 2(4\omega + 1)\mathbb{1}^T z.$$

Thus, we have proven Theorem 2.

References

1. Bateni, M., Hajiaghayi, M., Immorlica, N., Mahini, H.: The Cooperative Game Theory Foundations of Network Bargaining Games. In: Abramsky, S., Gavoille, C., Kirchner, C., Meyer auf der Heide, F., Spirakis, P.G. (eds.) ICALP 2010, Part I. LNCS, vol. 6198, pp. 67–78. Springer, Heidelberg (2010)
2. Biró, P., Kern, W., Paulusma, D.: On Solution Concepts for Matching Games. In: Kratochvíl, J., Li, A., Fiala, J., Kolman, P. (eds.) TAMC 2010. LNCS, vol. 6108, pp. 117–127. Springer, Heidelberg (2010)

3. Chalkiadakis, G., Elkind, E., Wooldridge, M.: Computational Aspects of Cooperative Game Theory. Synthesis Lectures on Artificial Intelligence and Machine Learning. Morgan & Claypool Publishers (2011)
4. Deng, X., Ibaraki, T., Nagamochi, H.: Algorithmic aspects of the core of combinatorial optimization games. Math. Oper. Res. 24(3), 751–766 (1999)
5. Faigle, U., Kern, W., Kuipers, J.: An efficient algorithm for nucleolus and prekernel computation in some classes of tu-games. Tech. Rep. 1464, U. of Twente (1998)
6. Kleinberg, J.M., Tardos, É.: Balanced outcomes in social exchange networks. In: Proceedings of ACM Symposium on Theory of Computing, pp. 295–304 (2008)
7. Könemann, J., Larson, K., Steiner, D.: Network bargaining: Using approximate blocking sets to stabilize unstable instances. Tech. Rep. submit/0522859, arXiV (full version, 2012)
8. Lau, L.C., Ravi, R., Singh, M.: Iterative Methods in Combinatorial Optimization. Cambridge University Press (2011)
9. Nash, J.: The bargaining problem. Econometrica 18, 155–162 (1950)
10. Peleg, B., Sudhölter, P.: Introduction to the Theory of Cooperative Games. Springer (2003)
11. Schrijver, A.: Theory of Linear and Integer Programming. Wiley, Chichester (1986)
12. Shapley, L.S., Shubik, M.: The assignment game: the core. International Journal of Game Theory 1(1), 111–130 (1971), `http://dx.doi.org/10.1007/BF01753437`

Uniform Price Auctions: Equilibria and Efficiency*

Evangelos Markakis[1] and Orestis Telelis[2]

[1] Dept. of Informatics, Athens University of Economics and Business, Greece
[2] Dept. of Computer Science, The University of Liverpool, United Kingdom
{markakis,telelis}@gmail.com

Abstract. We present our results on Uniform Price Auctions, one of the standard sealed-bid multi-unit auction formats, for selling multiple identical units of a single good to multi-demand bidders. Contrary to the truthful and economically efficient multi-unit Vickrey auction, the Uniform Price Auction encourages strategic bidding and is socially inefficient in general, partly due to a "Demand Reduction" effect; bidders tend to bid for fewer (identical) units, so as to receive them at a lower uniform price. Despite its inefficiency, the uniform pricing rule is widely popular by its appeal to the natural anticipation, that identical items should be identically priced. Application domains of its variants include sales of U.S. Treasury bonds to investors, trade exchanges over the internet facilitated by popular online brokers, allocation of radio spectrum licenses etc. In this work we study equilibria of the Uniform Price Auction in undominated strategies. We characterize a class of undominated pure Nash equilibria and quantify the social inefficiency of pure and (mixed) Bayes-Nash equilibria by means of bounds on the Price of Anarchy.

1 Introduction

We study *Uniform Price Auctions*, a standard *Multi-Unit Auction* format, for allocating multiple units of a single good to multi-demand bidders within a single auction process. Multi-unit auctions are deployed in a variety of diverse trade exchanges, including online sales over the internet held by various brokers [20], allocation of radio spectrum licenses [17], sales of U.S. Treasury bonds to investors [22], and allocation of advertisement slots on internet sites [8]. The particular feature of the Uniform Price Auction is a single price for every unit allocated to any bidder; this makes it a proper representative of a wider category of uniform pricing auctions, as opposed to *discriminatory pricing* ones, that sell identical units of a single item at different prices [20,13]). As observed by Milgrom in [17], resurgence of interest in auction design is owed to a large extent to the success of multi-unit and – particularly – uniform price auction

* Work partially supported by the project AGT of the action THALIS (co-financed by the EU and Greek national funds) and by EPSRC grant EP/F069502/1.

M. Serna (Ed.): SAGT 2012, LNCS 7615, pp. 227–238, 2012.

formats. Uniform pricing appeals to the intuitive anticipation of identical prices for identical items and eases proxy agents that bid on behalf of their employers; they do not have to explain why they payed more than their competitors.

The design of *mechanisms* for auctioning multiple units of a single good to multi-demand bidders dates back to the seminal work of Vickrey [23]. Since then three *standard* sealed-bid auction formats have been identified in Auction Theory [13]: the Multi-Unit Vickrey Auction, the Uniform Price Auction, and the *Discriminatory Price* Auction. A significant volume of research has been dedicated to identifying the properties of these standard formats [19,9,1,21,3]. All three auctions have the same bidding format and allocation rule, and have been studied extensively for bidders with *"downward sloping"* (*symmetric submodular* [14]) valuations; these prescribe that the *marginal* value that a bidder has for each additional unit is non-increasing. Each bidder is asked to issue such a non-increasing sequence of *marginal* bids for the k available units. The k highest marginal bids win the auction and each winning bid grants its issuing bidder a distinct unit. The Multi-Unit Vickrey auction charges according to an instance of the Clarke payment rule [6] and generalizes the celebrated single-item Second-Price Auction to the case of multiple units. The Discriminatory Price Auction charges the winning bids as payments thus generalizing the First-Price Auction. The Uniform Price Auction, which was proposed by Friedman [10], charges per allocated unit the highest rejected (losing) marginal bid. The multi-unit Vickrey Auction for submodular bidders optimizes the Social Welfare and is truthful (it is a –weakly – dominant strategy for every bidder to report his marginal values truthfully). Neither the Discriminatory nor the Uniform Price auctions are truthful; they encourage strategic bidding.

In fact, a particular form of strategic bidding in Uniform Price Auctions has been identified as the *Demand Reduction* effect, observed in [19,9] and formalized in a general model for multi-unit auctions by Ausubel and Cramton [1]. Bidders may shade their marginal bids for some units, only to win fewer ones at a lower uniform price. This leads to diminished revenue and inefficient allocations at equilibrium. In particular it is known that the socially optimal allocation cannot be generally implemented in an equilibrium in (weakly) *undominated strategies*. Despite this effect, variants of Uniform Price Auctions have seen extensive applications, contrary to the Vickrey auction, which has been largely overlooked in practice; implementations of variants of the standard format are offered by several online brokers [1] [20,12] and are also being used for sales of U.S. Treasury notes to investors since 1992 [22]. We note that the Uniform Price Auction does retain some interesting features: overbidding any marginal value is a weakly dominated strategy, and so is any misreport of the marginal bid for the *first* unit.

Contribution. We study pure Nash and (mixed) Bayes-Nash equilibria of the Uniform Price Auction in *undominated* strategies. We give a detailed description of (pure) undominated strategies in the standard model of Uniform Price Auctions for submodular bidders (Section 4) and demonstrate how their properties follow from a standard assumption, i.e., that bidders issue non-increasing

[1] Among them, eBay ceased its own variant in 2009.

marginal bids for additional units. Although these properties are mentioned or partially derived in previous works, our analysis aims at clarifying some ambiguity between assumptions and implications. Additionally, we give a proposition describing a subset of pure Nash equilibria in undominated strategies.

In Section 5 we study the inefficiency of pure Nash equilibria (PNE) of the Uniform Price Auction in undominated strategies, i.e., the Price of Anarchy (PoA) over the subset of such equilibria. We derive an upper bound of $\frac{e}{e-1}$ for submodular valuation functions. We note here that the auction does have a socially optimal equilibrium (discussed in Section 3, but not in undominated strategies; all undominated PNE are known to be socially inefficient). As noted earlier, this is largely due to the *Demand Reduction* effect [1], whereby a bidder shades his bids for additional units, so as to pay a lower price for the units he wins. Our analysis can be viewed as a quantification of this effect. For any number of units $k \geq 9$, we provide an almost matching lower bound, equal to $\left(1 - e^{-1} + \frac{2}{k}\right)^{-1}$. In Section 6 we consider (mixed) Bayes-Nash equilibria in the *incomplete information* model of Harsanyi. For Bayes-Nash equilibria that emerge from randomized bidding strategy profiles containing only undominated pure strategies in their support, we upper bound the Price of Anarchy by $O(\log k)$.

2 Related Work

Uniform Price Auctions have received extensive study within the economics community. Noussair [19] and Engelbrecht-Wiggans and Kahn [9] gave characterizations of pure Bayes-Nash equilibria under independent private values of bidders, drawn from continuous distributions. They made a first observation of the effect of demand reduction. Ausubel and Cramton formalized demand reduction for a more general model of multi-unit auctions in [1], that allows also interdependent private values. Bresky showed in [3] existence of pure Bayes-Nash equilibria in the independent private values model (with continuous valuation distributions) for several multi-unit auctions, including all three standard formats.

Partly dictated by the practice of auction design and in part because of the computational difficulty of satisfying truthfulness while approximating the social welfare efficiently, there has been a resurgence of interest in the computer science community in studying auction mechanisms that are not necessarily incentive compatible [5,2,11,15]. Our results also follow this line of work of analyzing non-truthful mechanisms. Christodoulou, Kovács and Schapira initialized the study of Combinatorial Auctions, where they proposed that each out of a universe of distinct goods is sold separately and simultaneously to all other goods, in a Second-Price auction. For bidders with fractionally subadditive valuations they proved that this scheme recovers at least $\frac{1}{2}$ of the optimal social welfare in Bayesian (mixed) Nash Equilibrium. Bhawalkar and Roughgarden showed a bound of $O(\log m)$ for the Bayesian Price of Anarchy for subadditive valuations and a bound of 2 for the PoA of pure Nash equilibria [2]. Hassidim *et al.* proved welfare guarantees for a similar scheme that incorporated simultaneous First-Price auctions instead. Very recently, Syrgkanis and Tardos studied in [15] sequential First- and Second-Price auctions, motivated by the practical issue that supply may not be available at once. Lucier and Borodin [16] analyzed the social inefficiency at (mixed) Bayes-Nash equilibrium of combinatorial auctions for multiple distinct goods, with *greedy* allocation algorithms. They proved Price of Anarchy bounds fairly comparable to the approximation factors of the greedy allocation algorithms, for the underlying welfare optimization problem.

From the mechanism design perspective, Vickrey designed in [23] the first truthful mechanism for auctioning multiple units "in one go", so as to maximize the social welfare. Since then, computationally efficient truthful approximation mechanisms for multi-unit auctions and multi-demand bidders were given by Mu'alem and Nisan in [18] and by Dobzinski and Nisan in [7], even for general valuation functions. Very recently, Vöcking gave a randomized universally truthful polynomial-time approximation scheme for bidders with general valuations [24] (a universally truthful mechanism is a probability distribution over deterministic truthful mechanisms), thus almost closing the problem. In these works, the bids are elicited by the allocation algorithms through polynomially many *value queries* to the bidders, for specific bundles (with the exception of k-minded bidders, whose valuation function has a succinct representation).

3 Model and Definitions

We consider auctioning k units of a single item to a set $\mathcal{N} = [n]$ of n bidders indexed by $i = 1, \ldots, n$. Every bidder $i \in \mathcal{N}$ has a private valuation defined over the quantity of units he receives i.e. $v_i : [k] \mapsto \Re^+$, where $v_i(0) = 0$ and each v_i is non-decreasing. In this work we consider submodular valuation functions:

Definition 1. *A valuation function $f : [k] \mapsto \Re^+$ is called (symmetric)* **submodular** *if for every $x < y$, $f(x) - f(x-1) \geq f(y) - f(y-1)$.*

The following is a well known fact concerning submodular valuations.

Proposition 1. *Given $x, y \in [k]$ with $x \leq y$, a submodular valuation function f satisfies $f(x)/x \geq f(y)/y$.*

A valuation function v_i can be specified by a vector $(m_i(1), ..., m_i(k))$ of the *marginal values* $m_i(j) = v_i(j) - v_i(j-1)$ incurred to bidder i, for each additional unit in his allocation (if v_i is submodular, $m_i(j) \geq m_i(j+1)$).

Uniform Price Auction. In the standard Uniform Price Auction, bidders are asked to submit non-increasing marginal bids. Every bidder i is expected to declare his whole valuation curve as a vector $\boldsymbol{b}_i = (b_i(1), b_i(2), \ldots, b_i(k))$, with $b_i(1) \geq b_i(2) \geq \cdots \geq b_i(k)$, where $b_i(j)$ is the declared marginal value of i for obtaining the j-th unit. A declared bid $b_i(j)$ may differ from the actual marginal value $m_i(j)$. Given a bidding configuration $\boldsymbol{b} = (\boldsymbol{b}_1, \ldots, \boldsymbol{b}_n)$, the allocation algorithm produces an allocation $\boldsymbol{x}(\boldsymbol{b}) = (x_1(\boldsymbol{b}), x_2(\boldsymbol{b}), \ldots, x_n(\boldsymbol{b}))$. The *Social Welfare* under configuration $\boldsymbol{b}$ equals the bidders' total value for $\boldsymbol{x}(\boldsymbol{b})$:

$$SW(\boldsymbol{b}) = \sum_{i=1}^{n} v_i(x_i(\boldsymbol{b}))$$

The allocation algorithm of the Uniform Price Auction is an instantiation of the greedy algorithm described in [14] and is shown in Figure 1. It allocates the next unit to the next highest bid. Every bidder i pays a uniform price $p(\boldsymbol{b})$ per received unit, which equals the highest rejected bid. If under configuration $\boldsymbol{b}$ bidder i is allocated $x_i(\boldsymbol{b})$ units and the uniform price is $p(\boldsymbol{b})$, i pays a total of $x_i(\boldsymbol{b}) \times p(\boldsymbol{b})$ and derives utility $u_i(\boldsymbol{b}) = v_i(x_i(\boldsymbol{b})) - x_i(\boldsymbol{b}) \times p(\boldsymbol{b})$.

This format is a generalization of the single-item Vickrey auction to the case of multiple units, but it does not retain strategyproofness. It always admits an efficient pure Nash equilibrium though: let $\boldsymbol{x}^* = (x_1^*, ..., x_n^*)$ be an optimal allocation[2] of units to the bidders. Consider the profile $\boldsymbol{b}$ with $\boldsymbol{b}_i = (m_i(1), ..., m_i(x_i^*), 0, ..., 0)$ if $x_i^* \geq 1$ and $\boldsymbol{b}_i = \boldsymbol{0}$ otherwise. It can be shown that this is a Nash equilibrium. However, $\boldsymbol{b}_i = \boldsymbol{0}$ is weakly dominated for bidders i with $x_i^* = 0$ (Nash equilibria in undominated strategies are also known to exist).

A *demand reduction* effect occurs in undominated equilibria of this auction format. Bidders may have an incentive to understate their marginal increase for the j-th unit onwards, for some $j > 1$ [1]. This induces economic inefficiency to equilibria in undominated strategies. Nonetheless, we show that Uniform Price Auctions approximate the optimal Social Welfare within a constant factor.

1. **Set** $x_i = 0$, **for** $i = 1, \ldots, n$.
2. **For** $j = 1, \ldots, k$ **do:**
   (a) $i^* \leftarrow \arg\max_i b_i(x_i + 1)$
   (b) $x_{i^*} \leftarrow x_{i^*} + 1$
3. **return** $\boldsymbol{x}$

Fig. 1. Allocation Algorithm

Incomplete Information Setting. Every bidder $i \in \mathcal{N}$ obtains his valuation function from a finite set V_i of valuation functions, through a discrete probability distribution $\pi_i : V_i \mapsto [0, 1]$ independently of the rest of the biddders; for any particular $v \in V_i$ we write $v \sim \pi_i$ to signify that it is drawn randomly from distribution π_i. The valuation function of every bidder is *private*. A valuation profile $\boldsymbol{v} = (v_1, \ldots, v_n) \in \mathcal{V} = \times_i V_i$ is drawn from a *publicly known distribution* $\pi = \times_i \pi_i$, $\pi : \mathcal{V} \mapsto [0, 1]$. We thus write accordingly $\boldsymbol{v} \sim \pi$.

Every bidder i knows his own valuation function v_i – drawn from V_i according to π_i, but does not know the valuation function $v_{i'}$ drawn by any other bidder $i' \neq i$. Bidder i may only use his knowledge of π to estimate $\boldsymbol{v}_{-i}$. Given the publicly known distribution π, the (possibly mixed) strategy of every bidder is a function of his own valuation v_i, denoted by $B_i(v_i)$. B_i maps a valuation function $v_i \in V_i$ to a *distribution* $B_i(v_i) = B_i^{v_i}$, over all possible bid vectors (strategies) for i. In this case we will write $\boldsymbol{b}_i \sim B_i^{v_i}$, for any particular bid vector $\boldsymbol{b}_i$ drawn from this distribution. We also use the notation $\boldsymbol{B}_{-i}^{\boldsymbol{v}_{-i}}$, to refer to the vector of randomized strategies of bidders other than i, under valuation profile $\boldsymbol{v}_{-i}$ for these bidders. A *Bayes-Nash equilibrium* (BNE) is a strategy profile $\boldsymbol{B} = (B_1, \ldots, B_n)$ such that for every bidder i and for every valuation v_i, $B_i(v_i)$ maximizes the utility of i in expectation, over the distribution of the other bidders' valuations $\boldsymbol{w}_{-i}$ *given* v_i, and over the distribution induced by the mixed strategies of the bidders. That is, for every pure strategy $\boldsymbol{c}_i$ of i:

[2] For symmetric submodular valuations the allocation algorithm of the Uniform Price Auction outputs an optimal allocation when bidders bid truthfully.

$$\mathbb{E}_{\substack{\boldsymbol{w}_{-i}|v_i,\\ \boldsymbol{b}\sim \boldsymbol{B}^{(v_i,\boldsymbol{w}_{-i})}}}\left[u_i(\boldsymbol{b})\right] \geq \mathbb{E}_{\substack{\boldsymbol{w}_{-i}|v_i,\\ \boldsymbol{b}_{-i}\sim \boldsymbol{B}^{\boldsymbol{w}_{-i}}}}\left[u_i(\boldsymbol{c}_i, \boldsymbol{b}_{-i})\right]$$

where we use notation $\mathbb{E}_{\boldsymbol{v}}$ and $\mathbb{E}_{\boldsymbol{w}_{-i}|v_i}$ to denote expectation over the distributions π and $\pi(\cdot|v_i)$ (*given* v_i) respectively. Fix a valuation profile $\boldsymbol{v} \in \mathcal{V}$ and consider a (mixed) bidding configuration $\boldsymbol{B}^{\boldsymbol{v}}$, under $\boldsymbol{v}$. The Social Welfare $SW(\boldsymbol{B}^{\boldsymbol{v}})$ under $\boldsymbol{B}^{\boldsymbol{v}}$ is defined in expectation over the bidding profiles chosen by the bidders from their randomized strategies. Then, $\mathbb{E}_{\boldsymbol{v}}[SW(\boldsymbol{B}^{\boldsymbol{v}})]$ is the *expected* Social Welfare in *Bayes-Nash Equilibrium*:

$$\mathbb{E}_{\boldsymbol{v}}[SW(\boldsymbol{B}^{\boldsymbol{v}})] = \mathbb{E}_{\substack{\boldsymbol{v}\sim\pi,\\ \boldsymbol{b}\sim \boldsymbol{B}^{\boldsymbol{v}}}}\left[\sum_i v_i(x_i(\boldsymbol{b}))\right]$$

We denote by $\boldsymbol{x}^{\boldsymbol{v}}$ the socially optimal assignment under valuation profile $\boldsymbol{v} \in \mathcal{V}$ and, by slight abuse of notation, $\mathbb{E}_{\boldsymbol{v}}[SW(\boldsymbol{x}^{\boldsymbol{v}})]$ is the expected optimal social welfare. We will study the *Bayesian Price of Anarchy*, i.e. the worst case ratio $\mathbb{E}_{\boldsymbol{v}}[SW(\boldsymbol{x}^{\boldsymbol{v}})]/\mathbb{E}_{\boldsymbol{v}}[SW(\boldsymbol{B}^{\boldsymbol{v}})]$ over all distributions π and Bayes-Nash equilibria $\boldsymbol{B}$.

4 Undominated Equilibria

We study bidders with submodular valuation functions. Following Krishna [13] and Milgrom [17], we consider the standard multi-unit auction format, where bidders submit a vector of non-increasing marginal bids, i.e., encode their actual valuation function in a submodular function[3]. A similar situation occurs in combinatorial auctions with item-bidding [5,2] wherein bidders encode their valuation functions with additive functions.

Assumption 1 *The strategy space of a bidder i consists of all bidding vectors $\boldsymbol{b}_i$ for which $b_i(1) \geq b_i(2) \geq ... \geq b_i(k)$.*

A direct consequence of Assumption 1 is that, under any strategy profile $\boldsymbol{b}$, the price $p(\boldsymbol{b})$ never exceeds any of the winning bids. Lemmas 1 and 2 below state two well known facts about the Uniform Price Auction with submodular bidders (see e.g. [13,17]). We state them here to signify that Lemma 1 follows from Assumption 1 and Lemma 2 follows from the assumption *and* from Lemma 1.

Lemma 1. *For bidders with submodular valuations, and for any $j \in [k]$, it is a weakly dominated strategy to declare a bid $b_i(j)$ with $b_i(j) > m_i(j)$.*

Remark 1. By Lemma 1, a weakly undominated strategy captures a stricter notion of conservative behavior, than the usual "no-overbidding" assumption [2,4,5]. In our setting, no-overbidding would mean $\sum_{j=1}^{r} b_i(j) \leq v_i(r)$ for any $r = 1, \ldots, k$.

[3] This requirement is implementable: the auctioneer can exclude non-conforming bidders. Also, simple examples exhibit its necessity for ensuring individual rationality.

To distinguish from the usual no-overbidding assumption, we call a bidder i who bids at most $m_i(j)$ for any $j \in [k]$ *conservative with respect to marginal bids.*

Lemma 2. *In an undominated strategy, a bidder with a submodular valuation never declares a bid* $b_i(1) \neq v_i(1)$.

We now give a characterization of a subset of undominated equilibria:

Proposition 2. *Let* $\boldsymbol{b}$ *be a pure Nash equilibrium strategy profile of the Uniform Price Auction in undominated strategies for submodular bidders, with uniform price* $p(\boldsymbol{b})$. *There always exists a pure Nash equilibrium* $\boldsymbol{b}'$ *in undominated strategies, satisfying* $\boldsymbol{x}(\boldsymbol{b}') = \boldsymbol{x}(\boldsymbol{b})$ *and:*

1. $b'_i(x) = m_i(x)$, *for every bidder* i *and every* $x \leq x_i(\boldsymbol{b})$.
2. $p(\boldsymbol{b}') \leq p(\boldsymbol{b})$ *and* $p(\boldsymbol{b}')$ *is either* 0 *or equal to* $v_i(1)$ *for some bidder* i.

5 Inefficiency of Pure Nash Equilibria

This section presents welfare guarantees for pure Nash equilibria of the standard form of the Uniform Price Auction, discussed in the previous section. First we are going to show a general result about upper bounding the Price of Anarchy of pure Nash equilibria. Given a configuration $\boldsymbol{b}$, we will be denoting by $\beta_j(\boldsymbol{b})$, $j = 1, \ldots, k$, the j-th lowest winning bid, so that $\beta_1(\boldsymbol{b}) \leq \beta_2(\boldsymbol{b}) \leq \cdots \leq \beta_k(\boldsymbol{b})$. In this section we will omit an explicit reference to $\boldsymbol{b}$ in this notation, as it will be clear from the context. Instead, we use simply β_j, $j = 1, \ldots, k$.

Lemma 3. *Let* $\boldsymbol{b}$ *denote an undominated pure Nash equilibrium of a Uniform Price Auction for* k *units and* $\boldsymbol{x}(\boldsymbol{b})$ *the corresponding allocation. Let* $\boldsymbol{x}^*$ *be an assignment that maximizes the social welfare. The Price of Anarchy is at most:*

$$PoA \leq \sup_{\boldsymbol{b}} \max_{i: x_i^* - x_i(\boldsymbol{b}) > 0} \left[v_i(x_i^*) \cdot \left(v_i\Big(x_i(\boldsymbol{b})\Big) + \sum_{j=1}^{x_i^* - x_i(\boldsymbol{b})} \beta_j \right)^{-1} \right] \quad (1)$$

The following result quantifies the inefficiency of the standard multi-unit Uniform Price auction for multi-demand bidders with symmetric submodular valuation functions and identifies the impact of *demand reduction* [1].

Theorem 1. *The Uniform Price Auction recovers in an undominated pure Nash equilibrium a fraction of at least* $1 - e^{-1}$ *of the optimal Social Welfare, for multi-demand bidders with symmetric submodular valuations.*

Proof. It suffices to upper bound the social inefficiency of undominated equilibria satisfying the properties of Proposition 2. Let $p(\boldsymbol{b})$ be the uniform price paid under equilibrium $\boldsymbol{b}$. To estimate a lower bound on the Social Welfare of $\boldsymbol{b}$, we consider possible deviations of bidders i with $x_i^* > x_i(\boldsymbol{b})$. At least one such bidder exists, otherwise, $x_i(\boldsymbol{b}) \geq x_i^*$ for every i implies that $\boldsymbol{b}$ is socially optimal.

For every bidder i with $x_i^* > x_i(\boldsymbol{b})$ define $r_i(\boldsymbol{b}) = x_i^* - x_i(\boldsymbol{b})$; for every value $j = 1, \dots, r_i(\boldsymbol{b})$ there exists a deviation that will grant him j additional units to the ones he already holds under $\boldsymbol{b}$; this is due to the fact that all bidders play marginal bids at most equal to their marginal valuations in $\boldsymbol{b}$. Since a sorting of the marginal values determines $\boldsymbol{x}^*$, every "socially optimal winner" i (with $x_i^* \geq 1$) can feasibly deviate under $\boldsymbol{b}$ so as to obtain at least x_i^* units. If $r_i(\boldsymbol{b}) > 0$, a deviation of i for obtaining any $j = 1, \dots, r_i(\boldsymbol{b})$ *additional* units will raise the uniform price to exactly β_j (using Proposition 2) and cannot be profitable for i:

$$v_i(x_i(\boldsymbol{b}) + j) - (x_i(\boldsymbol{b}) + j) \cdot \beta_j \leq v_i(x_i(\boldsymbol{b})) - x_i(\boldsymbol{b}) \cdot p(\boldsymbol{b})$$

To simplify notation, we use hereafter x_i for $x_i(\boldsymbol{b})$, p for $p(\boldsymbol{b})$ and r_i for $r_i(\boldsymbol{b})$. Then we deduce for every i with $r_i > 0$:

$$\beta_j \geq \frac{1}{j + x_i} \cdot \Big(v_i(x_i + j) - v_i(x_i)\Big), \text{ for } j = 1, \dots, r_i \tag{2}$$

We can now proceed to upper bound (1) from Lemma 3, using (2) as follows:

$$\begin{aligned}
v_i(x_i) + \sum_{j=1}^{r_i} \beta_j &\geq v_i(x_i) + \sum_{j=1}^{r_i} \frac{1}{j + x_i} \cdot \Big(v_i(x_i + j) - v_i(x_i)\Big) &(3)\\
&= v_i(x_i) + \sum_{j=1}^{r_i} \left(\frac{j}{j + x_i} \cdot \frac{v_i(x_i + j) - v_i(x_i)}{j} \right) \\
&\geq v_i(x_i) + \frac{v_i(x_i^*) - v_i(x_i)}{x_i^* - x_i} \cdot \sum_{j=1}^{r_i} \frac{j}{j + x_i} &(4)\\
&= v_i(x_i) + \frac{v_i(x_i^*) - v_i(x_i)}{x_i^* - x_i} \cdot \left(x_i^* - x_i - x_i \cdot \sum_{j=1}^{r_i} \frac{1}{j + x_i} \right) \\
&= v_i(x_i^*) - \frac{v_i(x_i^*) - v_i(x_i)}{x_i^* - x_i} \cdot x_i \cdot \sum_{j=1}^{r_i} \frac{1}{j + x_i} &(5)\\
&\geq \left(v_i(x_i^*) - \frac{x_i}{x_i^*} \sum_{j=1}^{r_i} \frac{v_i(x_i^*)}{j + x_i} \right) \geq \left(1 - \frac{x_i}{x_i^*} \int_{x_i}^{x_i^*} \frac{1}{y} dy \right) v_i(x_i^*) &(6)\\
&= \left(1 + \frac{x_i}{x_i^*} \cdot \ln \frac{x_i}{x_i^*} \right) \cdot v_i(x_i^*) \geq (1 - e^{-1}) \cdot v_i(x_i^*) &(7)
\end{aligned}$$

Here (3) occurs by substitution of β_j from (2). (4) follows by submodularity of the valuation functions, particularly that $\frac{v_i(x_i+j)-v_i(x_i)}{j} \geq \frac{v_i(x_i^*)-v_i(x_i)}{x_i^*-x_i}$, for any $j = 1, \dots, r_i$ where $r_i = x_i^* - x_i$. For (6) we used $\frac{v_i(x_i^*)-v_i(x_i)}{x_i^*-x_i} \leq \frac{v_i(x_i^*)}{x_i^*}$, given $v_i(0) = 0$; we bounded the sum of harmonic terms with the integral, using

$\sum_{k=m}^{n} f(k) \leq \int_{m-1}^{n} f(x)dx$, for a monotonically decreasing positive function. We obtain the final result by minimizing $f(y) = 1 + y \ln y$ over $(0, 1)$ for $y = e^{-1}$. The claimed bound for the *PoA* follows by Lemma 3. □

We will produce an almost matching lower bound for the result of theorem 1, which holds for any number of units $k \geq 9$. We note that for $k = 2, 3$ units, tight bounds of $\frac{4}{3}$ and $\frac{18}{13}$ can be derived by direct manipulation of (3).

Theorem 2. *For any $k \geq 9$, there exist instances where the Uniform Price Auction recovers in an undominated pure Nash equilibrium at most a factor of $(1 - e^{-1} + \frac{2}{k})$ of the optimal social welfare, even for 2 submodular bidders.*

Proof. Consider $k \geq 9$ units and 2 bidders. For $q = \lfloor e^{-1} \cdot k - 1 \rfloor$ (notice that $q \geq 1$) define the valuation functions to be:

$$v_1(x) = x \qquad \text{and} \qquad v_2(x) = \begin{cases} x - q \cdot (H_k - H_{k-x}) & x \leq k - q \\ k - q \cdot (1 + H_k - H_q) & x > k - q \end{cases}$$

where H_m is the m-th harmonic number. Notice that $m_2(x) = 0$ for $x > k - q$. It can be verified that v_2 is symmetric submodular in x; for $x \leq k - q$ we have:

$$v_2(x) = x - q \cdot \left(H_k - H_{k-x}\right) = \sum_{j=1}^{x} \left(1 - \frac{q}{k - j + 1}\right) = \sum_{j=1}^{x} \frac{r - j + 1}{k - j + 1}$$

where $r = k - q$. Then $\frac{r-j+1}{k-j+1} \leq \frac{r-j+2}{k-j+2} = \frac{r-(j-1)+1}{k-(j-1)+1}$, thus $v_2(x) - v_2(x-1) \leq v_2(x-1) - v_2(x-2)$, for $x \leq k - q$; for $x > k - q$, $v_2(x) = v_2(x-1)$, thus v_2 is submodular. For the socially optimal allocation we grant all units to bidder 1, i.e., $\boldsymbol{x}^* = (k, 0, \ldots, 0)$ and $SW(\boldsymbol{x}^*) = k$. Consider next the configuration $\boldsymbol{b}$ where:

$$b_1(j) = \begin{cases} 1, \text{ for } j \leq q \\ 0, \text{ for } j > q \end{cases} \qquad b_2(j) = \begin{cases} \frac{r-j+1}{k-j+1}, \text{ for } j \leq r = k - q \\ 0, \quad \text{ for } j > r \end{cases}$$

Thus, under $\boldsymbol{b}$, q units are obtained by bidder 1 and $k - q$ units by bidder 2. $\boldsymbol{b}$ is a pure Nash equilibrium; indeed, bidder 2 is essentially truthful and, with a uniform price of 0, obtains the maximum of his utility for the won units. Given that he plays undominated strategies, he may not raise any of his bids further. Player 1 also pays the uniform price of 0, so he does not have incentive to drop any of his units. Should player 1 retain any $j \leq r$ of the $r = k - q$ units held by bidder 2, he would hold a total of $k - r + j$ units at a uniform price $\frac{j}{k-r+j}$; the marginal value gain of j to bidder 1 from the extra units is cancelled out by a total payment equal to j. For the social welfare of $\boldsymbol{b}$ we have:

$$SW(\boldsymbol{b}) = v_1(q) + v_2(r) = k \cdot \left(1 - \frac{q}{k} \cdot (H_k - H_q)\right)$$

Then, the Price of Anarchy is at least $k/SW(\boldsymbol{b})$, i.e. at least:

$$\left(1-\frac{q}{k}\cdot\left(H_k-H_q\right)\right)^{-1} \geq \left(1-\frac{e^{-1}\cdot k-2}{k}\cdot\int_{q+1}^{k}\frac{1}{y}dy\right)^{-1}$$
$$=\left(1-\frac{e^{-1}\cdot k-2}{k}\cdot\ln\frac{k}{\lfloor e^{-1}k-1\rfloor+1}\right)^{-1}\geq\left(1-e^{-1}+\frac{2}{k}\right)^{-1}$$

where we used $H_k - H_q = \sum_{r=q+1}^{k}\frac{1}{r} \geq \int_{q+1}^{k+1}\frac{1}{y}dy \geq \int_{q+1}^{k}\frac{1}{y}dy$, for monotonically decreasing positive functions; the final derivation follows by $q+1 \leq e^{-1}\cdot k$ and $\lfloor e^{-1}k-1\rfloor+1 \geq e^{-1}k$ □

6 Inefficiency of Bayes-Nash Equilibria

In this section we investigate the social inefficiency of (mixed) Bayes-Nash equilibria. Following [5,2], to ensure the existence of mixed Bayes-Nash equilibria, we make the assumption of a finite bidding space for bidders, using Remark 1 combined with a sufficiently fine discretization. Just like for pure equilibria, we examine Bayes-Nash equilibria with undominated strategies in their support[4].

We introduce auxiliary notation for the analysis that follows. Recall that for any valuation profile $\boldsymbol{v} \in \mathcal{V}$, $\boldsymbol{x}^{\boldsymbol{v}} = (x_1^{\boldsymbol{v}}, \ldots, x_n^{\boldsymbol{v}})$ is the socially optimal assignment. For any bidder $i \in \mathcal{N}$ let $\mathcal{U}^i \subseteq \mathcal{V}$ denote the subset of valuation profiles $\boldsymbol{v} \in \mathcal{V}$ where $x_i^{\boldsymbol{v}} \geq 1$, i.e., $\mathcal{U}^i = \{\boldsymbol{v} \in \mathcal{V} | x_i^{\boldsymbol{v}} \geq 1\}$; these are the profiles under which i is a "socially optimal winner". Accordingly, define $\mathcal{W}^{\boldsymbol{v}} = \{i | x_i^{\boldsymbol{v}} \geq 1\}$. Given any (pure) bidding profile $\boldsymbol{b}$, we use the *"operator"* $\beta_j(\boldsymbol{b})$, to denote the j-th lowest winning bid in $\boldsymbol{b}$, as in section 5. The following Lemma facilitates the expression of BNE conditions regarding unilateral deviations; it has been proved in a different form and under a different context (for simultaneous single-unit auctions with combinatorial bidders) in [5,2].

Lemma 4. *For each bidder $i \in \mathcal{N}$ with symmetric submodular valuation v_i, define $\boldsymbol{m}_i^{[j]} = (m_i(1), m_i(2), \ldots, m_i(j), 0, 0, \ldots, 0)$. For any conservative bidding profile $\boldsymbol{b}_{-i}$, and for any number of units j: $u_i(\boldsymbol{m}_i^{[j]}, \boldsymbol{b}_{-i}) \geq v_i(j) - j\cdot\beta_j(\boldsymbol{b}_{-i})$.*

Theorem 3. *The Price of Anarchy of Bayes-Nash Equilibria in Uniform Price Auctions with symmemtric submodular bidders is at most $O(\log k)$.*

Proof. (Sketch) For any Bayes-Nash equilibrium $\boldsymbol{B}$, fix any valuation profile $\boldsymbol{v} \in \mathcal{V}$ and a bidder $i \in \mathcal{W}^{\boldsymbol{v}}$. For $j = 1, \ldots, x_i^{\boldsymbol{v}}$, for any valuation profile $\boldsymbol{w}_{-i} \in \mathcal{V}_{-i}$ and any strategy $\boldsymbol{b} \sim \boldsymbol{B}_{-i}^{\boldsymbol{w}_{-i}}$, apply Lemma 4. Then take expectation over $\boldsymbol{b}_{-i} \sim \boldsymbol{B}_{-i}^{\boldsymbol{w}_{-i}}$ and, subsequently, over all valuation profiles $\boldsymbol{w}_{-i} \in \mathcal{V}_{-i}$, to obtain:

[4] Such Bayes-Nash equilibria can be shown to exist; moreover the strategies in their support can be shown to be conservative with respect to marginal bids.

$$\mathbb{E}_{\boldsymbol{w}_{-i}|v_i}\left[\mathbb{E}_{\boldsymbol{b}_{-i}\sim \boldsymbol{B}_{-i}^{\boldsymbol{w}_{-i}}}[u_i(\boldsymbol{m}_i^{[j]},\boldsymbol{b}_{-i})]\right] \geq v_i(j) - j\cdot \mathbb{E}_{\boldsymbol{w}_{-i}|v_i}\left[\mathbb{E}_{\boldsymbol{b}_{-i}\sim \boldsymbol{B}_{-i}^{\boldsymbol{w}_{-i}}}[\beta_j(\boldsymbol{b}_{-i})]\right]$$

Because under BNE $\boldsymbol{B}$ bidder i does not have incentive to deviate:

$$\mathbb{E}_{\boldsymbol{w}_{-i}|v_i}\left[\mathbb{E}_{\boldsymbol{b}\sim \boldsymbol{B}^{(v_i,\boldsymbol{w}_{-i})}}[u_i(\boldsymbol{b})]\right] \geq \mathbb{E}_{\boldsymbol{w}_{-i}|v_i}\left[\mathbb{E}_{\boldsymbol{b}_{-i}\sim \boldsymbol{B}_{-i}^{\boldsymbol{w}_{-i}}}[u_i(\boldsymbol{m}_i^{[j]},\boldsymbol{b}_{-i})]\right]$$

Thus $\frac{1}{j}\mathbb{E}_{\boldsymbol{w}_{-i}|v_i}\left[\mathbb{E}_{\boldsymbol{b}\sim \boldsymbol{B}^{(v_i,\boldsymbol{w}_{-i})}}[v_i(x_i(\boldsymbol{b}))]\right]+\mathbb{E}_{\boldsymbol{w}_{-i}|v_i}\left[\mathbb{E}_{\boldsymbol{b}_{-i}\sim \boldsymbol{B}_{-i}^{\boldsymbol{w}_{-i}}}[\beta_j(\boldsymbol{b}_{-i})]\right] \geq \frac{v_i(j)}{j}$. For any pure strategy $\boldsymbol{c}_i$ of bidder i, $\beta_j(\boldsymbol{b}_{-i}) \leq \beta_j(\boldsymbol{c}_i, \boldsymbol{b}_{-i})$ since the presence of $\boldsymbol{c}_i$ means that more bids are competing to win. Also, by independence of π_i, we have that $\sum_{\boldsymbol{w}_{-i}} \pi(\boldsymbol{w}_{-i}|v_i) = 1$. By submodularity, $\frac{v_i(j)}{j} \geq \frac{v_i(x_i^{\boldsymbol{v}})}{x_i^{\boldsymbol{v}}}$. Then:

$$\frac{1}{j}\cdot \mathbb{E}_{\boldsymbol{w}_{-i}|v_i}\left[\mathbb{E}_{\boldsymbol{b}\sim \boldsymbol{B}^{(v_i,\boldsymbol{w}_{-i})}}[v_i(x_i(\boldsymbol{b}))]\right] + \mathbb{E}_{\boldsymbol{w}}\left[\mathbb{E}_{\boldsymbol{b}\sim \boldsymbol{B}^{\boldsymbol{w}}}[\beta_j(\boldsymbol{b})]\right] \geq \frac{v_i(x_i^{\boldsymbol{v}})}{x_i^{\boldsymbol{v}}}$$

Summing both sides over $j = 1, \ldots, x_i^{\boldsymbol{v}}$, then taking the expectation over the distribution of $\boldsymbol{v} \in \mathcal{U}^i$ and summing over $i \in \mathcal{N}$ yields:

$$\sum_i \sum_{\boldsymbol{v}\in\mathcal{U}^i} \pi(\boldsymbol{v}) \sum_{j=1}^{x_i^{\boldsymbol{v}}} \frac{1}{j}\cdot \mathbb{E}_{\substack{\boldsymbol{w}_{-i}|v_i,\\ \boldsymbol{b}\sim \boldsymbol{B}^{(v_i,\boldsymbol{w}_{-i})}}} \left[v_i(x_i(\boldsymbol{b}))\right] + \sum_i \sum_{\boldsymbol{v}\in\mathcal{U}^i} \pi(\boldsymbol{v}) \sum_{j=1}^{x_i^{\boldsymbol{v}}} \mathbb{E}_{\substack{\boldsymbol{w},\\ \boldsymbol{b}\sim \boldsymbol{B}^{\boldsymbol{w}}}} \left[\beta_j(\boldsymbol{b})\right]$$

$$\geq \sum_i \sum_{\boldsymbol{v}\in\mathcal{U}^i} \pi(\boldsymbol{v}) \sum_{j=1}^{x_i^{\boldsymbol{v}}} \frac{v_i(x_i^{\boldsymbol{v}})}{x_i^{\boldsymbol{v}}} = \sum_{\boldsymbol{v}\in\mathcal{V}} \pi(\boldsymbol{v}) \sum_{i\in\mathcal{W}^{\boldsymbol{v}}} v_i(x_i^{\boldsymbol{v}}) = \mathbb{E}_{\boldsymbol{v}}\left[SW(\boldsymbol{x}^{\boldsymbol{v}})\right] \quad (8)$$

The result follows by upper bounding the first and second summands of the first line of (8) by $(1+\ln k)\mathbb{E}_{\boldsymbol{v}}[SW(\boldsymbol{B}^{\boldsymbol{v}})]$ and $\mathbb{E}_{\boldsymbol{w}}[SW(\boldsymbol{B}^{\boldsymbol{w}})]$ respectively. The bounding of the second summand in particular can be carried out by usage of $\sum_i \sum_{j=1}^{x_i} \beta_j(\boldsymbol{b}) \leq SW(\boldsymbol{b})$, for any bidding configuration $\boldsymbol{b}$ that is conservative w.r.t. marginal bids and for any assignment $\boldsymbol{x}$ of all k units to n bidders. □

References

1. Ausubel, L., Cramton, P.: Demand Reduction and Inefficiency in Multi-Unit Auctions. Tech. rep., University of Maryland (2002)
2. Bhawalkar, K., Roughgarden, T.: Welfare Guarantees for Combinatorial Auctions with Item Bidding. In: Proceedings of the ACM-SIAM Symposium on Disctrete Algorithms, SODA, pp. 700–709 (2011)
3. Bresky, M.: Pure Equilibrium Strategies in Multi-unit Auctions with Private Value Bidders. Tech. Rep. 376, Center for Economic Research & Graduate Education - Economics Institute (CERGE-EI), Czech Republic (2008)
4. Caragiannis, I., Kaklamanis, K., Kanellopoulos, P., Kyropoulou, M., Lucier, B., Paes Leme, R., Tardos, E.: On the efficiency of equilibria in generalized second price auctions. arxiv:1201.6429 (2012)

5. Christodoulou, G., Kovács, A., Schapira, M.: Bayesian Combinatorial Auctions. In: Aceto, L., Damgård, I., Goldberg, L.A., Halldórsson, M.M., Ingólfsdóttir, A., Walukiewicz, I. (eds.) ICALP 2008, Part I. LNCS, vol. 5125, pp. 820–832. Springer, Heidelberg (2008)
6. Clarke, E.H.: Multipart pricing of public goods. Public Choice 11, 17–33 (1971)
7. Dobzinski, S., Nisan, N.: Mechanisms for Multi-Unit Auctions. Journal of Artificial Intelligence Research 37, 85–98 (2010)
8. Edelman, B., Ostrovsky, M., Schwartz, M.: Internet Advertising and the Generalized Second-Price Auction: Selling Billions of Dollars Worth of Keywords. The American Economic Review 97(1), 242–259 (2007)
9. Engelbrecht-Wiggans, R., Kahn, C.M.: Multi-unit auctions with uniform prices. Economic Theory 12(2), 227–258 (1998)
10. Friedman, M.: A Program for Monetary Stability. Fordham University Press, New York (1960)
11. Hassidim, A., Kaplan, H., Mansour, Y., Nisan, N.: Non-price equilibria in markets of discrete goods. In: Proceedings of the ACM Conference on Electronic Commerce, EC, pp. 295–296 (2011)
12. Kittsteiner, T., Ockenfels, A.: On the Design of Simple Multi-unit Online Auctions. In: Jennings, N., Kersten, G., Ockenfels, A., Weinhardt, C. (eds.) Negotiation and Market Engineering. Dagstuhl Seminar Proceedings, vol. 06461. Internationales Begegnungs- und Forschungszentrum für Informatik (IBFI), Schloss Dagstuhl, Germany (2007), `http://drops.dagstuhl.de/opus/volltexte/2007/1005`
13. Krishna, V.: Auction Theory. Academic Press (April 2002)
14. Lehmann, B., Lehmann, D.J., Nisan, N.: Combinatorial auctions with decreasing marginal utilities. Games and Economic Behavior 55(2), 270–296 (2006)
15. Leme, R.P., Syrgkanis, V., Tardos, E.: Sequential auctions and externalities. In: Proceedings of the ACM-SIAM Symposium on Discrete Algorithms, SODA, pp. 869–886 (2012)
16. Lucier, B., Borodin, A.: Price of Anarchy for Greedy Auctions. In: Proceedings of the ACM-SIAM Symposium on Discrete Algorithms, SODA, pp. 537–553 (2010)
17. Milgrom, P.: Putting Auction Theory to Work, Cambridge (2004)
18. Mu'alem, A., Nisan, N.: Truthful approximation mechanisms for restricted combinatorial auctions. Games and Economic Behavior 64(2), 612–631 (2008)
19. Noussair, C.: Equilibria in a multi-object uniform price sealed bid auction with multi-unit demands. Economic Theory 5, 337–351 (1995)
20. Ockenfels, A., Reiley, D.H., Sadrieh, A.: Economics and Information Systems. In: Handbooks in Information Systems, Online Actions, vol. 1, ch. 12, pp. 571–628. Elsevier Science (December 2006)
21. Reny, P.J.: On the existence of Pure and Mixed Strategy Nash Equilibria in Discontinuous Games. Econometrica 67, 1029–1056 (1999)
22. U.S. Dept. of Treasury: Uniform-price auctions: Update of the treasury experience, office of market finance (1998), `http://www.treasury.gov/domfin`
23. Vickrey, W.: Counterspeculation, auctions, and competitive sealed tenders. Journal of Finance 16(1), 8–37 (1961)
24. Vöcking, B.: A universally-truthful approximation scheme for multi-unit auctions. In: Proceedings of the ACM-SIAM Symposium on Discrete Algorithms, SODA, pp. 846–855 (2012)

Minimizing Expectation Plus Variance

Marios Mavronicolas[1] and Burkhard Monien[2]

[1] Department of Computer Science, University of Cyprus, Nicosia CY-1678, Cyprus
mavronic@cs.ucy.ac.cy
[2] Faculty of Computer Science, Electrical Engineering and Mathematics, University of Paderborn, 33102 Paderborn, Germany
bm@upb.de

Abstract. We consider *strategic games* in which each *player* seeks a *mixed strategy* to minimize her *cost* evaluated by a *concave valuation* V (mapping probability distributions to reals); such valuations are used to model *risk*. In contrast to games with *expectation-optimizer* players where *mixed equilibria* always exist [15,16], a mixed equilibrium for such games, called a V*-equilibrium,* may fail to exist, even though *pure equilibria* (if any) transfer over. *What is the impact of such valuations on the existence, structure and complexity of mixed equilibria?* We address this fundamental question for a particular concave valuation: ***expectation plus variance***, denoted as RA, which stands for *risk-averse*; so, *variance* enters as a measure of risk and it is used as an additive adjustment to *expectation.* We obtain the following results about RA-equilibria:

- A collection of general structural properties of RA-equilibria connecting to *(i)* E-equilibria and Var-equilibria, which correspond to the *expectation* and *variance* valuations E and Var, respectively, and to *(ii)* other weaker or incomparable equilibrium properties.
- A second collection of *(i)* existence, *(ii)* equivalence and separation (with respect to E-equilibria), and *(iii)* characterization results for RA-equilibria in the new class of *player-specific scheduling games.* Using examples, we provide the *first* demonstration that going from E to RA may as well create *new* mixed (RA-)equilibria.
- A *purification* technique to transform a player-specific scheduling game on identical links into a player-specific scheduling game so that all *non-pure* RA-equilibria are eliminated while new pure equilibria *cannot* be created; so, a *particular* game on two identical links yields one with *no* RA-equilibrium. As a by-product, the *first* $\mathcal{PLS}$-completeness result for the computation of RA-equilibria follows.

1 Introduction

In a *strategic game*, each *player* is choosing a *strategy,* and her *utility* (resp., *cost*) depends on the choices of all players in the game. The *player* is allowed to use a *mixed strategy,* a probability distribution over her strategies. (A *pure strategy* is the case where the player is choosing a certain strategy with probability 1.) Much of *Non-Cooperative Game Theory* has been built on the fundamental assumption that players are *expectation-optimizers*: each player maximizes

M. Serna (Ed.): SAGT 2012, LNCS 7615, pp. 239–250, 2012.

(resp., minimizes) the *expectation* of her utility (resp., cost); the earliest account of *Expected Utility Theory*, or EUT for short, is the historical book of von Neumann and Morgenstern [17]. In his ground-breaking result, John F. Nash [15,16] used the *linearity* of expectation (in the *probabilities*) to prove the existence of a *Nash equilibrium*, where no player can improve her expectation by switching to another mixed strategy. Existence extends beyond expectation to *concave* (resp. *convex*) ***valuations*** (functions from probability distributions to reals) for games where players *maximize* (resp., *minimize*) their utilities (resp., costs) [4]. The assumption that players are expectation-optimizers may not fit well into the context of *risk* (cf. [1]), where economic agents may risk for a mixed strategy yielding uncertain utility over one with more certain, but possibly higher, expectation. This is because valuation functions that *cannot* be cast as expectations remain outside the framework of EUT. Even worse, when trading expectation for arbitrary valuations, mixed equilibria are no *a priori* preserved even though pure equilibria *are*. So, *concave* (resp., *convex*) valuations may even fail to guarantee existence of mixed equilibria when players minimize (resp., maximize) valuations.

Modelling risk has been a very active research topic (see, e.g., [1,11,12,18,19]). Rabin [18] underlines the inadequacy of expectation to model risk. Already in 1906, Fisher [1] proposed *standard deviation* (the square root of ***variance***) as a measure of riskiness which should be added to expectation. Standard deviation of the return on investment is the standard measure of risk in modern *Portfolio Theory* (cf. [11]). Markowitz [11], Nobel Laureate of 1990, advocates that investors should care about both the *risk* **and** the (expected) return of their investment; he posed the problem of ***minimizing the variance*** of a portfolio taking as a constraint a required return in ***expectation***. To the best of our knowledge, Crawford [3] was the *first* to ask how non-EUT valuations impact the ***existence*** and ***structure*** of mixed equilibria; he presented [3] a simple game with players maximizing a *non-concave* valuation as the *first* counter-example to the existence of mixed equilibria outside EUT. So, mixed equilibria get *endangered* and their decision problem becomes non-trivial when players *minimize* (resp., *maximize*) a *non-convex* (resp., *non-concave*) valuation. *The possibility that new mixed equilibria be created was left open.*

Fiat and Papadimitriou [6] initiated the study of the complexity of mixed equilibria in contexts where players *maximize* a *non-concave* valuation, thus bringing a major challenge into the *Algorithmic Game Theory* community; we continue this study. We shall consider games with players **minimizing** a *concave* valuation since we are interested in congestion-like games [20] where players seek to *minimize* their delay costs. In this vein, we shall present a collection of structural and complexity results for mixed equilibria incurred by a particular *concave* valuation, namely ***expectation plus variance***, denoted as RA, in both general games and in ***player-specific scheduling games***, a new class of congestion-like games we introduce.

To model concave valuations, we adopt a definition of concavity which applies to a function over the *convex closure* of a finite *set* and is more general than the standard. Roughly speaking, such a function is ***concave*** if its values on the convex closure *cannot* go below any of its values on the set, but *must* go above some of its values on the set *in case those are not all equal.* We chose to use this definition since it is the most general definition allowing to prove our results. We prove that RA is concave in the probabilities of the corresponding player under the adopted definition of concavity (Proposition 5).

A player i is interested in *minimizing* her ***valuation***: the value of her ***valuation function*** V_i on the tuple of the mixed strategies of players. A collection of valuation functions V induces a V***-equilibrium*** where no player could decrease her valuation by unilaterally switching to another mixed strategy; so, each player is playing a *best-response* mixed strategy with respect to V. We shall also treat two concepts similar to but different than V-equilibrium. The ***Strong Equilibrium*** property requires that for each player, each pure strategy chosen with non-zero probability is a *best-response* (with respect to V) to the other players' mixed strategies. The ***Weak Equilibrium*** property loosens *Strong Equilibrium* by only requiring that for a player, all pure strategies chosen with non-zero probability incur *the same* value of V conditioned on the other players' mixed strategies. So, *Strong Equilibrium* implies *Weak Equilibrium* (but not vice versa). Interestingly, we show that for a *concave* V, the *Strong Equilibrium* property holds for each player in a V-equilibrium (Proposition 1).

We shall focus on the structure of RA-equilibrium, in relation to E-equilibrium and Var-equilibrium corresponding to the sibling ***expectation*** and ***variance*** valuations, respectively, denoted as E and Var. Note that E-equilibrium coincides with both Nash equilibrium and the *Strong Equilibrium* (for E). We shall consider the new class of ***player-specific scheduling games*** where the cost (that is, *delay*) of a player on the *link* it chooses is a sum of *weights,* one for each player choosing that link and each depending on the two players involved and the link; so, each player has her own way of evaluating the influence of others on her. Player-specific scheduling games are different than *congestion games with player-specific payoff functions* [14] where *(i)* the weight of player depends neither on other players nor on the link, and *(ii)* each player uses her *player-specific latency function* to evaluate her delay on the link it chooses. Player-specific scheduling games *simultaneously capture (i)* the ***unrelated links game*** [10], which generalizes the (extensively studied) ***related links game*** [9], and *(ii)* the ***max-cut game*** [2,5] inspired by the MAX-CUT problem [21].

We first prove that if for a player *(i)* the *Strong Equilibrium* property holds with respect to RA and *(ii)* the *Weak Equilibrium* property holds with respect to E, then the player is playing a best-response with respect to RA (Theorem 6). In a sense, this result provides a converse to Proposition 1 for a particular concave valuation RA: the *Weak Equilibrium* property with respect to E suffices for the *Strong Equilibrium* property with respect to RA to imply best-responses with respect to RA. More interestingly, we show that RA-equilibrium implies *Weak Equilibrium (i)* with respect to E in both the game and the ***square game*** where utilities are squared, and *(ii)* with respect to Var (Theorem 7). The necessary conditions on RA-equilibria established in Theorem 7 take the form of a system of non-linear equations. So, Theorem 7 can be seen as suggesting the *first* theoretical explanation behind any *general* inexistence result of (*mixed*) V-equilibria when (players minimize and) V is concave: it may be due to the general unsolvability of non-linear systems. We shall treat the necessary conditions on RA-equilibria (in Theorem 7) as a tool for characterizing their *(in)existence* in concrete cases. We continue with results specific to particular games.

The fully mixed case: (I) We prove that in the identical links game [9], we get an RA-equilibrium if every player chooses every link with equal probability (Theorem 11). We continue to demonstrate **limitations** on the existence of *fully mixed* equilibria in the form of characterizations; these follow from the necessary *Weak Equilibrium* properties (Theorem 7). (II) In the related links game, even for two players, no single player could randomize over three links with distinct capacities in an RA-equilibrium (Theorem 12); this excludes fully mixed RA-equilibria in the presence of such a triple of links. This implies a **separation** between fully mixed RA-equilibria and fully mixed E-equilibria [13] for the case of two players in the related links game. (III) For the case of three players on three links in the player-specific scheduling game, there is a fully mixed Var-equilibrium if and only if a certain *triangle inequality* holds among the weights (Proposition 14). (IV) Finally, we look at ***bimatrix games*** on two strategies. We derive a combinatorial characterization (Theorem 15) of the two (2×2) matrices for which a fully mixed RA-equilibrium may exist.

Two links: (I) RA-equilibria and E-equilibria *coincide* in the identical links game with two links (Theorem 16). Using an example of an identical links game on *three links* (Example 1), the coincidence could not extend beyond two (identical) links. (II) We next assume that the two links are *ordered*: link 1 incurs less weight to a *fixed* player due to some other (also *fixed*) player choosing the same link than link 2; this is fulfilled in the related links game with two links of different capacities. We show that, in an RA-equilibrium, each player either *(i)* is *pure* or *(ii)* other players influencing her on link 2 are all *pure* (Theorem 17). So, there is no fully mixed RA-equilibrium, implying a **separation** between fully mixed RA-equilibria and E-equilibria in the related links game with two links.

Two players: We consider a generalization of player-specific scheduling games, called *player-specific scheduling games with monotone latencies,* where the incurred cost to a player is the value of some *monotone* function (specific to the link she chooses) on the sum of weights of players choosing the same link. We show that, in an RA-equilibrium, either *(i)* there is a pure player, or *(ii)* the two players are choosing from *disjoint* sets of links, *(iii)* they are choosing from the same set of links (Theorem 18). We explore the potential of these necessary conditions to suffice for an RA-equilibrium or for an E-equilibrium. We present two suitable related links games (Example 2) to demonstrate that (I/a) Condition *(iii)* is *not* sufficient for an RA-equilibrium even for a fully mixed profile that is an E-equilibrium; (I/b) Condition *(ii)* is *not* sufficient for an E-equilibrium even for a fully mixed profile that is an RA-equilibrium. Even more so, there is a game with an E-equilibrium violating both Conditions *(ii)* and *(iii)* (Example 3).

We present a new *purification* technique to **eliminate** all mixed RA-equilibria from a given player-specific scheduling game on two *identical links*: given such a game, we transform it to a *modified* player-specific scheduling game with no *mixed* RA-equilibrium whose pure equilibria are also pure equilibria of the original game

(Theorem 20). (The transformation is surprisingly simple: we simply multiply the incurred weight of each player choosing link 2 by a large factor and add 1 to it.) So, any RA-equilibrium in the modified game is pure. This is the *first* concrete instance of a specific class of games and a specific valuation (RA) such that there may only be pure equilibria with respect to the valuation. The proof of Theorem 20 utilizes Theorem 17 to conclude that for a player in the modified player-specific scheduling game, either *(i)* she is pure or *(ii)* her neighbors are pure; we argue that case *(ii)* is excluded, so that each player is pure. A particular subclass of player-specific scheduling games are the ***max-cut games*** where all incurred weights are *symmetric* in the two players. This subclass may be viewed either as a class of games or as a $\mathcal{PLS}$*-problem* corresponding to the minimization of *non-cut* edges in the weighted graph induced by the weights, known to be $\mathcal{PLS}$-complete [21]. We observe that local optima for the $\mathcal{PLS}$-problem exactly correspond to pure equilibria in the modified max-cut game (Lemma 21). Hence, computing an RA-equilibrium for the modified max-cut game is $\mathcal{PLS}$-complete (Theorem 22). This is the *first* $\mathcal{PLS}$-completeness result about computing an RA-equilibrium. Finally, we present an example of a player-specific scheduling game on two identical links with no pure equilibrium. By Theorem 20, this game is transformed into a player-specific scheduling game on two links with no RA-equilibrium (Corollary 23). Hence, restricting to player-specific scheduling games *cannot* outlaw the inexistence of RA-equilibria.

We remark that the standard definition of *strict concavity* (cf. Section 2), which is a restriction of concavity, does *not* cover RA while it excludes mixed equilibria (cf. [19]). In contrast, the definition of concavity we adopted, which encompasses concavity, covers RA while it does *not* exclude mixed equilibria. Fiat and Papadimitriou [6, Section 2] introduced a definition of strict concavity which refers to a *pair* of a valuation function *and* a game. It *generalizes* the standard definition by requiring strict concavity to hold *only if* the game is in ***general position***; games *not* in general position have measure 0. Whether the definition of strict concavity in [6, Section 2] covers RA was not considered in [6]. Fiat and Papadimitriou [6] proved that games in general position with a strictly concave valuation may *not* have mixed quilibria. Hence, the definition of strict concavity in [6, Section 2] does *not* exclude mixed equilibria, but they may exist for a class of games with measure 0. Fiat and Papadimitriou [6, Proposition 2] prove that RA is concave. We use a simple proof to prove that RA is concave under the more general definition of concavity (Proposition 5). Fiat and Papadimitriou [6, Theorem 5] provide a proof sketch to claim that it is $\mathcal{NP}$-complete to decide if a game with two players has an RA-equilibrium.

2 Framework and Preliminaries

Consider a finite set $\mathsf{T} = \{t_1, \ldots, t_n\}$ and denote as τ the ***convex closure*** of T: $\tau = \left\{ x = \sum_{i=1}^{n} x_i t_i \mid x_i \geq 0 \text{ for each } i \in [n] \text{ and } \sum_{i \in [n]} x_i = 1 \right\}$. For a point $x \in \tau$, the ***support*** of x is $\sigma(x) = \{i \in [n] \mid x_i > 0\}$. A function $\mathsf{V} : \tau \to \mathsf{R}$. is ***concave*** if the following two conditions hold for each point $x \in \tau$: (C1) If

$\mathsf{V}(t_j) = \mathsf{V}(t_k)$ for all indices $j, k \in \sigma(x)$, then $\mathsf{V}(x) \geq \mathsf{V}(t_j)$ for each $j \in \sigma(x)$. (C2) If there are indices $j, k \in \sigma(x)$ with $\mathsf{V}(t_j) \neq \mathsf{V}(t_k)$, then $\mathsf{V}(x) > \mathsf{V}(t_\ell)$ for some $\ell \in \sigma(x)$. This is different from the two standard definitions: (1) A real-valued function $\mathsf{V} : \mathsf{T} \rightarrow \mathsf{R}$ on a convex set T is ***concave*** if for any two points $t_1, t_2 \in \mathsf{T}$ and any number $\delta \in [0,1]$, $\mathsf{V}(\delta\, t_1 + (1-\delta)\, t_2) \geq \delta\, \mathsf{V}(t_1) + (1-\delta)\, \mathsf{V}(t_2)$. (2) The function V is ***strictly concave*** if for any two points $t_1, t_2 \in \mathsf{T}$ with $t_1 \neq t_2$ and any number $\delta \in (0,1)$, $\mathsf{V}(\delta\, t_1 + (1-\delta)\, t_2) > \delta\, \mathsf{V}(t_1) + (1-\delta)\, \mathsf{V}(t_2)$. Our definition is *more general* than the standard definition of concavity.

For an integer $n \geq 2$, an ***n-players game*** G, or ***game*** for short, consists of *(i)* n finite sets $\{S_k\}_{k \in [n]}$ of ***strategies***, and *(ii)* n ***cost functions*** $\{\mu_k\}_{k \in [n]}$, each mapping $S = \prod_{k \in [n]} S_k$ to R. So $\mu_i(\mathbf{s})$ is the ***cost*** of player i on $\mathbf{s}$. The ***square game*** G^2 results from G by substituting in *(ii)* the squares of the cost functions for G. A ***profile*** is a tuple $\mathbf{s} = \langle s_1, \ldots, s_n \rangle$ of strategies, one per player. For a player $i \in [n]$, the ***partial profile*** $\mathbf{s}_{-i}$ results by eliminating s_i from $\mathbf{s}$; denote S_{-i} the set of partial profiles $\mathbf{s}_{-i}$. A ***mixed strategy*** for player i is a probability distribution p_i on her strategy set S_i. The support of player i in p_i is the set $\sigma(p_i) = \{\ell \in S_i \mid p_i(\ell) > 0\}$. Denote as p_i^ℓ the ***pure strategy*** of player i choosing the strategy $\ell \in S_i$ with probability 1. Denote as $\Delta(S_i)$ the set of mixed strategies for player i. A ***mixed profile*** is a tuple $\mathbf{p} = \langle p_1, \ldots, p_n \rangle$ of n mixed strategies, one per player; denote as Δ the set of mixed profiles. For a player i, the ***partial mixed profile*** $\mathbf{p}_{-i}$ results by eliminating p_i from $\mathbf{p}$. A mixed profile $\mathbf{p}$ induces probabilities $\mathbf{p}(\mathbf{s})$ and $\mathbf{p}(\mathbf{s}_{-i})$ for each profile $\mathbf{s}$ and partial profile $\mathbf{s}_{-i}$. A player i is ***fully mixed*** in $\mathbf{p}$ if for each strategy $s_i \in S_i$, $0 < p_i(s_i) < 1$; $\mathbf{p}$ is ***fully mixed*** if each player is fully mixed. Fix a mixed profile $\mathbf{p}$. For a player $i \in [n]$ and a strategy $\ell \in S_i$, denote as A_i^ℓ and B_i^ℓ the conditional expectations of her cost and square of the cost *had she chosen strategy* ℓ.

For each player $i \in [n]$, a ***valuation function*** V_i is a mapping from $\Delta(S)$ to R, which yields a ***valuation*** $\mathsf{V}_i(\mathbf{p})$ to each mixed profile $\mathbf{p} \in \Delta(S)$ for player i. A ***valuation*** $\mathsf{V} = \langle \mathsf{V}_1, \ldots, \mathsf{V}_n \rangle$ is a tuple of valuation functions, one per player. Denote as G^V the game G together with the tuple V of valuation functions; so, each player aims at minimizing her valuation in the game G^V. Fix a player $i \in [n]$. Say that the mixed strategy p_i is a V_i***-best response*** to the partial mixed profile $\mathbf{p}_{-i}$ if $\mathsf{V}_i\,(p_i, \mathbf{p}_{-i}) = \min \{\mathsf{V}_i\,(p_i', \mathbf{p}_{-i}) \min p_i' \in \Delta(S_i)\}$; so, the mixed strategy p_i minimizes the valuation $\mathsf{V}_i(\mathbf{p})$ of player i. The mixed profile $\mathbf{p}$ is a V***-equilibrium*** (for the game G^V) if for each player $i \in [n]$ the mixed strategy p_i is a V_i-best response to the partial mixed profile $\mathbf{p}_{-i}$; so, no player could unilaterally deviate to another mixed strategy p_i' to decrease her valuation $\mathsf{V}_i(\mathbf{p})$. *Nash's Theorem* [15,16] establishes that the game G has at least one mixed equilibrium with respect to expectation; it was extended by Debreu [4] to establish the existence of at least one V-equilibrium in case all valuation functions V are convex (resp., concave) and players are minimizers (resp., maximizers). If the valuation functions V are *concave* (resp., convex) and players are minimizers (resp., maximizers), then the existence of a V-equilibrium is not guaranteed.

The mixed profile $\mathbf{p}$ has the ***Strong Equilibrium*** property for player $i \in [n]$ in G^V if for each strategy $\ell \in \sigma(p_i)$, $\mathsf{V}_i\,(p_i^\ell, \mathbf{p}_{-i}) = \min \left\{ \mathsf{V}_i \left(p_i^{\ell'}, \mathbf{p}_{-i} \right) \mid \ell' \in S_i \right\}$;

so, each pure strategy in the support of player i is a V_i-best-response to the partial mixed profile $\mathbf{p}_{-i}$. The mixed profile $\mathbf{p}$ has the ***Strong Equilibrium*** property if it has the *Strong Equilibrium* property for each player. The mixed profile $\mathbf{p}$ has the ***Weak Equilibrium*** property for player $i \in [n]$ in the game G^{V} if for each pair of strategies $\ell, \ell' \in \sigma(p_i)$, $\mathsf{V}_i\left(p_i^{\ell}, \mathbf{p}_{-i}\right) = \mathsf{V}_i\left(p_i^{\ell'}, \mathbf{p}_{-i}\right)$. The mixed profile $\mathbf{p}$ has the ***Weak Equilibrium*** property in the game G^{V} if it has the *Weak Equilibrium* property for each player. A fully mixed profile with the *Weak Equilibrium* property, has the *Strong Equilibrium* property. We show:

Proposition 1. *For G^{V}, fix player $i \in [n]$. Assume for each partial mixed profile $\mathbf{p}_{-i} \in \Delta(S_{-i})$, the function $\mathsf{V}_i(p_i, \mathbf{p}_{-i})$ is concave in p_i. If p_i is a V_i-best-response to $\mathbf{p}_{-i}$, then $\mathbf{p}$ has the Strong Equilibrium property for i.*

A ***player-specific scheduling game*** is equipped with a ***weight*** $\omega(i,j,\ell)$ for each triple of a player $i \in [n]$, a player $j \in [n]$ and a strategy $\ell \in S_i$, with $S_1 = \ldots = S_n = [m]$; $\omega(i,j,\ell)$ represents the load due to player j incurred to player i on strategy ℓ. The m strategies are also called ***links***. Given the collection of weights $\{\omega(i,j,r)\}_{i,j\in[n],r\in S_i}$, the cost function μ_{i} is defined by $\mu_{\mathsf{i}}(\mathbf{s}) = \sum_{j \mid s_j = s_i} \omega(i,j,s_i)$. A ***player-specific scheduling game on identical links*** is the special case of a player-specific scheduling game where for each pair of players $i, i' \in [n]$, for each pair of links ℓ, ℓ', $\omega(i,j,\ell) = \omega(i,j,\ell')$. The following are important special cases of player-specific scheduling games: (1) The ***unrelated links game*** [10] where for each player $i \in [n]$, for each strategy $\ell \in [m]$, and for each player $j \in [n]$, $\omega(i,j,\ell) = \omega(j,\ell)$; so, there is incurred the *same* load $\omega(j,\ell)$ due to player j on strategy ℓ to all players $i \in [n]$. (This special case is *more general* than the well-studied ***related links game*** [9] where $\omega(j,\ell) = \frac{w_j}{c_\ell}$ for a collection of ***weights*** $\{w_i\}_{i\in[n]}$ and a collection of ***capacities*** $\{c_\ell\}_{\ell\in[m]}$.) (2) The ***max-cut game*** or ***party affiliation game*** [2,5], denoted as MCG, is the **symmetric** special case of a player-specific game on two identical links where for each pair of players $i, j \in [n]$, $\omega(i,j) = \omega(j,i)$. The non-zero weights in a max-cut game induces an undirected (edge-weighted) graph $G = \langle V, E\rangle$ with $V = [n]$. Each player is identified with a vertex $v \in V$; for a player $i \in [n]$, denote as $\mathsf{N}(i) = \{j \in [n] \mid \{i,j\} \in E\}$ the ***neighborhood*** of i in $[n]$. Clearly, $\mu_i(\mathbf{s}) = \sum_{j\in\mathsf{N}(i) \mid s_j = s_i} \omega(i,j)$. So, each player is minimizing the sum of weights on edges to neighbors choosing the same strategy, called ***non-cut edges***. Thus, a pure equilibrium for the max-cut game is a *local minimum* with respect to the sum of weights on *neighboring* non-cut edges; the well-known MAX-CUT problem asks for the *global minimum* (resp., *global maximum*) with respect to the *total* sum of weights on non-cut edges (resp., cut edges). We now derive expressions for A_i^{ℓ} and B_i^{ℓ}, with $i \in [n]$ and $\ell \in [m]$, to be used later:

Lemma 2. *Consider a player-specific scheduling game. Fix a mixed profile $\mathbf{p}$. Then, for each pair of a player $i \in [n]$ and a link $\ell \in [m]$, $\mathsf{A}_i^{\ell} = \omega(i,i,\ell) + \sum_{k\in[n]\setminus\{i\}} p_k(\ell) \cdot \omega(i,k,\ell)$ and*

$$\mathsf{B}_i^{\ell} = \omega^2(i,i,\ell) + 2\,\omega(i,i,\ell) \sum_{k\in[n]\setminus\{i\}} p_k(\ell) \cdot \omega(i,k,\ell) + \sum_{k\in[n]\setminus\{i\}} p_k(\ell) \cdot \omega^2(i,k,\ell)$$

$$+2 \sum_{j,k \in [n] \setminus \{i\}} p_j(\ell)\, p_k(\ell)\, \omega(i,j,\ell)\, \omega(i,k,\ell)\,.$$

We shall treat *(i)* the ***expectation*** $\mathsf{E}_i(\mathbf{p}) = \mathsf{E}_{\mathbf{s}\sim\mathbf{p}}(\mu_i(\mathbf{s})) = \sum_{\mathbf{s}\in S} \mathbf{p}(\mathbf{s})\, \mu_i(\mathbf{s})$, *(ii)* the ***variance*** $\mathsf{Var}_i(\mathbf{p}) = \sum_{\mathbf{s}\in S} \mathbf{p}(\mathbf{s})\, \mu_i^2(\mathbf{s}) - \left(\sum_{\mathbf{s}\in S} \mathbf{p}(\mathbf{s})\, \mu_i(\mathbf{s})\right)^2$, so that for a strategy $\ell \in S_i$, $\mathsf{Var}_i\left(\mathbf{p}_{-i}, p_i^\ell\right) = \mathsf{B}_i^\ell - \left(\mathsf{A}_i^\ell\right)^2$, and *(iii)* the ***expectation plus variance*** $\mathsf{RA}_i(\mathbf{p})$ given as $\mathsf{RA}_i(\mathbf{p}) = \sum_{\mathbf{s}\in S} \mathbf{p}(\mathbf{s})\, \mu_i(\mathbf{s}) + \sum_{\mathbf{s}\in S} \mathbf{p}(\mathbf{s})\, \mu_i^2(\mathbf{s}) - \left(\sum_{\mathbf{s}\in S} \mathbf{p}(\mathbf{s})\, \mu_i(\mathbf{s})\right)^2$; so, $\mathsf{RA}_i(\mathbf{p}) = \sum_{\ell\in S_i} p_i(\ell) \cdot \left(\mathsf{A}_i^\ell + \mathsf{B}_i^\ell\right) - \left(\sum_{\ell\in S_i} p_i(\ell) \cdot \mathsf{A}_i^\ell\right)^2$. Here, RA stands for *risk-averse.* With $\mathsf{E} = \langle \mathsf{E}_1, \ldots, \mathsf{E}_n \rangle$, $\mathsf{Var} = \langle \mathsf{Var}_1, \ldots, \mathsf{Var}_n \rangle$ and $\mathsf{RA} = \langle \mathsf{RA}_1, \ldots, \mathsf{RA}_n \rangle$, the games G^{E}, $\mathsf{G}^{\mathsf{Var}}$ and G^{RA} result.

3 Structural Results

Since RA is the sum of E and Var, it immediately follows:

Lemma 3. *Assume that* $\mathbf{p}$ *is (i) an* E*-equilibrium (for* G^{E}*) and (ii) a* Var*-equilibrium (for* $\mathsf{G}^{\mathsf{Var}}$*). Then,* $\mathbf{p}$ *is an* RA*-equilibrium (for* G^{RA}*).*

We also observe:

Lemma 4. *Consider an* E*-equilibrium* $\mathbf{p}$ *with* $\sigma(p_i) \cap \sigma(p_k) = \emptyset$ *for each pair of players* $i, j \in [n]$*. Then, for each player* $i \in [n]$, $\mathsf{Var}_i(\mathbf{p}) = 0$, *so that* $\mathbf{p}$ *is an* RA*-equilibrium.*

We continue to show:

Proposition 5. *Fix a player* $i \in [n]$ *and a partial mixed profile* $\mathbf{p}_{-i} \in \Delta(S_{-i})$*. Then, the valuation function* $\mathsf{RA}_i(p_i, \mathbf{p}_{-i})$ *is concave in* p_i*.*

We now show:

Theorem 6. *Fix a player* $i \in [n]$ *and assume that the mixed profile* $\mathbf{p}$ *has* (A1) *the Strong Equilibrium property for player* i *in the game* G^{RA}, *and* (A2) *the Weak Equilibrium Property for player* i *in the game* G^{E}*. Then, the mixed strategy* p_i *is an* RA*-best response to* $\mathbf{p}_{-i}$.

We continue to show:

Theorem 7. *Fix a player* $i \in [n]$ *and assume that the mixed strategy* p_i *is an* RA*-best-response to the partial mixed profile* $\mathbf{p}_i$ *in the game* G^{RA}*. Then,* $\mathbf{p}$ *has the Weak Equilibrium property for player* i *in the games* G^{E}, $(\mathsf{G}^2)^{\mathsf{E}}$ *and* $\mathsf{G}^{\mathsf{Var}}$.

We finally show:

Proposition 8. *A fully mixed profile is an* RA*-equilibrium if and only if it is both an* E*-equilibrium and a* Var*-equilibrium.*

The following observation follows immediately from Theorem 7.

Lemma 9. *Consider a game* G *with two players, for which* $\langle p, q \rangle$ *is an* RA*-equilibrium with* $\sigma(p) = \sigma(q)$*. Then,* $\langle p, q \rangle$ *is an* E*-equilibrium.*

4 Existence, Equivalence, Separation and Characterization Results

We partition into (I) results about ***the fully mixed case***; (II) results for the case of ***two links***; and (III) results for the case of ***two players***. We shall use an implication of Lemma 2, providing expressions for Var in two special cases.

Corollary 10. *Consider the player-specific scheduling game. Fix a profile* $\mathbf{p}$. *Then, for each pair of a player* $i \in [n]$ *and a link* $\ell \in [m]$, $\mathsf{Var}_i\left(p_i^\ell, \mathbf{p}_{-i}\right) = \mathsf{B}_i^\ell - \left(\mathsf{A}_i^\ell\right)^2 = \sum_{k \in [n] \setminus \{i\}} p_k^\ell \left(1 - p_k^\ell\right) \omega^2(i, k, \ell)$; *for games on two links,* $\mathsf{Var}_i\left(p_i^\ell, \mathbf{p}_{-i}\right) = \sum_{k \in [n] \setminus \{i\}} p_k(1)\, p_k(2)\, \omega^2(i, k, \ell)$; *for games with two players,* $\mathsf{Var}_i\left(p_i^\ell, \mathbf{p}_{-i}\right) = p_{\bar{i}}^\ell \left(1 - p_{\bar{i}}^\ell\right) \omega^2\left(i, \bar{i}, \ell\right)$, *where* $\bar{i}$ *denotes the player opponent to* i.

For the fully mixed case, we first show:

Theorem 11. *Consider the identical links game. Then, the fully mixed profile* $\mathbf{p}$ *with all probabilities equal is an* RA*-equilibrium.*

We continue to show:

Theorem 12. *Consider the related links game* G *with two players on* $m \geq 3$ *links. Fix a triple of links* $1, 2, 3$ *with different capacities. Then, there is no* RA*-equilibrium with a mixed player over links* $1, 2, 3$.

By Theorem 7, Corollary 10 immediately implies:

Lemma 13. *Consider an* RA*-equilibrium* $\mathbf{p}$ *for the player-specific scheduling game* G *with two players. Then, for each player* $i \in [2]$, $p_{\bar{i}}^\ell \left(1 - p_{\bar{i}}^\ell\right) \omega^2\left(\bar{i}, \bar{i}, \ell\right)$ *is constant over all links* $\ell \in \sigma(i)$.

The strictly positive numbers $\{\alpha_1, \alpha_2, \alpha_3\}$ satisfy the ***strict triangle inequality*** if *(i)* $\alpha_1 < \alpha_2 + \alpha_3$, *(ii)* $\alpha_2 < \alpha_1 + \alpha_3$, and *(iii)* $\alpha_3 < \alpha_1 + \alpha_2$. We show:

Proposition 14. *The player-specific scheduling game* G *with two players on three links has a fully mixed* Var*-equilibrium if and only if for each player* $i \in [2]$, *the numbers* $\left\{ \frac{1}{\omega^2\left(i, \bar{i}, \ell\right)} \mid \ell \in [3] \right\}$ *satisfy the strict triangle inequality.*

A ***bimatrix game*** is a two-players game; it is represented as the pair of $r \times c$ matrices $\langle \mathsf{A}, \mathsf{B} \rangle$ such that for each pair of indices $\ell \in [r], \ell' \in [c]$, $a_{\ell\ell'} = \mu_1(\ell, \ell')$ and $b_{\ell\ell'} = \mu_2(\ell, \ell')$. A 2×2 matrix $\mathsf{A} = \begin{pmatrix} a_{11}\, a_{12} \\ a_{21}\, a_{22} \end{pmatrix}$. is ***column-nice*** if there holds one of conditions (C1) and (C2): (C1): $a_{11} = a_{21}$ and $a_{22} = a_{12}$. (C2) (C2/a) $a_{11} \neq a_{21}$ and $a_{12} \neq a_{22}$; (C2/b) $a_{11} > a_{21}$ if and only if $a_{22} > a_{12}$; (C2/c) $a_{11} + a_{21} = a_{22} + a_{12}$. We show:

Proposition 15. *A bimatrix game* $\mathsf{SG} = \langle \mathsf{A}, \mathsf{B} \rangle$ *on two strategies has a fully mixed* RA*-equilibrium if and only if both matrices* A *and* B *are column-nice.*

We now present results for the case of two links. We first show:

Theorem 16. *For the identical links game with two links, a mixed profile is an* RA*-equilibrium if and only if it is an* E*-equilibrium.*

We now show:

Example 1. Consider the identical links game with three identical players on three links. Then. there are: (1) An RA-equilibrium which is not an E-equilibrium. (2) An E-equilibrium which is not an RA-equilibrium.

We finally show:

Theorem 17. *Consider a player-specific scheduling game* G *on two links such that for each pair of players* $i, j \in [n]$, *either* (A1) $\omega(i, j, 1) = \omega(i, j, 2) = 0$ *or* (A2) $\omega(i, j, 1) < \omega(i, j, 2)$. *Consider an* RA*-equilibrium* $\mathbf{p}$ *(for the game* G^{RA}*). Then, for each player* $i \in [n]$, *either* (C1) p_i *is a pure strategy, or* (C2) p_k *is a pure strategy for each player* $k \in [n] \setminus \{i\}$ *with* $\omega(i, k, 2) \neq 0$.

We now present results for the case of two players. We first consider a generalization of a player-specific scheduling game, called a ***player-specific scheduling game with strictly monotone latencies***, which comes with a collection of strictly monotone ***latencies*** $\{\mathsf{f}_j : \mathsf{N} \rightarrow \mathsf{N}\}_{j \in [m]}$ so that for each player $i \in [n]$ and a profile $\mathbf{s} \in S$, $\mu_{\mathsf{i}}(\mathbf{s}) = \mathsf{f}_{s_i}\left(\sum_{k \in [m] \mid s_k = s_i} \omega(i, k, s_i)\right)$. A ***weighted scheduling game with strictly monotone latencies*** is the special case where for all triples of players $i, j \in [n]$ and strategy $\ell \in [m]$, $\omega(i, j, \ell) = w_j$, for a collection of weights $\{w_k\}_{k \in [n]}$. So, for a player $i \in [n]$ and a profile $\mathbf{p} \in S$, $\mu_{\mathsf{i}}(\mathbf{s}) = \mathsf{f}_{s_i}\left(\sum_{k \in [m] \mid s_k = s_i} w_k\right)$. We show:

Theorem 18. *Consider a weighted scheduling game with strictly monotone latencies* G, *with two players. Then, for an* RA*-equilibrium* $\langle p, q \rangle$ *of* G, *either* (C1) *at least one player is pure, or* (C2) $\sigma(p) \cap \sigma(q) = \emptyset$, *or* (C3) $\sigma(p) = \sigma(q)$.

We continue with two examples:

Example 2. Consider the related links game with two players. Then, there are: (1) An E-equilibrium $\langle p, q \rangle$ with $\sigma(p) = \sigma(q)$ which is not an RA-equilibrium. (2) An RA-equilibrium $\langle p, q \rangle$ with $\sigma(p) \cap \sigma(q) = \emptyset$ which is not an E-equilibrium.

Example 3. Consider the related links game with two players. Then, there is an E-equilibrium $\langle p, q \rangle$ such that neither (C2) $\sigma(p) \cap \sigma(q) = \emptyset$ nor (C3) $\sigma(p) = \sigma(q)$.

Finally, we observe:

Lemma 19. *A player-specific scheduling game with two players on three links has a pure equilibrium.*

5 Inexistence and Complexity Results

Consider an instance of a player-specific scheduling game G on ***identical links***, defined by the weights $\{\omega(i,j)\}_{i,j\in[n]}$. Assume there are only two links. Denote as $\mathcal{G}$ the corresponding class of player-specific scheduling games on two identical links. Induced by the game G is the ***modified game*** $\widehat{\mathsf{G}}$ defined by the weights $\widehat{\omega}(i,j,\ell) = \mathsf{D} \cdot \omega(i,j)$ if $\ell = 1$, or $\mathsf{D} \cdot \omega(i,j) + 1$ if $\ell = 2$, for each pair of players $i,j \in [n]$. where $\mathsf{D} = n+1$. Clearly, the modified game G is a player-specific scheduling game. Denote as $\widehat{\mathcal{G}}$ the class of games resulting from the transformation of games in $\mathcal{G}$. Note that $\widehat{\mu_i}(\mathbf{s}) = \mu_i(\mathbf{s})$ if $s_i = 1$, or $\mu_i(\mathbf{s}) + \delta_i(\mathbf{s})$ if $s_i = 2$, where $\delta_i(\mathbf{s}) = |\{j \in [n] \mid s_j = s_i = 2\}|$. So, $1 \leq \delta_i(\mathbf{s}) \leq n$. We show:

Theorem 20. *$\widehat{\mathsf{G}}$ has no mixed* RA*-equilibrium, and every pure equilibrium of $\widehat{\mathsf{G}}$ is a pure equilibrium of* G.

Denote as $\mathcal{MCG}$ the subclass of $\mathcal{G}$ consisting of games where the non-zero weights are *symmetric*; so, for each pair of players $i,j \in [n]$, $\omega(i,j) = \omega(j,i)$. Each game $\mathsf{MCG} \in \mathcal{MCG}$ is called a ***max-cut game*** [5,5]; its underlying (symmetric) graph $G = \langle V, E\rangle$ is defined by the non-zero weights of the game MCG. So, $\mathcal{MCG}$ is a class of games, but it can also be viewed as a $\mathcal{PLS}$*-problem* [8] where the global function (from profiles to integers) to be (locally) minimized is $\mathsf{W}(\mathbf{s}) = \sum_{\{i,j\}\in E \mid s_i = s_j} \omega(i,j)$, and the *neighborhood* of each player i is defined as $[2] \setminus \{s_i\}$; so it is the link to which player i can switch. In a similar way, we associate with the class $\mathcal{MCG}$ the function $\widehat{\mathsf{W}}(\mathbf{s}) = \mathsf{D} \cdot \mathsf{W}(\mathbf{s}) + \eta(\mathbf{s})$, where $\eta(\mathbf{s}) = |i \in [n] \mid s_i = 2|$. We abuse notation to denote as $\mathcal{MCG}$ and $\widehat{\mathcal{MCG}}$ these $\mathcal{PLS}$-problems. We prove:

Lemma 21. *A profile* $\mathbf{s}$ *is a local optimum in the* $\mathcal{PLS}$*-problem* $\widehat{\mathcal{MCG}}$ *if and only if* $\mathbf{s}$ *is a pure equilibrium in the game* $\widehat{\mathsf{MCG}}$.

$\widehat{\mathcal{MCG}}$, viewed as a $\mathcal{PLS}$-problem, has a local optimum. So, by Lemma 21 the game $\widehat{\mathsf{MCG}}$ has a pure equilibrium and $\widehat{\mathcal{MCG}}$ is a $\mathcal{PLS}$-problem. By [21], $\mathcal{MCG}$ is $\mathcal{PLS}$-complete. Theorem 20 implies that $\widehat{\mathsf{MCG}}$ has no mixed RA-equilibrium. Theorem 20 and Lemma 21 establish a $\mathcal{PLS}$-reduction from $\mathcal{MCG}$ to $\widehat{\mathcal{MCG}}$. So:

Theorem 22. *It is* $\mathcal{PLS}$*-complete. to compute an* RA*-equilibrium for the modified max-cut game* $\widehat{\mathsf{MCG}}$.

We continue with a player-specific scheduling game G_1 with three players on two identical links with no pure equilibrium. By Theorem 20, this implies:

Corollary 23. *There is a player-specific scheduling game* $\widehat{\mathsf{G}_1}$ *with three players on two links which has no* RA*-equilibrium.*

References

1. Arrow, K.J.: The Theory of Risk-Aversion. In: Aspects of the Theory of Risk Bearing. Y. J. Assatio, Helsinki
2. Christodoulou, G., Mirrokni, V.S., Sidiropoulos, A.: Convergence and Approximation in Potential Games. In: Durand, B., Thomas, W. (eds.) STACS 2006. LNCS, vol. 3884, pp. 349–360. Springer, Heidelberg (2006)
3. Crawford, V.P.: Equilibrium Without Independence. Journal of Economic Theory 50, 127–154 (1990)
4. Debreu, G.: A Social Equilibrium Existence Theorem. Proceedings of the National Academy of Sciences of the United States of America 38, 886–893 (1952)
5. Fabrikant, A., Papadimitriou, C., Talwar, K.: The Complexity of Pure Nash Equilibria. In: Proceedings of the 36th Annual ACM Symposium on Theory of Computing, pp. 604–612 (June 2004)
6. Fiat, A., Papadimitriou, C.: When the Players Are Not Expectation Maximizers. In: Kontogiannis, S., Koutsoupias, E., Spirakis, P.G. (eds.) SAGT 2010. LNCS, vol. 6386, pp. 1–14. Springer, Heidelberg (2010)
7. Fisher, I.: The Nature of Capital and Income. The Macmillan Company (1906)
8. Johnson, D., Papadimitriou, C., Yannakakis, M.: How Easy is Local Search? Journal of Computer and System Sciences 37, 79–100 (1988)
9. Koutsoupias, E., Papadimitriou, C.: Worst-Case Equilibria. Computer Science Review 3, 65–69 (2009)
10. Lücking, T., Mavronicolas, M., Monien, B., Rode, M., Spirakis, P.G., Vrťo, I.: Which Is the Worst-Case Nash Equilibrium? In: Rovan, B., Vojtáš, P. (eds.) MFCS 2003. LNCS, vol. 2747, pp. 551–561. Springer, Heidelberg (2003)
11. Markowitz, H.: Portfolio Selection. Journal of Finance 7, 77–91 (1952)
12. Marschak, J.: Rational Behavior, Uncertain Prospects, and Measurable Utility. Econometrica 18, 111–141 (1950)
13. Mavronicolas, M., Spirakis, P.G.: The Price of Selfish Routing. Algorithmica 48, 91–126 (2007)
14. Milchtaich, I.: Congestion Games with Player-Specific Payoff Functions. Games and Economic Behavior 13, 111–124 (1996)
15. Nash, J.F.: Equilibrium Points in n-Person Games. Proceedings of the National Academy of Sciences of the United States of America 36, 48–49 (1950)
16. Nash, J.F.: Non-Cooperative Games. Annals of Mathematics 54, 286–295 (1951)
17. von Neumann, J., Morgenstern, O.: Theory of Games and Economic Behavior. Princeton University Press (1944)
18. Rabin, M.: Diminishing Marginal Utility of Wealth Cannot Explain Risk Aversion. In: Kahneman, D., Tversky, A. (eds.) Choices, Values, and Frames, pp. 202–208. Cambridge University Press (2000)
19. Ritzberger, K.: On Games Under Expected Utility with Rank Dependent Probabilitis. Theory and Decision 40, 1–27 (1996)
20. Rosenthal, R.W.: A Class of Games Possessing Pure Strategy Nash Equilibria. International Journal of Game Theory 2, 65–67 (1973)
21. Schäffer, A.A., Yannakakis, M.: Simple Local Search Problems That are Hard to Solve. SIAM Journal on Computing 20, 56–87 (1991)

A Theoretical Examination of Practical Game Playing: Lookahead Search

Vahab Mirrokni[1], Nithum Thain[2], and Adrian Vetta[3,*]

[1] Google Research, New York
[2] Department of Mathematics and Statistics, McGill University.
[3] Department of Mathematics and Statistics and School of Computer Science, McGill University.

Abstract. Lookahead search is perhaps the most natural and widely used game playing strategy. Given the practical importance of the method, the aim of this paper is to provide a theoretical performance examination of lookahead search in a wide variety of applications. To determine a strategy play using *lookahead search*, each agent predicts multiple levels of possible re-actions to her move (via the use of a search tree), and then chooses the play that optimizes her future payoff accounting for these re-actions. There are several choices of optimization function the agents can choose, where the most appropriate choice of function will depend on the specifics of the actual game - we illustrate this in our examples. Furthermore, the type of search tree chosen by computationally-constrained agent can vary. We focus on the case where agents can evaluate only a bounded number, k, of moves into the future. That is, we use depth k search trees and call this approach *k-lookahead search*. We apply our method in five well-known settings: industrial organization (Cournot's model); AdWord auctions; congestion games; valid-utility games and basic-utility games; cost-sharing network design games. We consider two questions. First, what is the expected social quality of outcome when agents apply lookahead search? Second, what interactive behaviours can be exhibited when players use lookahead search? We demonstrate how the answer depends on the game played.

Keywords: game theory, market games, valid utility games, cournot, stackelberg, adwords, network design, bounded rationality.

1 Introduction

Our goal here is not to prescribe how games should be played. Rather, we wish to analyse how games actually are played. To wit we consider the strategy of lookahead search, described by Pearl [27] in his classical book on heuristic search as being used by "almost all game-playing programs". To understand the lookahead method and the reasons for its ubiquity in practice, consider an agent trying to decide upon a move in a game. Essentially, her task is to evaluate each of her

* Supported in part by NSERC grants 288334 and 429598.

M. Serna (Ed.): SAGT 2012, LNCS 7615, pp. 251–262, 2012.

possible moves (and then select the best one). Equivalently, if she knows the values of each child node in the game tree then she can calculate the value of the current node. However, the values of the child nodes may also be unknown! Recall two prominent ways to deal with this. Firstly, crude estimates based upon local information could be used to assign values to the children; this is the approach taken by *best response dynamics*. Secondly, the values of the children can be determined recursively by finding the values of the grandchildren. At its computational extreme, this latter approach in a finite game is Zermelo's algorithm - assign values to the leaf nodes[1] of the game tree and apply backwards induction to find the value of the current node.

Both these approaches are special cases of *lookahead search*: choose a local search tree T rooted at the current node in the game tree; valuations (or estimates thereof) are given to leaf nodes of T; valuations for internal tree nodes are then derived using the values of a node's immediate descendants via backwards induction; a move is then selected corresponding to the value assigned the root. For best response dynamics the search tree is simply the star graph consisting of the root node and its children. With unbounded computational power, the search tree becomes the complete (remaining) game tree used by Zermelo's algorithm.

In practice the actual shape of the search tree T is chosen *dynamically*. For example, if local information is sufficient to provide a reliable estimate for a current leaf node w then there is no need to grow T beyond w. If not, longer branches rooted at w need to be added to T. Thus, despite our description in terms of "backwards induction", lookahead search is a very forward looking procedure. Subject to our computational abilities, we search further forward only if we think it will help evaluate a game node. Indeed, in our opinion, it is this forward looking aspect that makes lookahead search such a natural method, especially for humans and for dynamic (or repeated) games.[2]

Interestingly, the lookahead method was formally proposed as long ago as 1950 by Shannon [31], who considered it a practical way for machines to tackle complex problems that require "general principles, something of the nature of judgement, and considerable trial and error, rather than a strict, unalterable computing process". To illustrate the method, Shannon described in detail how it could be applied by a computer to play chess. The choice of chess as an example is not a surprise: as described the lookahead approach is particularly suited to game-playing. It should be emphasised again, however, that this approach is natural for all computationally constrained agents, not just for computers. Lookahead search is an instinctive strategic method utilised by human beings as well. For example, Shannon's work was in part inspired by De Groot's influential psychology thesis [16] on human chess players. De Groot found that all players (of whatever standard) used essentially the same thought process - one based upon a lookahead heuristic. Stronger players were better at evaluating positions

[1] Often the values of the leaf nodes will be true values rather than estimates, for example when they correspond to end positions in a game.

[2] In contrast, strategies that are prescribed by axiomatic principles, equilibrium constraints, or notions of regret are much less natural for dynamic game players.

and at deciding how to grow (prune or extend) the search tree but the underlying approach was always the same.

Despite its widespread application, there has been little theoretical examination of the consequences of decision making determined by the use of local search trees. The goal of this paper is to begin such a theoretical analysis. Specifically, what are the quantitative outcomes and dynamics in various games when players use lookahead search?

1.1 Lookahead Search: The Model

Having given an informal presentation, let's now formally describe the lookahead method. Here we consider games with sequential moves that have complete information. These assumptions will help simplify some of the underlying issues, but the lookahead approach can easily be applied to games without these properties.

We have a strategic game $G(\mathcal{P}, \mathcal{S}, \{\alpha_i : i \in \mathcal{P}\})$. Here $\mathcal{P}$ is the set of n players, S_i is the set of possible strategies for $i \in \mathcal{P}$, $\mathcal{S} = (S_1 \times S_2 \ldots \times S_n)$ is the strategy space, and $\alpha_i : \mathcal{S} \to R$ is the payoff function for player $i \in \mathcal{P}$. A *state* $\bar{s} = (s_1, s_2, \ldots, s_n)$ is a vector of strategies $s_i \in S_i$ for each player $i \in \mathcal{P}$.

Suppose player $i \in \mathcal{P}$ is about to decide upon a move. With lookahead search she wishes to assign a value to her current state node $\bar{s} \in \mathcal{S}$ that corresponds to the highest value of a child node. To do this she selects a search tree T_i over the set of states of the game rooted at $\bar{s}$. For each leaf node $\bar{l}$ in T_i, player i then assigns a valuation $\Pi_{j,\bar{l}} = \alpha_j(\bar{l})$ for each player j. Valuations for internal nodes in T_i are then calculated by induction as follows: if player p is destined to move at game node $\bar{v}$ then his valuation of the node is given by $\Pi_{p,\bar{v}} = \max_{\bar{u} \in \mathcal{C}(\bar{v})}[r_{p,\bar{v}} + \Pi_{p,\bar{u}}]$. Here, $\mathcal{C}(\bar{v})$ denotes the set of children of $\bar{v}$ in T_i, and $r_{p,\bar{v}}$ is some additional payoff received by player p at node $\bar{v}$. Should p choose the child $\bar{u}^* \in \mathcal{C}(\bar{v})$ then assume any non-moving player $j \neq p$ places a value of $\Pi_{j,\bar{v}} = r_{j,\bar{v}} + \Pi_{j,\bar{u}^*}$ on node $\bar{v}$. Then given values for children of the root node $\bar{s}$ of T_i, player i is thus able to compute the lookahead payoff $\Pi_{i,\bar{s}}$ which she uses to select a move to play at $\bar{s}$. [The method is defined in an analogous manner if players seek to minimise rather than maximise their "payoffs".]

After i has moved, suppose player j is then called upon to move. He applies the same procedure but on a local search tree T_j rooted at the new game node. Note that j's move may **not** be the move anticipated by i in her analysis. For example, suppose all the players use 2-lookahead search. Then player i calculates on the basis that player j will use a 1-lookahead search tree T'_j when he moves – because for computational purposes it is necessary that $T'_j \subseteq T_i$. But when he moves player j actually uses the 2-lookahead search tree T_j and this tree goes beyond the limits of T_i.

1.2 Lookahead Search: The Practicalities

There is still a great deal of flexibility in how the players implement the model. For example

• **Dynamic Search Trees.** Recall that search trees may be constructed dynamically. Thus, the exact shape of the search tree utilized will be heavily influenced by the current game node, and the experience and learning abilities of the players. Whilst clearly important in determining gameplay and outcomes, these influences are a distraction from our focal point, namely, computation and dynamics in games in which players use lookahead search strategies. Therefore, we will simply assume here that each T_i is a breadth first search tree of depth k_i. Implicitly, k_i is dependent on the computational facilities of player i.
• **Evaluation Functions.** Different players may evaluate leaf nodes in different ways. To evaluate internal nodes, as described above, we make the standard assumption that they use a max (or min) function. This need not be the case. For example, a risk-averse player may give a higher value to a node (that it does not own) with many high value children than to a node with few high value children – we do not consider such players here.
• **Internal Rewards or Not: Path Model vs. Leaf Model.** We distinguish between two broad classes of game that fit in this framework but are conceptually quite different. In the first category, payoffs are determined only by outcomes at the end of game. Valuations at leaf nodes in the local search trees are then just estimates of the what the final outcome will be if the game reaches that point. Clearly chess falls into this category. In the second category, payoffs can be accumulated over time - thus different paths with the same endpoints may give different payoffs to each player. Repeated games, such as industrial games over multiple time periods, can be modelled as a single game in this category. The first category is modelled by setting all internal rewards $r_{p,\bar{v}} = 0$. Thus what matters in decision making is simply the initial (estimated) valuations a player puts on the leaf nodes. We call this the *leaf (payoff) model* as an agent then strives to reach a leaf of T_i with as high a value as possible. The second category arises when the internal rewards, $r_{p,\bar{v}}$, can be non-zero. Each agent then wishes to traverse paths that allow for high rewards along the way. More specifically, in this model, called the *path (payoff) model*, the internal reward is $r_{p,\bar{v}} = \alpha_p(\bar{v})$.
• **Order of Moves: Worst-Case vs. Average-Case.** In multiplayer games, the order in which the players move may not be fixed. This adds additional complexity to the decision making process, as the local search tree will change depending upon the order in which players move. Here, we will examine two natural approaches a player may use in this situation: *worst case lookahead* and *average case lookahead*. In the former situation, when making a move, a risk-averse player will assume that the subsequent moves are made by different players chosen by an adversary to minimize that player's payoff. In the latter case, the player will assume that each subsequent move is made by a player chosen uniformly at random; we allow players to make consecutive moves. In both cases, to implement the method the player must perform calculations for multiple search trees. This is necessary to either find the worst-case or perform expectation calculations.

In practice, such versatility is a major strength and a key reason underlying the ubiquity of lookahead search in game-playing. For example, it accords well with Simon's belief, discussed in Section 1.4, that behaviours should be adaptable. For theoreticians, however, this versatility is problematic because it necessitates application-specific analyses. This will be apparent as we present our applications; we will examine what we consider to be the most natural implementation(s) of lookahead search for each game, but these implementations may vary each time!

1.3 Techniques and Results

We want to understand the social quality of outcomes that arise when computationally bounded agents use k-lookahead search to optimise their *expected* or *worst-case* payoff over the next k moves. Two natural ways we do this are via **equilibria** and via the study of **game dynamics**. To explain these approaches, consider the following definition. Given a lookahead payoff function, $\Pi_{i,\bar{s}}$, a *lookahead best-response* move for player i, at a state $\bar{s} \in \mathcal{S}$, is a strategy s_i maximising her lookahead payoff, that is, $\forall s'_i \in S_i$: $\Pi_{i,\bar{s}} \geq \Pi_{i,(\bar{s}_{-i},s'_i)}$. [A move s'_i for player i, at a state $\bar{s} \in \mathcal{S}$, is *lookahead improving* if $\Pi_{i,\bar{s}} \leq \Pi_{i,(\bar{s}_{-i},s'_i)}$.] A *lookahead equilibrium* is then a collection of strategies such that each player is playing her lookahead best-response move for that collection of strategies. Our focus here is on pure strategies. Then, given a social value for each state, the *coordination ratio* (or price of anarchy) *of lookahead equilibria* is the worst possible ratio between the social value of a lookahead equilibrium and the optimal global social value.

To analyse the dynamics of lookahead best-response moves, we examine the expected social value of states on polynomial length random walks on the *lookahead state graph*, $\mathcal{G}$. This graph has a node for each state $s \in \mathcal{S}$ and an edge from $\bar{s}$ to a state $\bar{t}$ with a label $i \in \mathcal{P}$ if the only difference between $\bar{s}$ and $\bar{t}$ is that player i changes strategy from s_i to t_i, where t_i is the lookahead best response move at $\bar{s}$. The *coordination ratio of lookahead dynamics* is the worst possible ratio between the expected social value of states on a polynomially long random walk on $\mathcal{G}$ and the optimal global social value.

For practical reasons, we are usually more interested in the dynamics of lookahead best-response moves than in equilibria. For example, as with other equilibrium concepts, lookahead best-response moves may not lead to lookahead equilibria. Indeed, such equilibria may not even exist. Typically, though, the methods used to bound the coordination ratio for k-lookahead equilibria can be combined with other techniques to bound the coordination ratio for k-lookahead dynamics. Consequently, for both simplicity and brevity, most of the results we give here concern the coordination ratio for lookahead equilibria. We are particularly interested in discovering when lookahead equilibria guarantee good social solutions, and how outcomes vary with different levels of foresight (k). We perform our analyses for an assortment of games including an AdWord auction game, the Cournot game, congestion games, valid-utility games, and a cost-sharing network design game.

We begin, in Section 2, with the Cournot duopoly game. Here two firms compete in producing a good consumed by a set of buyers via the choice of production quantities. We study equilibria in these simple games resulting from k-lookahead search. The equilibria for myopic game playing, $k = 1$, are well-understood in Cournot games. For $k > 1$, however, firms produce over 10% more than if they were competing myopically; this is better for society as it leads to around a 5% increase in social surplus. Surprisingly, the optimal level of foresight for society is $k = 2$. Furthermore, we show that Stackelberg behaviours arise as a special case of lookahead search where the firms have asymmetric computational abilities.

Next, in Section 3, we examine strategic bidding in an AdWord generalised second-price auction, and studying the social values of the allocations in the resulting equilibria. In particular, we show that 2-lookahead game playing results in the optimal outcome or a constant-factor approximate outcome under the leaf and path models, respectively. This is in contrast to 1-lookahead (myopic) game playing which can result in arbitrarily poor equilibrium outcomes, and shows that more forward-thinking bidders would produce efficient outcomes.

Third, we examine congestion games with linear latency functions, and study the average of delay of players in those games. We show that 2-lookahead game playing results in constant-factor approximate solutions. In particular, the coordination ratio of lookahead dynamics is a constant. These guarantees are similar to those obtained via 1-lookahead.

Fourth, we consider two classes of resource sharing games, known as valid-utility and basic-utility games. For both of these games, we show that lookahead game playing may result in very poor solutions. For valid-utility games, we show k-lookahead can give a coordination ratio for lookahead dynamics of $\Theta(\sqrt{n})$, where n is the number of players. Myopic game play can also give very poor solutions [15], but additional foresight does not significantly improve outcomes in the worst case. For basic-utility games, however, myopic game dynamics give a constant coordination ratio [15] whereas we show that 2-lookahead game playing may result in $o(1)$-approximate social welfare with the leaf model. Thus, additional foresight in games need not lead to better outcomes, as is traditionally assumed in decision theory.

Finally, we present a simple example of a cost-sharing network design game that illustrates how the use of lookahead search can encourage cooperative behaviour (and better outcomes) *without* a coordination mechanism.

Due to space constraints, all the proofs as well as our results for congestion games, valid and basic-utility games, and cost-sharing networks are omitted from this proceedings version but can be found in the full paper.

1.4 Background and Related Work

This work is best viewed within the setting of *bounded rationality* pioneered by Herb Simon [32]. An extensive discussion on this relationship is given in the full paper. The value of lookahead search in decision-making has been examined by the artificial intelligence community [25]; for examples in effective diagnostics and real-time planning see [17] and [29]. Lookahead search is also related to the sequential thinking framework in game theory [23,34]. Compared to such works, our focus is more theoretical and less experimental and psychological. Specifically, we desire quantitative performance guarantees for our heuristics.

Our study also relates to the price of anarchy in a game, and to the convergence of game dynamics to approximately optimal solutions [22,15] and to sink equilibria [15,10]. Numerous articles study the convergence rate of best-response dynamics to approximately optimal solutions [8,12,3,4]. For example, polynomial-time bounds has been proven for the speed of convergence to approximately optimal solutions for approximate Nash dynamics in a large class of potential games [3], and for learning-based regret-minimisation dynamics for valid-utility games [4]. Our work differs from all the above as none of them capture lookahead dynamics. In another line of work, convergence of best-response dynamics to (approximate) equilibria and the complexity of game dynamics and sink equilibria have been studied [11,1,7,33,10,21], but our paper does not focus on these types of dynamics or convergence to equilibria.

A much broader discussion of other concepts of equilibria and game dynamics that have been studied in the economics and computer science literatures can be found in the full paper.

2 Industrial Organisation: Cournot Competition

For our first example, we consider the classical game theoretic topic of duopolistic competition. Economists have considered a number of alternative models for market competition [35], prominent amongst them is the Cournot model [9]. Our main result is that the social surplus increases when firms are not myopic; surprisingly, social welfare is actually maximized when firms use 2-lookahead.

The Cournot model assumes players sell identical, nondifferentiated goods, and studies competition in terms of quantity (rather than price). Each player takes turns choosing some quantity of good to produce, q_i, and pays some marginal cost to produce it, c. The price for the good is then set as a function of the quantities produced by both players, $P(q_i+q_j) = (a-q_i-q_j)$, for some constant $a > c$. On turn l, each player i makes profit: $\Pi_i^l(q_i, q_j) = q_i(a-q_i-q_j-c)$. In this form, the model then only has one equilibrium, called the *Cournot equilibrium*, where $q_i = (a-c)/3$ for each player. We may assume that $a = 1$ and $c = 0$. Then, at equilibrium, each player makes a profit of $\Pi^i(q_i, q_j) = q_i(1-2q_i)$. The consumer surplus is $2q_i^2$ and the social surplus (the sum of the firms profits and the consumer surplus) is $2q_i(1-q_i)$.

We analyse this game when players apply k-lookahead search. In industrial settings it is natural to assume that payoffs are collected over time (as in a repeated game); thus, we focus upon the path model. We define this model inductively. In a k-step lookahead path model, each player i's utility is the sum of his utilities in the current turn and the $k-1$ subsequent turns. He models the quantities chosen in the subsequent turns as though the player acting during those turns were playing the game with a smaller lookahead. More specifically, he assumes that the player acting in the t'th subsequent turn chooses their quantity to maximise their utility under a $k-t$ lookahead model. In order to rewrite this rigorously, let π^i_l be the contribution to his utility that player i expects on the lth subsequent turn (and π^i_0 be the contribution to his utility that player i expects on his current turn), let π^j_l be the contribution to player j's utility that player i expects on the l'th subsequent turn, and let q^i_l (respectively, q^j_l) be the quantity that player i expects to choose (respectively, expects his opponent to choose) under this model.

Then in the path model, player i's expected utility function is $\Pi^i = \sum_{t=0}^{k-1} \pi^i_t$. Player j's expected utility function on player i's turn is $\Pi^j = \sum_{t=0}^{k-1} \pi^j_t$. Our aim now is to determine the quantities that player i expects to be chosen by both players in the subsequent turns and, thereby, determine the quantity he chooses this turn and the utility he expects to garner. To facilitate the discussion, it should be noted that unless noted otherwise, any reference to a "turn" refers to a turn during player i's calculation and not an actual game turn.

To simplify our analysis, we will define q_l to be the quantity chosen on turn l by whichever player is acting and Π_l to be the expected utility that that player garners from turn l to turn k. So $\Pi_0 = \Pi^i$, $\Pi_1 = \sum_{t=1}^{k-1} \pi^j_t$, etc. We define $\overline{\Pi}_l$ to be the utility garnered from turn l to turn k by the player who does not act during turn l. So $\overline{\Pi}_0 = \Pi^j$, $\overline{\Pi}_1 = \sum_{t=1}^{k-1} \pi^i_t$, etc. It is clear that on each turn l, the active player is trying to maximise Π_l.

We are now ready to compute these quantities and utilities recursively. From our definition above, we obtain $\Pi_k = q_k(1 - q_k - q_{k-1})$ and $\overline{\Pi}_k = q_{k-1}(1 - q_k - q_{k-1})$, and the recursive formula for $l < k$ that $\Pi_l = q_l(1 - q_l - q_{l-1}) + \overline{\Pi}_{l+1}$ and $\overline{\Pi}_l = q_{l-1}(1 - q_l - q_{l-1}) + \Pi_{l+1}$. Note that in each of these formulas, Π_l and $\overline{\Pi}_l$ are each functions of q_t for $t \geq l$; q_{l-1} is in fact fixed on the previous turn and is, therefore, not a variable in Π_l. It is then possible to calculate q_l recursively.

Lemma 1. *It holds that q_l is $\beta_l - \alpha_l q_{l-1}$, where $\beta_k = \alpha_k = \beta_{k-1} = \frac{1}{2}, \alpha_{k-1} = \frac{1}{3}$ and, for $l < k - 1$,*

$$\beta_l = \frac{2 - \beta_{l+1} + \alpha_{l+1}\beta_{l+2} - \alpha_{l+1}\alpha_{l+2}\beta_{l+1}}{4 - 2\alpha_{l+1} - \alpha_{l+1}^2\alpha_{l+2}}, \quad \alpha_l = \frac{1}{4 - 2\alpha_{l+1} - \alpha_{l+1}^2\alpha_{l+2}}$$

Our goal is now to calculate q_0 as this will tell us the quantity that player i actually chooses on his turn. From the above lemma, we can calculate q_0 if we can determine α_0 and β_0. Using numerical methods on the above recursive formula, we see that as $k \to \infty$, α_0 decreases towards a limit of $0.2955977\ldots$ and β_0 approaches a limit of $0.4790699\ldots$. These values also converge quite quickly; they both converge to within 0.0001 of the limiting value for $k \geq 10$. Thus, at a lookahead equilibrium, player i will choose $q_i \approx .0.4790699 - 0.2955977 q_j$ and player j, symmetrically, will choose $q_j \approx 0.4790699 - 0.2955977 q_i$. So each player will choose a quantity $q \approx 0.369767$. which is more than in the myopic equilibrium. Indeed, it is easy to show that for every $k \geq 2$, each player will produce more than the myopic equilibrium. We observe the quantity produced does not change monotonically with the length of foresight k, but it does increase significantly if non-myopic lookahead is applied at all. Consequently, in the path model looking ahead is better for society overall but worse for each individual firm's profitability (as the increase in sales is outweighed by the consequent reduction in price).

Theorem 1. *For Cournot games under the path model, output at a k-lookahead equilibrium peaks at* $k = 2$ *with output* 12.5% *larger than at a myopic equilibrium* ($k = 1$). *As foresight increases, output is* 10.9% *larger in the limit. The associated rises in social surplus are* 5.5% *and* 4.9%, *respectively.* □

`Stackelberg Behaviours`: We could also analyse this game under the leaf model, but this model is both less realistic here and trivial to analyse. However, it is interesting to note that for the leaf model with asymmetric lookahead, where player i has 2-lookahead and player j has 1-lookahead, we get the same equilibrium as the classic Stackelberg model for competition. Thus, the use of lookahead search can generate leader-follower behaviours.

3 Generalised Second-Price Auctions

For our second example, we apply the lookahead model to generalised second-price (GSP) auctions. Our main results are that outcomes are provably good when agents use additional foresight; in contrast, myopic behaviour can produce very poor outcomes.

The auction set-up is as follows. There are T slots with click-through rates $c_1 > c_2 > ... > c_T > 0$, that is, higher indexed slots have lower click-through rates. There are n players bidding for these slots, each with a private valuation v^i. Each player i makes a bid b^i. Slots are then allocated via a *generalised second price auction*. Denote the jth highest bid in the descending bid sequence by b_j, with corresponding valuation v_j. The jth best slot, for $j \leq T$, is assigned to the jth highest bidder who is charged a price equal to b_{j+1}. The T highest bidders are called the "winners". According to the pricing mechanism, if bidder i were to get slot t in the final assignment, then he would get utility $u_t^i = (v^i - b_{t+1})c_t$. We denote a player i's utility if he bids b^i by $u^i(b^i)$ (the other players bids are implicit inputs for u^i).

This auction is used in the context of keyword ad auctions (e.g, Google AdWords) for sponsored search. Given the continuous nature of bids in the GSP auction, the best response of each bidder i for any vector of bids by other bidders corresponds to a range of bid values that will result in the same outcome from i's perspective. Among these set of bid values, we focus on a specific bid value b^i, called the *balanced* bid [6]. The balanced bid b^i is a best-response bid that is as high as possible such that player i cannot be harmed by a player with a better slot undercutting him, i.e. bidding just below him. It is easy to calculate that for player i in slot t, $1 \leq t < T$, the only balanced bid is $b^i = (1 - \frac{c_t}{c_{t-1}})v^i + \frac{c_t}{c_{t-1}} b_{t+1}$.

An important property of balanced bidding is that each "losing" player i (one not assigned a slot) should bid truthfully, that is $b^i = v^i$. To see this add dummy slots with $c_t = 0$ if $t > T$. The player who wins the top slot should also bid truthfully under balanced bidding. Balanced bidding is the most commonly used bidding strategy [6,20]. For some intuition behind this, note that balanced bidding has several desirable properties. For a competitive firm, bidding high obviously increases the chance of obtaining a good slot. Within a slot this also has the benefit of pushing up the price a competitor pays without affecting the price paid by the firm. On the other hand, bidding high increases the upper bound on the price the firm may pay, leading to the possibility that the firm may end up paying a high price for one of the less desirable slots. Balanced bidding eliminates the possibility that a change in bid from a higher bidder can hurt the firm. (Clearly, it is impossible to obtain such a guarantee with respect to a lower bidder.) Thus, balanced bidding provides some of the benefits of high bidding at less risk. Balanced bidding naturally converges to Nash equilibria unlike other bidding strategies such as altruistic bidding or competitor busting [6]. Moreover, the other bidding strategies would require some discretization of players' strategy space in order to analyse the best response dynamics [6,20]. Consequently, balanced bidding is the most natural strategy choice for our analysis.

For this auction, we consider only the leaf model. This model seems more natural than the path model for a single auction as players are interested in the final allocation output by the auction (there are no intermediary payoffs). We analyse both worst-case and average-case lookahead; depending upon the level of risk-aversion of the agents both cases seem natural in auction settings.

Let player i's lookahead payoff (or utility) at bid b^i with respect to player j, denoted by $u^{ij}(b^i)$, be player i's payoff (or utility) after player j makes a best-response move. In the worst-case lookahead model, we define player i's lookahead payoff for a vector $\bar{b}$ of bids as $\Pi_{i,\bar{b}} = \tilde{u}^i(b^i) = \min_j u^{ij}(b^i)$. In the average-case lookahead model, player i's lookahead payoff $\Pi_{i,\bar{b}}$ for a bid vector $\bar{b}$ is $\Pi_{i,\bar{b}} = \bar{u}^i(b^i) = \frac{1}{n}\sum_j u^{ij}(b^i)$. Changing strategy from bid b^i to bid $\bar{b}^i$ is a *lookahead improving* move if lookahead utility increases, i.e., $\bar{u}^i(\bar{b}^i) > \bar{u}^i(b^i)$. We are at a *lookahead equilibrium* if no player has a lookahead improving move.

It is known that the social welfare of Nash equilibria for myopic game playing can be arbitrarily bad [6] unless we disallow over-bidding [19]. Here, we prove the advantage of additional foresight by showing that 2-lookahead equilibria have much better social welfare. In particular, we show that all such equilibria are optimal in the worst-case lookahead model, and all such equilibria are constant-factor approximate solutions in the average-case lookahead model.

3.1 Worst-Case Lookahead

Our proof for the worst-case lookahead model can be seen as a generalisation of the proof of [5] for a slightly different model. A useful lemma in this context is

Lemma 2. *Consider the worst-case lookahead model with the leaf model. Label the players so that player i is in slot i, and suppose there is a player t such that $v^t < v^{t+1}$. Then player t myopically prefers slot $t+1$ to slot t.*

An equilibrium is *output truthful* if the slots are assigned to the same bidders as they would be if bidders were to bid truthfully. It is easy to verify that an allocation optimizes social welfare if and only if it is output truthful. Thus to prove 2-lookahead equilibria are socially optimal it suffices to show they are output truthful.

Theorem 2. *For GSP auctions, any 2-lookahead equilibrium gives optimal social welfare in the worst-case, leaf model.*

3.2 Average Case Lookahead

For the average-case lookahead model, optimality is not guaranteed at equilibria.

Theorem 3. *In GSP auctions, there exist 2-lookahead equilibria that are not output-truthful in the average-case, leaf model.*

Despite this negative result, 2-lookahead equilibria cannot have arbitrarily bad social welfare.

Theorem 4. *In GSP auctions, the coordination ratio of 2-lookahead equilibria is constant in the average-case, leaf model.*

Acknowledgements. The authors would like to thank Kevin Leyton-Brown and Tim Roughgarden for interesting discussions on this topic.

References

1. Ackermann, H., Röglin, H., Vöcking, B.: On the impact of combinatorial structure on congestion games. Journal of the ACM 55(6) (2008)
2. Awerbuch, B., Azar, Y., Epstein, A.: The price of routing unsplittable flow. In: STOC (2005)
3. Awerbuch, B., Azar, Y., Epstein, A., Mirrokni, V., Skopalik, A.: Fast convergence to nearly-optimal solutions in potential games. In: EC, pp. 264–273 (2008)
4. Blum, A., Hajiaghayi, M., Ligett, K., Roth, A.: Regret minimization and the price of total anarchy. In: STOC, pp. 373–382 (2008)
5. Bu, T., Deng, X., Qi, Q.: Forward looking Nash equilibrium for keyword auction. Information Processing Letters 105(2), 41–46 (2008)
6. Cary, M., Das, A., Edelman, B., Giotis, I., Heimerl, K., Karlin, A., Mathieu, C., Schwarz, M.: Greedy bidding strategies for keyword auctions. In: EC (2007)
7. Chien, S., Sinclair, A.: Convergence to approximate Nash equilibria in congestion games. Games and Economic Behavior 71(2), 315–327 (2011)
8. Christodoulou, G., Mirrokni, V.S., Sidiropoulos, A.: Convergence and Approximation in Potential Games. In: Durand, B., Thomas, W. (eds.) STACS 2006. LNCS, vol. 3884, pp. 349–360. Springer, Heidelberg (2006)
9. Cournot, A.: Recherces sur les Principes Mathématiques de la Théorie des Richesse, Paris (1838)
10. Fabrikant, A., Papadimitriou, C.: The complexity of game dynamics: BGP oscillations, sink equlibria, and beyond. In: SODA, pp. 844–853 (2008)
11. Fabrikant, A., Papadimitriou, C., Talwar, K.: The complexity of pure Nash equilibria. In: STOC, pp. 604–612 (2004)
12. Fanelli, A., Flammini, M., Moscardelli, L.: The Speed of Convergence in Congestion Games under Best-Response Dynamics. In: Aceto, L., Damgård, I., Goldberg, L.A., Halldórsson, M.M., Ingólfsdóttir, A., Walukiewicz, I. (eds.) ICALP 2008, Part I. LNCS, vol. 5125, pp. 796–807. Springer, Heidelberg (2008)

13. Friedman, M.: The methodology of positive economics. In: Friedman, M. (ed.) Essays in Positive Economics, pp. 3–43. University of Chicago Press (1953)
14. Goemans, M., Li, L., Mirrokni, V., Thottan, M.: Market sharing games applied to content distribution in ad-hoc networks. In: MOBIHOC (2004)
15. Goemans, M., Mirrokni, V., Vetta, A.: Sink equilibria and convergence. In: FOCS, pp. 142–154 (2005)
16. de Groot, A.: Thought and Choice in Chess, 2nd edn. Mouton (1978), Original Version: Het denken van den Schaker, een experimenteel-psychologische studie, Ph.D. thesis, University of Amsterdam (1946)
17. de Kleer, J., Raiman, O., Shirley, M.: One step lookahead is pretty good. In: Hamscher, W., Console, L., de Kleer, J. (eds.) Readings in Model-Based Diagnosis, pp. 138–142. Morgan Kaufmann (1992)
18. Koutsoupias, E., Papadimitriou, C.: Worst-Case Equilibria. In: Meinel, C., Tison, S. (eds.) STACS 1999. LNCS, vol. 1563, pp. 404–413. Springer, Heidelberg (1999)
19. Paes Leme, R., Tardos, E.: Pure and Bayes-Nash price of anarchy for generalized second price auction. In: FOCS (2010)
20. Markakis, E., Telelis, O.: Discrete Strategies in Keyword Auctions and Their Inefficiency for Locally Aware Bidders. In: Saberi, A. (ed.) WINE 2010. LNCS, vol. 6484, pp. 523–530. Springer, Heidelberg (2010)
21. Mirrokni, V., Skopalik, A.: On the complexity of Nash dynamics and sink equilibria. In: EC (2009)
22. Mirrokni, V.S., Vetta, A.: Convergence Issues in Competitive Games. In: Jansen, K., Khanna, S., Rolim, J.D.P., Ron, D. (eds.) APPROX and RANDOM 2004. LNCS, vol. 3122, pp. 183–194. Springer, Heidelberg (2004)
23. Nagel, R.: Unraveling in guessing games: an experimental study. The American Economic Review 85(5), 1313–1326 (1995)
24. Nau, D.: An investigation of the causes of pathology in games. Artificial Intelligence 19, 257–278 (1982)
25. Nau, D.: Decision quality as a function of search depth on game trees. Journal of the ACM 30(4), 687–708 (1983)
26. Newell, A., Simon, H.: Human Problem Solving. Prentice-Hall (1972)
27. Pearl, J.: Heuristics: Intelligent Search Strategies for Computer Problem Solving. Addison-Wesley (1984)
28. Rubenstein, A.: Modeling Bounded Rationality. MIT Press (1998)
29. Sefer, E., Kuter, U., Nau, D.: Real-time A* Search with Depth-k Lookahead. In: Proceedings of the International Symposium on Combinatorial Search (2009)
30. Selten, R.: What is bounded rationality? In: Gigerenzer, G., Selten, R. (eds.) Bounded Rationality: the Adaptive Toolbox, pp. 13–36. MIT Press (2001)
31. Shannon, C.: Programming a computer for playing chess. Philosophical Magazine, Series 7 41(314), 256–275 (1950)
32. Simon, H.: A behavioral model of rational choice. Psychological Review 63, 129–138 (1955)
33. Skopalik, A., Vöcking, B.: Inapproximability of pure Nash equilibria. In: STOC, pp. 355–364 (2008)
34. Stahl, D., Wilson, P.: Experimental evidence on players' models of other players. Journal of Economic Behavior and Organization 25(3), 309–327 (1994)
35. Tirole, J.: The Theory of Industrial Organization. MIT Press (1988)
36. Varian, H.: Position auctions. International Journal of Industrial Organization 25, 1163–1178 (2007)
37. Vetta, A.: Nash equilibria in competitive societies, with applications to facility location, traffic routing and auctions. In: FOCS, pp. 416–425 (2002)

Author Index